MAN
and his
GEOLOGIC
ENVIRONMENT

MAN
and his
GEOLOGIC
ENVIRONMENT

DAVID N. CARGO
BOB F. MALLORY
Northwest Missouri State University

ADDISON-WESLEY
PUBLISHING COMPANY
Reading, Massachusetts
Menlo Park, California • London • Don Mills, Ontario

ISBN 0-201-00892-0
BCDEFGHIJ-HA-7987654

Dedicated to man's constant quest for knowledge and to his perception of new ways to approach old problems.

PREFACE

The authors have long believed that the traditional introductory geology course was not adequate to meet the needs of the liberal arts student who was not majoring in geology. Our discussions with these students, both in and out of class, often centered on such topics as the use and availability of water, mineral and energy resources, effect of the natural environment on health, the impact of predicted population increases on geologic resources, and the role the geologist will play in the future. We concluded that there is a need for geology courses that emphasize the environmental aspects of geology and that give students outside our field some insight into problems they will be called upon to face in the future, such as shortages in water, mineral resources, and energy. This book emerged from that need.

In the fall of 1970 we decided to offer an alternative to the traditional, general geology-historical geology sequence which has been, more or less, the standard way of fulfilling a science requirement in most schools for longer than we or our colleagues could remember. We would offer a course entitled "Environmental Geology" to be taken in lieu of the Historical Geology course, at the option of the student. Having made our decision, we began a search for a suitable text. None was available! We spent most of the spring and summer of 1970 frantically putting together our notes from widely scattered and varied sources and building our course from the ground up. Since there was no established format, we were free to explore areas which were not normally frequented by geologists and were unheard of at the freshman level. The results of our efforts met with a great deal of enthusiasm from the students.

We began to introduce more and more of the material from the "Environmental" course into the "General" course, and today our introductory course has evolved to the point that it is close to what our "Environmental" course was originally. The latter course has evolved also, so that today we offer what could be considered a two-semester course in "studies in the environmental aspects of geology."

This book is the result of our building this sequence. Recognizing that many, like ourselves, have begun an evolution of their introductory course in geology to better accommodate their students, we have tried to write this book so that it is usable in either a beginning course or as a text for a second course following a traditional physical or general geology course. We have used the material in both and are confident it can be used as the basis for meeting both needs.

World population is used as the framework within which our subjects are viewed, for it is within this framework that people must function in the future. Water, energy, and mineral resources are presented in relation to geology and to present and future needs. We have explored such geologic

hazards as earthquakes, landslides, floods, and volcanic eruptions in view of
their relation to man. We have entered the new area of medical geology by
looking at the natural geochemical situation and its relation to health problems. We hope we have been successful in relating, in a new way, the study
of geology to man and his daily living.

We would like to thank the hundreds of geology instructors all over the
country who took the time to respond to a questionnaire asking about the
nature of environmental geology courses they teach. The results were most
revealing. It was particularly interesting to learn of the great diversity in
courses in terms of philosophy content and level. The results suggested that
students do want to learn about the earth, particularly insofar as the relation
of the earth to people is concerned.

Maryville, Missouri   D. N. C.
December 1973   B. F. M.

CONTENTS

4. Soils

5. Mineral Resources

6. Minerals for the Future

7. Energy—The Fossil Fuels

"The earth's natural processes, and not man alone, are the principal agents of modifying our environment.

"This is not to excuse or to put aside what man does, but rather to put man's actions in proper natural perspective. Those individuals who speak about restoring our inherited environment of pure air, pure rain, pure water, pure lakes, and pure coastlines ignore the inevitability of nature."

"Although earth processes dwarf the actions of man, man can still become a major geologic agent. And this interaction of man with nature is without question a most important issue for future years."

From a commencement address at Colorado School of Mines, 1970, by the late William Thomas Pecora, former Director of the United States Geological Survey.

Man has a profound impact on his geologic environment. (Photograph by George Sheng)

INTRODUCTION

It was only a few years ago that the word "environment" was little known to the general public. In recent times, however, the word has invaded the vocabularies of a vast number of individuals, as have such other words as "ecology," "population explosion," and "pollution." There has been a proliferation of new publications, college courses, political maneuverings, and a lot of talk (some of it important) about the environment. The reason for all this is quite real: the environment has become a matter of greatest concern to all who inhabit this planet. Not that the environment has not always been a matter of concern, but never until now have so many people put so many demands on the earth and its physical and biological environments.

But what is the *environment*? The term implies a relationship between something and its surroundings. The relationship may be active or passive, dynamic or static. The *total surroundings* constitute the environment of any given entity. Any condition, state, or circumstance which

1

surrounds an entity is abutted against by still other conditions, states, or circumstances. In terms of space, therefore, the environment is infinite. At the same time, however, no two environments can be precisely alike, for each entity must necessarily occupy its own central point in its environment. If all environments, of all the respective entities, are infinite, however, they must overlap. Even in the immediate or nearby surroundings of an entity, we may find overlap if there is a nearby entity.

Although we have spoken of environments, or *the* environment if you will, as being infinite, we may for some practical purposes place a limit on it. We might for example, consider a lake or a city as constituting an environment. And finally, we come to the point where we may consider the earth itself to embody a single, total environment. (Obviously, we can find fault with this consideration, too, for how can we pluck the earth out of a gravitational field or energy field of the solar system, and the solar system out of its place in the galaxy, *ad infinitum*?)

Since we have to stop somewhere, let us stop at this point: the earth is a finite body, although under the influence of external things. The earth has on it a vast number of entities. Each of these entities occupies a position in the total environment of the earth and also constitutes the environment of each other entity to greater or lesser degrees.

What are these entities? Entities of the environment are in the broadest sense *things,* any things. They may be organic or inorganic, animate or inanimate. They may be the living organisms of the earth's community or simply the physical or chemical substances that make up the world and all therein. They may be material, like rock, forces, like gravity, or even senses, like love and fear. All these things of the earth comprise the environment.

One notable entity in the earth's environment is man. There was a time, perhaps, when man could scarcely have been considered notable, but today he has become notable, and particularly so for two outstanding features about him. The one which comes immediately to mind, and which we all like to pride ourselves upon, is his brain. Man's brain has given him enormous power in becoming a significant environmental entity. And the other notable feature? His numbers. His abundance and ability to proliferate and survive have set man forth in the environment by the billions. These billions of men, each with his own highly developed brain, today constitute a factor never before present in the earth's environment. If the earth is exactly 4,500,000,000 years old, then for 4,499,999,900 years, give or take a few, the environment did not include man in great numbers, for it has only been during the last hundred or so years that intelligent man with his vast numbers has become the environmental factor that he has. And for all that, we can even quite reasonably make that the last ten years.

Brain + numbers: this is indeed a deadly combination. There is not any other conceivable combination, in today's earth at least, that could possibly have the impact on the environment that this one is having. Man has become a central figure in the environment. No other creature has the ability to produce or consume, create or destroy, beautify or foul, or otherwise modify, the total earth environment that man has. With this vastness of intelligence and numbers, it is bound to happen that man has a resounding impact on his (and every other living creature's) environment, and likewise that the environment has a resounding impact on man.

It has been argued, and perhaps rightly so, that the term *environmental geology* is redundant because geology is the study of the earth, and the earth *is* the environment. But as long as more and more emphasis is being placed on "the environment," a bit of redundancy may be excused for the sake of emphasis. There are numerous branches of geology, some of which are mineralogy, stratigraphy, and paleontology. It would be illogical, indeed, to regard environmental geology as a branch of geology; rather, the term implies a certain way of focusing on the science or, for that matter, on the environment.

Some rules are needed in playing any game. Therefore, we should state a few for environmental geology. The principles should be few, broad, and simple, because after all, when you deal with an environment as broad as the geoenvironment, too many specifics can turn the environment into a swamp. Then, too, not very many people have had in the past much clear notion of just what to include in these principles.

There are a number of things that ought to be stressed. First, the earth is a finite body. It has fixed dimensions and consists of virtually unchanging quantities of given substances. As such, it is a closed system, the amount of its substance lost to or added from outer space being insignificant. The distribution, use, and redistribution of the various materials of the earth are matters of some concern, since the supply of these materials is finite. Second, the various materials of the earth are not only present in certain places in finite amounts, but they also participate in various cycles and reactions, and they possess unique properties. Third, the earth is a dynamic body. The rocks, the waters, and the air are in continuous motion. The motion may be swift and occasionally violent, or it may be glacially slow. For that matter, some glaciers figuratively "gallop" in comparison to the drift of continents. Fourth, the earth and the events which take place on it lie in the dimension of time. There is nothing that remains static; given sufficient time, all things change, however rapidly or slowly. Finally, man himself is a participant in his environment. He is not apart from the earth, nor can he greatly alter the natural processes which occur in the earth. He does, however, seem to possess limitless capacity for meddling in the dynamism of the earth, which easily leads to his being poisoned, buried, washed out, sliding off or falling in, drowned, burned, starved, frozen, or at best his going thirsty or running out of fuel or building blocks. When man was a primitive being, which was essentially until he developed modern technology, he had little capacity to clash with his environment except to get mauled by a bear or to drown. But when technocracy, the bomb, and the population explosion arrived, man achieved a capacity to clash with the earth on a grand scale. Hopefully, man also has the capacity to live in harmony as an integral part of the earth-environment. It is a capacity he simply must develop, for it is either that or perish.

It is our intention to examine a few broad aspects of the earth's environment. We will do this by describing them and showing how man, ever more numerous and ever more demanding, relates to these aspects.

Increases in population lead to increased demands on man's geologic environment. (Photograph by Bruce Anderson)

POPULATION

INTRODUCTION

Why have a population chapter in an introductory geology text? Very simply, we feel that any discussion of the role geology will play in the future of man must necessarily be within the framework of what demand the population will make on our geological resources. This demand will stem from two sources: the *total number* of people drawing on these resources, and the drain on the resources caused by the *level of affluence* at which these people live. This chapter is designed to give an overview of the current population picture in terms of numbers of people and some present demands, as well as of the forecast for the number of people to be expected in the future. Hopefully, this will provide a convenient framework which you can use throughout the remainder of the text.

7

Exponential and Linear Growth

In dealing with many of the subjects covered in this text, we will make reference to exponential and linear growth rates. It might be profitable to review the difference between these two. Many of the things of interest to us, such as population growth, energy and mineral consumption, water use, and generation of waste products follow exponential growth. Therefore, it is very important to have a clear understanding of this concept.

Linear growth can be defined as the increase gained by adding a fixed amount to a total at the end of successive units of time. The growth is by fixed increments. For instance, suppose a man has buried $500 in a can in his backyard, but digs it up at the end of every year in order to add an additional $500. His savings grow by increments of $500 for every time period; it is growing in a linear fashion.

Exponential growth can be defined as the addition of a certain *percent of the total* to the total at the end of successive units of time. For each successive unit of time, then, the amount added is larger because the total is larger. This is analogous to a man's putting his money in the bank at a certain rate of interest annually and allowing the interest to accumulate with the original money. At the end of each successive year the interest payment will be larger than the year before because interest is being paid on the original amount *plus* prior interest payments.

When dealing with growth of any kind, there comes a point at which the original quantity will have doubled in size because of the added increments. This "doubling time" is easy to determine when dealing with linear growth, but is a little more difficult when dealing with exponential growth. It is often vital to know what the "doubling time" is under certain conditions. To calculate the "doubling time" when exponential growth is involved, simply divide seventy (70) by the percent of growth per year; the quotient is the *number of years* it takes to double the original quantity at that rate of growth. Table 1–1 gives the "doubling time" for several rates of growth. The world population is increasing at about 2 percent annually, and we see by this table that the world population will double in about 35 years.

Table 1–1 Growth rate and corresponding
doubling time in years, using exponential
growth

Growth rate (% per year)	Doubling time in years (rounded to nearest year)
1.0	70
2.0	35
3.0	23
4.0	18
5.0	14
6.0	12
7.0	10
10.0	7

Present Metal Usage

The world, especially the developed countries, is using metals at fantastic
rates. Lovering has estimated that "the amount of metal consumed in about
30 years (by the United States) at the current rate of increase in consumption
approximates the total amount of metal used in all previous times." In 1967
the United States' consumption of copper was 18 pounds per person,
compared to a worldwide consumption of about 3.2 pounds per person.
The consumption of lead was 12 pounds per person in the United States
and 1.5 pounds per person on a worldwide basis.

The per capita usage of iron and steel in the United States was about
0.5 ton, and worldwide usage was about 0.1 ton per capita. Figures 1–1,
1–2, and 1–3 are of man's usage of these three metals in the past, plotted
against the population curve from the year 1000 to the year 2000. These
three cases are representative of the status of much of our metallic resources today. Today, when consumption of most metallic resources is
plotted against time, the plot is almost a straight, *nearly vertical,* line.

The question for which we must try to determine an answer is: How
much of each of these metals is available to us on the earth, and how long
will they last, assuming present trends continue? Any attempt to answer
this question will be complicated by a further unknown variable, the level
of affluence. The higher the level of affluence, the greater the consumption
of materials necessary to support this affluence. If we are to project reserves in the future and make estimates about the length of time our reserves will last, we must also attempt to project some figure for affluence
level. We will concern ourselves with this problem in a later chapter.

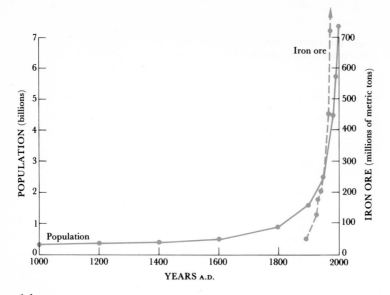

1-1

Iron ore consumption plotted against population growth. **Estimates by the Bureau of Mines** *of world iron ore demand for the year 2000 range from a low of 12.6 billion metric tons to a high of 15.2 billion metric tons. These figures represent an increase of 17–21 times over the amount used in 1969. (From the U.S. Bureau of Mines and the 1972* **World Population Data Sheet** *with the permission of the Population Reference Bureau, Inc.)*

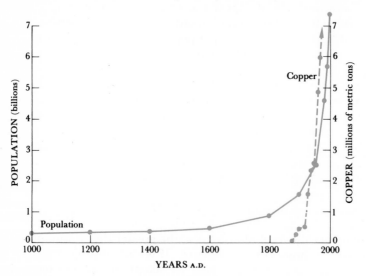

1-2

Copper consumption plotted against population growth. **Estimates by the Bureau of Mines** *of world copper demand for the year 2000 range from a low of 19.68 million metric tons to a high of 38.78 million metric tons. (From the U.S. Bureau of Mines and the 1972* **World Population Data Sheet** *with the permission of the Population Reference Bureau, Inc.)*

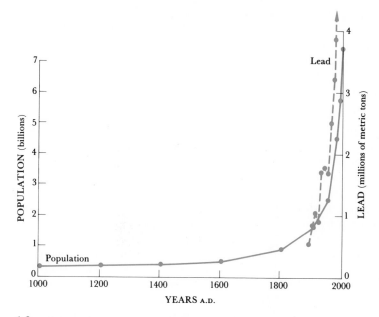

1-3
Lead consumption plotted against population growth. Estimates by the Bureau of Mines of world lead demand for the year 2000 range from a low of 5.0 million metric tons to a high of 7.4 million metric tons. (From the U. S. Bureau of Mines and the 1972 World Population Data Sheet *with the permission of the Population Reference Bureau, Inc.)*

Present Use of Water

The United States used approximately 270 billion gallons of water per day in 1960. This represented approximately 22½ percent of the average aggregate streamflow, that is, the available water runoff, of 1200 billion gallons per day. Projections for 1980 average approximately 559 billion gallons per day, or about 46 percent of the available runoff. Water usage, like our usage of mineral resources, shows a marked increase over the past few decades.

As our population increases and our level of affluence rises, we will make greater and greater demands on our water resources. In some areas of the country these demands have already reached the limit of the area's ability to supply. Extended droughts are bringing this to our attention with great impact. These areas are currently facing severe problems. Our water supply, like our supply of other resources, is limited, and these limits can be estimated. Unlike the situation with many other resources, there is the possibility of removing or easing the limits currently imposed on the

amount of water available. There is also the possibility that through desalination the oceans could become a source of fresh water, but the pros-
pect seems unlikely on a large scale at present. We must learn to live within the water budget we now have.

Present Use of Energy

The total worldwide production of energy for 1970 was estimated to be
between 45,000 and 50,000 billion kilowatt hours. Of this, approximately
76 percent was produced by the petroleum group of fuels, 20 percent by
coal and lignite, and about 3.8 percent by water. The United States alone
used 1600 billion kilowatt hours of electricity in 1970. It has been estimated that the next ten years will bring demands totaling 18,000 billion
kilowatt hours for the United States. This figure represents more electrical
energy than that used by the United States in the preceding 90 years.

The consumption of energy from fossil fuels during the last century
has increased on the average of about 4 percent per year on a worldwide
basis. During this same period the world's population grew at an average
rate of 2 percent per year. Thus, the energy demands are related not only
to the increase in the number of people using energy, but also, apparently,
to an increase in affluence of at least some of these people. If standards of
living remained fairly constant, then one would expect the use of power to
remain reasonably constant. The increased demands for power, above that
of normal increase due to population, can only be attributed to an associated improvement in the standard of living. The obvious implications of
both increased population and increased affluence are that we will eventually demand more energy than we can produce. This situation has already been reached in some areas of our East Coast.

WORLD POPULATION FIGURES

Past Populations

No information exists on sizes or distribution patterns of man's earliest
populations. However, some estimates have been made, and one of the
most acceptable, prepared by Fletcher Wellemeyer in conjunction with
Frank Lorimer, will be summarized here. Additional information on this
subject is given in the Population Reference Bureau's *Population Bulletin,*
18, 1.

Wellemeyer divided his estimate into three broad time spans (Fig.
1-4): Period I extends from 600,000 B.C. to 6000 B.C.; Period II extends
from 6000 B.C. to A.D. 1650; and Period III extends from 1650 to

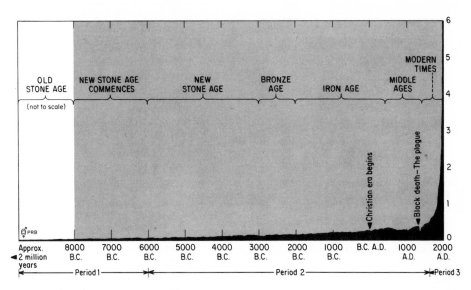

1-4
Growth of human numbers in the past. This illustrates the way in which man's population has grown. If the Old Stone Age were in scale, its base line would extend 35 feet to the left. ("How Many People Have Ever Lived on Earth," Population Bulletin 18, 1, February 1962. Reprinted with the permission of the Population Reference Bureau, Inc.)

1952, the date his work was published. He makes several major assumptions which should be listed so that you may assess the limits of the estimates for yourself. His first assumption was that man first appeared around 600,000 years ago. This choice was a compromise between extremes and probably is quite conservative. At the time Wellemeyer did this work, L. S. B. Leakey in Tanzania was using one million years or more, a more realistic figure. Wellemeyer was aware of this, but felt it inadvisable to "assume so early a beginning for purposes of estimating human population growth."

He also assumed that the human population was five million at the beginning of the New Stone Age (that is, 6000 B.C.) and that the annual birth rate at that time was 50 per thousand. His estimates assumed a smooth increase from that point on, and he felt that the estimates were representative of the net increase, even though the growth was undoubtedly irregular. Figure 1–4 shows these estimates schematically, and Tables 1–2 and 1–3 tabulate the various estimates.

Table 1-2 How many people have ever lived

Period	Number of years in the period	Number of births in the period
I. 1.6 x 10^6 B.C.–600,000 B.C.	1,000,000	20 billion
II. 600,000 B.C.–6000 B.C.	594,000	12 billion
III. 6000 B. C.	7,650	42 billion
IV. 1650 A.D.–	312	23 billion
		97 billion

"How Many People Have Ever Lived on Earth," *Population Bulletin*, 18, 1 (February 1962). Reprinted with the permission of the Population Reference Bureau, Inc.

Table 1-3 Approximate population of the world (in millions)

Year	Population	Year	Population
1000	275	1500	446
1100	306	1600	486
1200	348	1700	– –
1300	384	1800	906
1400	373	1900	1608

"How Many People Have Ever Lived on Earth," *Population Bulletin* 18, 1 (February 1962). Reprinted with the permission of the Population Reference Bureau, Inc.

Using Wellemeyer's conservative figure, today's population of 3.7 billion people represents only about 5 percent of all the people who have ever lived. On the other hand, if we use the less conservative estimate for the "beginning" of man at 1.6 million years ago, the estimated total number of persons who have ever lived becomes about 97 billion. Our present population is only slightly less than 4 percent of that figure. More recent work by Richard Leakey in East Africa would indicate that the figure of 1.6 million years may itself be a conservative estimate and that perhaps 2.5 to 3.0 million years would be a better estimate. This would reduce the percentages even more. However, if you remain alert to the fact that the estimates presented are subject to alteration with the gathering of new information, the figures in the table may prove to be a useful framework within which to work.

Table 1-4 World population—mid-1972

Region	Population estimate (1971) (in millions)	Current rate of annual population growth (percentage)	Years to double population
World	3,782	2.0	35
Africa	364	2.6	27
Northern Africa	92	3.0	23
Western Africa	107	2.5	28
Eastern Africa	103	2.5	28
Middle Africa	38	2.1	33
Southern Africa	24	2.4	29
Asia	2,154	2.3	30
Southwest Asia	82	2.8	25
Middle South Asia	806	2.6	27
Southeast Asia	304	2.8	25
East Asia	962	1.7	41
North America	231	1.1	63
Canada	22	1.7	41
United States	209	1.0	70
Latin America	300	2.8	25
Middle America	72	3.2	22
Caribbean	27	2.2	32
Tropical South America	160	3.0	23
Temperate South America	41	1.7	41
Europe	469	0.7	99
Northern Europe	82	0.5	139
Western Europe	151	0.5	139
Eastern Europe	106	0.7	99
Southern Europe	131	0.9	77
U. S. S. R.	248	0.9	77
Oceania	20	2.0	35

From the 1972 *World Population Data Sheet* with the permission of the Population Reference Bureau, Inc.

Present Population

The 1971 "World Population Data Sheet" prepared by the Population Reference Bureau estimates the mid-1971 world population as 3,706,000,000 people. Table 1-4, condensed from this data sheet, lists the population of major areas of the world, their rate of growth for 1971, and the estimated number of years necessary to double the population of each area,

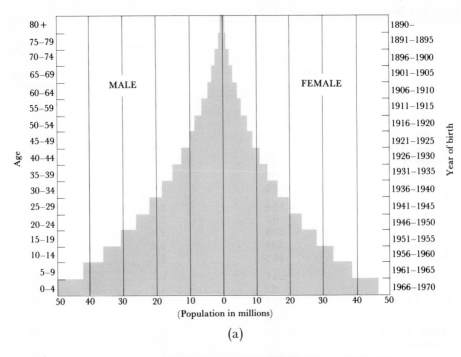

(a)

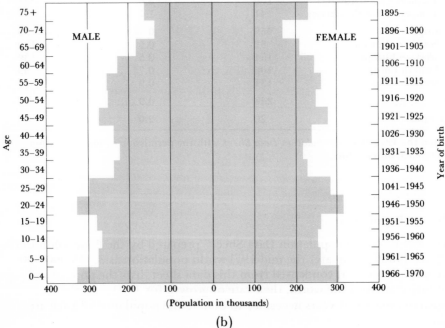

(b)

◀ *1-5*
*Comparison 1970 of age-sex pyramids for (a) India, whose population is doubling in about
27 years and (b) Sweden, whose population is currently doubling in about 140 years.
("India: Ready or Not, Here They Come," Population Bulletin* **26**, *5, November 1970. Reprinted with the permission of the Population Reference Bureau, Inc.)*

based on the current rate of population growth. By looking at the "Current Rate of Growth" column, one can fairly easily rank the major areas
according to their relative stability in terms of population. The least stable
area is Latin America, followed by Asia, Africa, Oceania, the U. S. S. R.,
and North America. Europe is the most stable.

One problem associated with current populations which is not evident from the data in Table 1-4 is the age distribution of the people in
these areas. This is shown graphically in Fig. 1-5 by contrasting the age
pyramids of India, which has a high percentage of young, and Sweden, a
country with a more even age distribution. Specifically, what percent of
the population is young or below the reproductive age? These data are
tabulated in Table 1-5. When they start reproducing, the potential impact of this group on the economy of their respective countries is formi-
dable. They will overwhelm all present facilities (schools, public utilities,
transportation, and so forth) with their numbers and will have a resounding
impact upon the population growth, if they reproduce at rates equal to
those of their predecessors. And they likely will!

Table 1-5 Percent of population below 15 years of age

World	37	Latin America	42
Africa	44	Europe	25
Asia	40	USSR	28
North America	30	Oceania	32

Projections for Future Populations

The Population Reference Bureau, Inc., has projected population figures
to the year 2000, using the United Nations' estimates. They have suggested
a probable population of 7.4 billion by that time, barring nuclear war, famine, or some new and effective form of fertility control. The United Nations' low projection indicates that we may fall short of this figure by
approximately 2 billion, but in the light of present conditions the indications are that the higher figure will be the more realistic. In Table 1-6
projections are listed along with the projected annual rate of increase. The
figures given are "Continued Trends" projections based on 1965 growth
rates. In 1965 the world population was estimated to be growing at a rate

Table 1-6 Population estimates to the year 2000 (in millions)

Region	1980 Population estimate	1980 Rate of increase	1990 Population estimate	1990 Rate of increase	2000 Population estimate	2000 Rate of increase
World	4,486	2.2	5,703	2.4	7,413	2.5
Africa	457	3.0	619	3.4	816	3.7
Northern Africa	117	3.5	161	3.8	228	4.1
Western Africa	156	3.7	220	4.1	318	4.5
Eastern Africa	114	2.6	147	2.9	193	3.1
Middle Africa	40	2.1	50	2.5	64	2.8
Southern Africa	30	3.3	41	3.6	58	3.9
Asia	2,557	3.1	3,317	3.4	4,402	3.6
Southwest Asia	103	3.4	140	3.6	193	3.8
Middle South Asia	945	2.9	1,255	3.3	1,709	3.6
Southeast Asia	370	3.1	503	3.6	697	3.9
East Asia	1,139	2.8	1,419	3.0	1,803	3.1
North America	272	1.8	325	1.9	388	2.0
Canada	31	1.8	37	2.0	50	2.1
United States*	241	1.8	288	1.8	338	1.9
Latin America	387	3.3	537	3.5	756	3.6
Middle America	92	4.2	133	4.5	195	4.6
Caribbean	34	3.1	45	3.4	62	3.6
Tropical S. Am.	214	3.9	302	4.1	431	4.3
Temperate S. Am.	47	2.0	57	2.0	68	2.0
Europe	496	0.8	533	0.8	571	0.7
Northern Europe	87	0.7	92	0.6	98	0.7
Western Europe	152	0.6	161	0.6	172	0.6
Eastern Europe	116	1.0	126	0.9	136	0.8
Southern Europe	141	0.9	154	0.9	165	0.8
U. S. S. R.	295	1.7	345	1.7	402	1.7
Oceania	22	1.9	27	2.1	33	2.2

*For the United States, "continued trends" are based on high post-World War II fertility patterns. Recently, fertility in the United States has been declining slightly. Since a continuation of this trend approximates the median projection, the median projection was used here. "World Population Projections 1965-2000," *Population Bulletin* 21, 4 (October 1965). Reprinted with the permission of the Population Reference Bureau, Inc.

of 2.1 percent per year. This results in a present annual net increase of approximately 70 million persons. A continuation of this trend would result in *adding* 200 million people to the population *yearly* by the year 2000. That is, by the year 2000 we would be adding a population *nearly equal to that of the present population of the United States each and every year!* (See Fig. 1-6.) Indications are that the rate of growth has

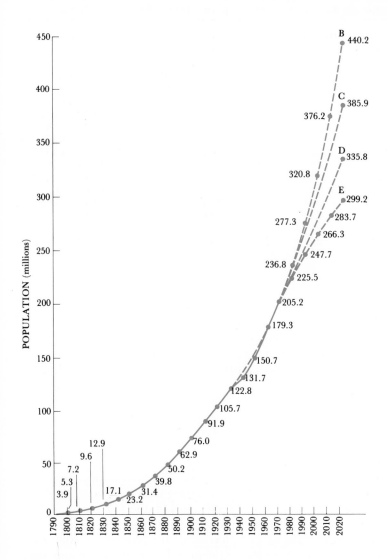

1-6
Population of the United States from 1790 to 2020, based on data from the Bureau of the Census. The first three projections were prepared in 1967; the fourth, in 1970. The assumption underlying all four projections was that the completed fertility would range as follows: B−3.10 children per woman; C−2.78; D−2.34; E−2.11. Series E would eventually result (without immigration) in zero population growth. Series A (3.35 children per woman) was dropped because it no longer seemed a reasonable possibility.

Table 1-7 Population size for successive doublings (initial population of 3.5 billion)

Doubling number	Population (billions)	Doubling number	Population (billions)	Doubling number	Population (billions)
1	7	16	229,376	31	7,516,192,768
2	14	17	458,752	32	15,032,385,536
3	28	18	917,504	33	30,064,771,072
4	56	19	1,835,008	34	60,129,542,144
5	112	20	3,670,016	35	120,259,084,288
6	224	21	7,340,032	36	240,518,168,576
7	448	22	14,680,064	37	481,036,337,152
8	896	23	29,360,128	38	962,072,674,304
9	1,792	24	58,720,256	39	1,924,145,348,608
10	3,584	25	117,440,512	40	3,848,290,697,216
11	7,168	26	234,881,024	41	7,696,581,394,432
12	14,336	27	469,762,048	42	15,393,162,788,864
13	28,672	28	930,524,096	43	30,786,325,577,728
14	57,344	29	1,879,048,192	44	61,592,651,155,456
15	114,688	30	3,758,096,384	45	123,145,302,310,912

declined to approximately 2 percent per year, shifting the estimates downward slightly. However, at this new rate we would still be adding a yearly increment of some 150 million people by the year 2000.

Projections of future populations often have little meaning to most of us when presented as raw figures. At the same time we often feel that enormous populations are in the far-distant future and are really not worth worrying about. To give you a better feeling for both of these areas, we will present some situations which we call "Population Absurdities." These situations are absurd in the sense that they can never occur, for reasons which will become obvious to you; however, they are worth considering simply to enhance our understanding of the population numbers and timetables we may be dealing with in the near future. The basic datum used here is population size for successive doublings. Starting with a population of 3.5 billion, we have simply doubled it several times (Table 1-7).

In order to make our calculations and create our estimates of future situations, we must consider the rate of increase of the population (that is, how many additional people are added to the population in a given amount of time). We have chosen to look at the size of future populations within the framework of two different assumptions: (1) the growth of the population based on a static rate of increase, using a doubling time of 35 years, which is the current estimated rate of increase, and (2) the growth of the population based on a doubling time which becomes less with each successive doubling.

Population Growth—Constant Doubling Time of 35 Years

We will use the currently accepted doubling time of 35 years and project various population growth landmarks on this basis. The following calculations use the data which have been tabulated in Tables 1–7 and 1–8.

The weight of the earth is approximately 6.6×10^{21} tons. If we use the figure that the weight of 15 people equals one ton (average weight of 150 lb each), a little arithmetic will show that approximately 98.9×10^{21} people would have a weight equal to that of the earth. A quick reference to Table 1–8 shows that our population would have to double between 44 and 45 times to reach this number of people. We see also that this would happen under our assumption of a constant 35-year doubling rate by the year 3545, according to Table 1–8. This is 1575 years in the future, but is not as far in the future as the birth of Christ was in the past.

We might carry our calculations further and ask: When, at this rate, would we achieve a population equal to one person per square mile, square yard, or square foot of space? We see from Table 1–9 that we have already exceeded the "one-person-per-square-mile" figure, even if we include the surface of the oceans. Currently, we have slightly less than 19 persons per square mile of the total earth surface. If we use only the land surface, the figure is 65 persons per square mile. We now have only about 0.000020 persons per square yard of land area and only about 0.0000022 persons per square foot of land area. A quick reference to the above tables indicates

Table 1–8 Doubling times for world population (based on constant rate of 35 years doubling time)

Doubling number	Complete by year	Doubling number	Complete by year	Doubling number	Complete by year
1	2005	16	2530	31	3055
2	2040	17	2565	32	3090
3	2075	18	2600	33	3125
4	2110	19	2634	34	3160
5	2145	20	2670	35	3195
6	2180	21	2705	36	3230
7	2215	22	2740	37	3265
8	2250	23	2775	38	3300
9	2285	24	2810	39	3335
10	2320	25	2845	40	3370
11	2355	26	2880	41	3405
12	2390	27	2915	42	3440
13	2425	28	2950	43	3475
14	2460	29	2985	44	3510
15	2495	30	3020	45	3545

Table 1-9 Surface area of the earth (in billions)

	Square miles	Square yards	Square feet
Ocean	0.139	431,691.0	3,886,399.3
Land	0.057	178,071.7	1,602,644.6
Total	0.196	609,892.7	5,489,033.9

that by the year 2330 we will have one person per square yard of land surface, and by the year 2600 we will have one person per square yard of total
earth surface. We also see that by the year 2635 we would have one person
per square foot of land surface, and by the year 2705 we would have one
person per square foot of total earth space. That is, *at our current rate,*
in a little over 700 years we will have such a population size that there
will be only one square foot of space per person. This will include oceans,
deserts, mountains, and the ice fields of Greenland, as well as the more
hospitable places on the earth.

   We have all heard suggestions that with our space capabilities, we
should be able to send our excess population to the other planets in our
solar system to begin their colonization. Table 1-10 shows the total area
in square miles of the other planets in our solar system (to the best of our
present knowledge) and the number of persons necessary to populate each to
produce a density of 60 persons per square mile, which is approximately the
current density of people on our land surface. We are assuming that we have
the capability (which we don't) of taking approximately 275,000 people
(the current rate of increase) into space each day and that the planets are
habitable (which they aren't). If these assumptions were valid, we would
find the entire solar system as overcrowded by the year 2370, or in just 600
years, as the earth is today!

Table 1-10 Surface area of planets and population necessary to equal current
land density on earth

Planet	Total area in square miles	Population necessary for density of 60 per square mile
Mercury	28,880,804	143,184,824
Venus	178,191,567	10,691,494,020
Mars	56,191,567	336,377,138
Jupiter	24,804,153,250	1,488,149,195,000
Saturn	17,759,153,250	1,065,555,555,980
Uranus	2,679,387,752	160,763,265,120
Neptune	2,456,333,612	147,380,016,720
Pluto	43,661,906	2,619,712,360

Population Growth With Ever-Decreasing Doubling Time

If we look at the length of time necessary to have doubled the population in the past, we see that each successive doubling time has been less than the previous. Starting with the time of Christ, it has been estimated that it took approximately 1450 years to double the population. However, the next doubling came in 400 years, the next in 75, the next in 50, and the current rate is every 35 years. For the calculations in this section we have used the following assumptions:

1. The world population in 1970 equals 3.5 billion.
2. Doubling time decreases with each successive doubling.
3. Based on the shortened period of the last two doubling times, the length of each successive doubling time is only 68 percent of the previous period. Therefore, doubling time decreases, on the average, by 32 percent each time.

Using these assumptions, we find a drastically different timetable for our population growth landmarks. With this rate of population increase, we find that we can expect the earth to reach its own weight in people (that is, 6.6×10^{21} tons), shortly after 9:00 A.M. on December 6, 2062, only 92 years in the future! We would expect to reach a density of one person per square yard of *total* earth surface on November 1, 2062, and a density of one person per square foot of total earth surface on November 25 of that year (Happy Thanksgiving!). If we started today, we could expect to have the solar system populated to a density of 60 persons per square mile by late May 2061.

Others have calculated "population absurdities" (although they didn't call them that) to point out the seriousness of this problem (Fig. 1-7). An article entitled "Doomsday: Friday, 13 November A. D. 2026," written by H. von Foerster and others, appeared in *Science* in 1960. They calculated that on that date the population of the world "will approach infinity if it grows as it has grown in the last two millennia."

H. H. Fremlin calculates that, everything else aside, we will be unable to continue at our present rate for more than 800–1000 years. At that time the heat generated by the population and by the support processes necessary for that population (10^{16} to 10^{18} people, or approximately 120 persons per square meter of total earth surface space) would be impossible to dispose of. "Such an increase would need an outer skin temperature of 5000° C. (comparable with the surface of the sun) to radiate away the body heat, which would seem to be well beyond the possible limits." That is, if we were able to overcome all other population-limiting factors (food, space, materials, and so forth), we would still reach this "absolute limiting

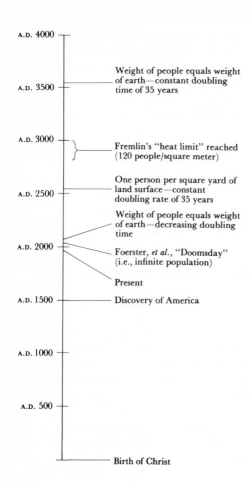

A.D. 4000

A.D. 3500 — Weight of people equals weight
of earth—constant doubling
time of 35 years

A.D. 3000 — Fremlin's "heat limit" reached
(120 people/square meter)

One person per square yard of
land surface—constant
A.D. 2500 — doubling rate of 35 years

Weight of people equals weight
of earth—decreasing doubling
time

A.D. 2000 — Foerster, *et al.*, "Doomsday"
(i.e., infinite population)

Present

A.D. 1500 — Discovery of America

A.D. 1000

A.D. 500

— Birth of Christ

1-7
Relative positions of population milestones on a time scale.

factor" in less than a thousand years and at that time would become incandescent and perish from our own heat.

One additional thought should be considered under the general head-
ing of "future populations," and that is the possibility of reaching the point
at which couples just replace themselves and the question of what will
happen to the population at that point. Most people seem to feel that if
we were to reach that point today, our problems would be solved, for after
that we would have a stable population size. Not so, say the demographers!

Population Reference Bureau Selection 33 states in a footnote on page 2 that:

> *If, for instance, by the year 2000, the developed countries were to reach the point at which couples only replace themselves, and the developing countries were to reach that point by the year 2050—and both these achievements appear unlikely—the world's present population of 3.5 billion would not become stationary before the year 2120, and would then stand at 15 billion.* *

The final stabilization level of a population is far greater than the level of the population at the time the desired rate of increase is reached. That is, there is a "coasting" effect that takes place, and when a figure is set as the level of population to be maintained, it is imperative that a zero rate of increase be reached before the target number is reached. If this is not done, it is impossible, because of the "coasting" effect, to keep from exceeding the target number.

Possible Models of Future Populations

As we mull over the projections that have just been presented, we cannot possibly overlook the fact that our present population growth cannot continue for an indefinite time. If you become concerned with which projection is "right" or which disaster date is "correct," you will have missed the entire point of the presentation, which is that *man's future is finite under present conditions,* and that the time at which he will will reach impossible situations may not be as far in the future as the discovery of America by Columbus is in the past. Since the consensus seems to be that the situations just discussed are impossible and can never be reached, the corollary would be that "something" must happen between now and the time at which the "impossible situation" would become theoretically possible. The "something" is the unknown factor in this equation.

Demographers suggest that there are "essentially three principal ways by which men may come into more stable balance with their environment." These methods are illustrated and discussed briefly in the sections below. For a more detailed look at these models and other aspects of the population problem, refer to *Population Reference Bulletin,* 27, No. 2.

---

*Population Reference Bureau, Inc., "The World Bank and Population," *PRB Selection No. 33* (October 1970), p. 2n. Reprinted by permission.

Population in billions

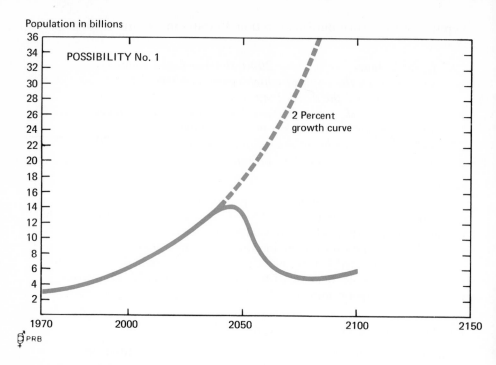

1-8
Schematic of the continuation of a 2 percent growth rate and the possible calamitous popu-
lation crash associated with this growth. ("Man's Population Predicament," Population
Bulletin 27, 2, April 1971. *Reprinted with the permission of the Population Reference*
Bureau, Inc.)

Model No. 1. The "Population Crash Curve." In this model (Fig. 1–8),
the population is allowed to grow unchecked until the carrying capacity
of the earth is exceeded.

> *Many delayed adverse effects upon the environment would cumu-*
> *late and culminate in a precipitous decline through extreme famine,*
> *excessive pollution, social chaos, high death rates from communi-*
> *cable and chronic diseases, wars to hold or seize the ever-scarcer*
> *resources of the earth, very low birth rates and other factors. In*
> *the process of severely overtaxing his environment, man would also*
> *undoubtedly continue the process — a process which is much further*
> *along than many now realize — of denuding the earth of numerous*
> *other species of animals. These animals have contributed to an*

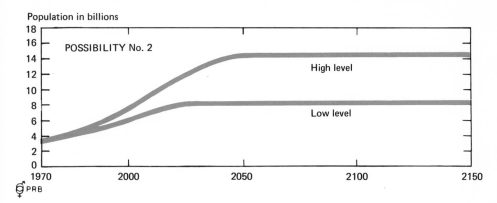

Population in billions

1-9
Schematic of zero population growth being achieved at some time in the twenty-first century. ("Man's Population Predicament," Population Bulletin 27, 2, April 1971. Reprinted with the permission of the Population Reference Bureau, Inc.)

> *ecological balance which has sustained man through hundreds of millennia. Without them, we do not know what kind of an ecological balance would be possible for a very much smaller number of men in future centuries.**

This model is not a new one. We have seen it actually occur in other groups of animals (especially mammals) when their numbers overshoot the capacity of their environment to sustain them. Will man be an exception to the rule?

Model No. 2. The "Gradual Approach to Zero Population Growth." To arrive at this curve the following assumptions were used: (1) we have not exceeded the population level at which the earth can continue to sustain mankind at a reasonable level of health and culture; (2) the growth rate can be decreased, and a population within the limits of what the earth's ecosystem can sustain can be maintained; (3) at the stable population reached, there will be resources and human capability sufficient to achieve a state in which the number of births and deaths are equal.

The lower line in Fig. 1–9 is set at a population level of 7 billion. It is based on the assumption that the worldwide growth rate will be zero (at the replacement level) by 1985. This means that birth rates must be cut in half, or even more, in a matter of 15 years or so. Even if this

---

*"Man's Population Predicament," *Population Bulletin* 27, 2, April 1971, p. 33. Reprinted with the permission of the Population Reference Bureau, Inc.

Population in billions

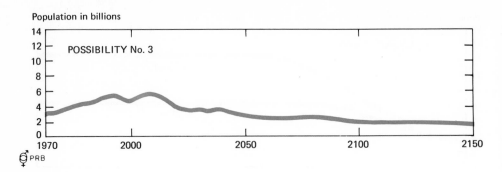

1-10
*Schematic of the "Modified Irish Curve," caused by a series of disasters which resulted in
temporary population declines and a consensus to keep the population level down. ("Man's
Population Predicament,"* Population Bulletin **27, 2, April 1971.** *Reprinted with the
permission of the Population Reference Bureau, Inc.)*

occurs, we see that the great number of those approaching reproductive
age (remember that about 37 percent of the world's population is under
15 years of age, and in many underdeveloped countries the number approaches 50 percent) will bring about the "coasting" phenomenon mentioned earlier. The momentum of this factor is so great that even under
the most favorable conditions, a leveling off could not occur in less than
about 70 years and at less than a population level of about 7 billion.

A less optimistic, but perhaps more realistic, view is represented by
the upper line of the graph. This line represents a steady state being
reached at a population level of approximately 14 billion and assumes
one or more doubling times after the 7 billion population figure.

Model No. 3. The "Modified Irish Curve." This model (Fig. 1–10) takes
its name from the experience of Ireland in the disastrous potato famine
in the 1840s. In a period of ten years during and after the famine, the
population was reduced 24 percent by death and emigration. There followed a seemingly unstated consensus that the Irish would not allow
overpopulation to occur again; that they would modify their family
structure toward fewer and smaller families; and that they would diver-
sify their production to avoid dependency on a single crop, as they had
depended on the potato prior to the famine.

CONCLUDING REMARKS

Regardless of how the picture is viewed, the geologic implication is that
we will need a great deal more in the way of natural resources in the near

future. The professional geologist will be called upon to find more mineral wealth, to rework material that was previously unprofitable, and to keep reevaluating our position with regard to our remaining mineral resources.

The social implication is that we, as a world people, must alter our reproductive habits. We are reproducing ourselves out of existence and are fouling our nest while we do so. The best solution, of course, is that each citizen of the world should recognize the problem and voluntarily curb the reproduction rate. This seems a remote possibility, for this solution assumes that we will overcome, on a worldwide basis, religious concepts ("contraception is a sin"; "go forth and multiply"), illiteracy (the most effective way to promulgate an idea is by the written word), and hunger (the immediate problem of where the next meal will come from is far more pressing to a majority of the world's population than how many people will be here 30 years from now). This seems an overwhelming task.

There are two alternatives to a voluntary reduction of the growth rate at the personal level: (1) government intervention to control reproductive habits, and (2) intervention of nature to reduce population to a manageable size. Both of these alternatives portend dire consequences. The first involves further loss of personal freedom and choice and relegates a very powerful weapon to the government, while the latter presents us with a picture of wholesale death, suffering, and agony. The choice is ours to make. Either we will control our population, or it will be done for us!

A word at this point to those who feel that zero population growth is the ultimate goal. The best evidence at present seems to indicate that this is only part of the goal. We should look, perhaps, to a population growth *based on a smaller stable population than we have now.* Many studies have been done to attempt to determine what population level our presently known reserves and technology are capable of sustaining. The standard of living in the United States is usually used as a target by the rest of the world. It has been suggested that to support 5 billion people at the 1970 United States' standard of living would require a worldwide consumption of resources nearly ten times as great as the 1970 world level of consumption actually was. If, as suggested by the line in Fig. 1–9 representing the most optimistic view, we actually level off at a population of 7 billion, it seems likely that most of these people will be accommodated only under the most difficult and unacceptable conditions. If a population level of only 5 billion would require a tenfold increase in resource consumption, what would a population level of 14 billion bring about?

REFERENCES

Beaton, J. R. and Alexander R. Doberenz, eds. *Proceedings of the Symposium, Population Growth: Crisis and Challenge*, January 9 and 10, 1970, at the College of Human Biology, University of Wisconsin-Green Bay.

Foerster, Heinz von, Patricia M. Mora, and Lawrance W. Amiot. "Doomsday: Friday, 13 November, A.D. 2026," *Science* 132 (1960): 1291-1295.

Fremlin, H. H. "How many people can the world support?" in *Population, Evolution, and Birth Control — A Collage of Controversial Ideas*, ed. Garrett Hardin, San Francisco: Freeman, 1964.

Hubbert, M. K. "Energy Resources," in *Resources and Man*, National Academy of Sciences, National Research Council, San Francisco: Freeman, 1969, pp. 157-241.

Lovering, T. S. "Mineral resources from the land," in *Resources and Man*, pp. 109-134.

Park, Charles F., Jr. *Affluence in Jeopardy: Minerals and the Political Economy*, San Francisco: Freeman, Cooper, 1968.

Population Reference Bureau. "U.S.A. population in 2050 one billion?" *Population Bulletin* XVI, 5 (1960): 89-106.

————. "How many people have ever lived on earth?" *Population Bulletin* XVIII, 1 (1962): 1-19.

————. "U.S. population growth — the dilemma of the fractional child," *Population Bulletin* XX, 1 (1964): 1-27.

————. "World population projections 1965-2000," *Population Bulletin* XXI, 4 (1965): 73-99.

————. "The food-population dilemma," *Population Bulletin* XXIV, 4 (1968): 81-99.

————. "The thin slice of life," *Population Bulletin* XXIV, 5 (1968): 101-123.

————. "Spaceship Earth in peril," *Population Bulletin* XXV, 1 (1969): 1-23.

————. "India: ready or not, here they come," *Population Bulletin* XXVI, 5 (1970): 1-32.

————. *The World Bank and Population*, Population Reference Bureau Selection No. 33, 1970.

————. "Man's population predicament," *Population Bulletin* XXVII, 2 (1971): 1-39.

————. "World population data sheet," 1972.

U.S. Bureau of the Census. *Population Estimates — Projections of the Populations of the United States, by Age and Sex: 1964 to 1985*, Bureau of the Census Series P–24, no. 286, 1964.

————. *Projections of the Population of the United States, by Age and Sex (Interim Rev.): 1970 to 2020*, Bureau of the Census Series P–25, no. 448, 1970.

Man and all other forms of life depend on water, which is distributed over the earth through the hydrologic cycle. (Photograph by George Sheng)

WATER 2 HYDROLOGY

INTRODUCTION

Water is an extremely important substance. Not only do all forms of
life require it, but so do a great number of physical, chemical, and geological processes. As a geological agent, water is unmatched in its abundance and its unique chemical properties. It is a powerful solvent; its
solid state (ice) is less dense than its liquid state; and it has a remarkably
high heat capacity. It is important as a dynamic agent of weathering,
erosion, and the transportation and deposition of earth materials. In
addition, water's action on the surface of the land, on the rocks and
soil, and in the air and sea has significant and continuing effects on the
activities of man. He uses it to irrigate his crops, to generate electrical
power, to carry away his waste products, to cool his manufacturing processes and power plants, and to provide him with sport and recreation.

2-1
Water wagon used by the U.S. Army Cavalry to haul water to Fort Robinson, Nebraska.
(Photograph courtesy USDA–Soil Conservation Service)

In this and the next chapter we will investigate the significance of water in geology and to humans. This chapter will be concerned largely with laying a foundation for understanding the natural occurrence of water, that is, hydrology. In Chapter 3 we will look at the present use of water and our future needs of it. We will also explore some of the sources of water, such as desalination, wastewater reclamation, water importation, and plant control, all of which can be used to supplement our present water supply.

DISTRIBUTION OF WATER

The waters of the earth, other than that locked in mineral matter or trapped in deep rock layers, form the *hydrosphere*. The hydrosphere consists of oceans, atmospheric water, ground water, surface lakes and streams, and ice. In Table 2-1, you can see that the ocean contains by far the greatest percentage of water and that what little water is not in the ocean is unevenly distributed among icecaps, lakes, streams, and the atmosphere.

It can be seen from this table that very little of the earth's water is of actual use to man at any one time. Sea water is not generally available to us for our direct use, which reminds us of the problem of the Ancient Mariner. Only about 3 percent of the hydrosphere consists of fresh water. Therefore, the sources for ready use are surface and ground water, ice being quantitatively important but presently economically unobtainable. Of the surface water, saline waters are of virtually no use. Fresh-water lakes and streams furnish nearly all of our surface water supplies. Most of the water obtained from the ground is from wells that are less than about one-half mile deep, much of the water in deeper zones being either of lower quality or too expensive to obtain by drilling. Soil moisture above the water table is of benefit to man, but is not withdrawn in bulk for use. (The *water table* is the level below which the ground is saturated with water.) The total water present in lakes, streams, and shallow ground water reservoirs is about 1,030,000 cubic miles. Figure 2-2 shows the sources for the major water uses in the United States.

Table 2-1 Distribution of water in the earth

	Cubic miles	Percentage of total
Surface waters		
Lakes	30,000	0.009
Saline lakes and inland seas	25,000	0.008
Streams	300	0.0001
Subsurface waters		
Soil moisture and vadose water	16,000	0.005
Ground water		
Shallow	1,000,000	0.310
Deep (½ mile)	1,000,000	0.310
Total water in land areas	8,630,000	0.635
Ice	7,000,000	2.150
Atmosphere	3,100	0.001
The oceans	317,000,000	97.200
	326,000,000	100.000

(Source: U. S. Geological Survey)

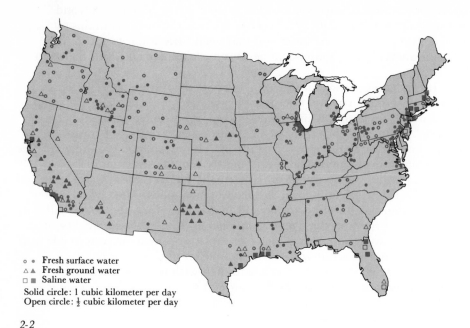

o • Fresh surface water
△ ▲ Fresh ground water
□ ■ Saline water
Solid circle: 1 cubic kilometer per day
Open circle: ½ cubic kilometer per day

2-2
Sources and approximate rate of use of water supply in the United States for public supply,
irrigation, fuel-electric power generation, and industrial uses. (Source: The National
Atlas of the United States of America, *1970.)*

THE HYDROLOGIC CYCLE

All water of the hydrosphere participates in the *hydrologic cycle,* a complex circulation of water between the ocean, atmosphere, and land (Fig.
2–3). From the vast reservoir of the ocean, water evaporates into the air.
It is estimated that about 80 percent of the water vapor in the air finds its
way there in this manner. The evaporation of water from the ocean depends
to a great extent on the temperature of the air and of the surface water of the
ocean, on how much vapor already is in the air, and on the movement of
the air in the form of winds. Thus, evaporation rates are highest in the belts
of trade winds north and south of the Equator, somewhat less at the Equator,
and minimal in high latitudes. G. L. Pickard presents evaporation values of
up to 146 cm per year in the trade wind belts, 110 cm per year at the Equator,
and 8 cm per year in high latitudes. S. N. Davis and J. M. DeWiest give a
value of 50 cm per year in the high latitudes.

    After entering the atmosphere, water is distributed by air masses to
all parts of the globe. This distribution is far from uniform and depends
on many factors, such as size and distribution of land and water bodies,

2-3
The hydrologic cycle.

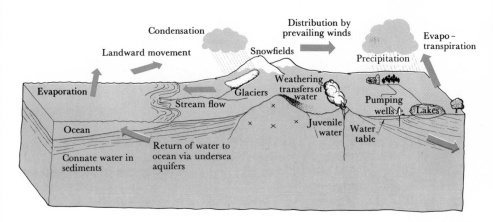

topography, elevation, and latitude. When weather conditions are right, the water is distributed to the land or returned directly to the ocean as precipitation.

Water that is present in, or that falls from, the atmosphere as precipitation is called *meteoric water.* Water that reaches the land as rain, snow, sleet, or hail may evaporate, run off, or soak into the ground. About 70 percent of the precipitation is lost back to the atmosphere as evaporation from streams and lakes, evaporation from water left in puddles on the ground after a rain, or sublimation from snow and ice. Loss of water also includes the transpiration of water to the atmosphere from plants. These kinds of water loss are collectively called *evapotranspiration.* In one very real sense, however, transpiration is not a loss, because it is a necessary function of plant life.

About 20 percent of the precipitation flows over the surface of the land, first as sheet wash or in small riverlets, then in larger streams. Such surface flow forms part of the *runoff.* Runoff, however, also includes the flow of water through the ground, as certainly it must, because surface and ground water form an integrated system. Some of the surface flow may evaporate, some may seep into the ground from influent streams, and some may be temporarily impounded in lakes or reservoirs. The ultimate destination of runoff, however, is the ocean. Nearly all of the streams of the world reach this destination, except for those in certain areas of wholly interior drainage, e.g., the Great Basin of the western United States. Yet, ever since man became a major force in the environment the streams have been increasingly robbed of their water for domestic, industrial, and agricultural purposes. Accordingly, much of the runoff is temporarily sidetracked from its route to the sea.

2-4
Stream-gauging station, Marion County, Oregon. (Photograph by K. N. Phillips, U.S. Geological Survey)

If meteoric water does not evaporate or run overland, it may seep directly into the ground. Such *infiltration*, which includes seepage from streams and lakes, contributes to the *ground water* supply. Ground water is simply water that occupies the open spaces in the soil and rock. However, most geologists restrict the term to mean only the water in the saturated zone below the water table.

SURFACE WATERS

The surface waters of the earth consist of oceans, lakes, and streams. The oceans are vast surface features from which we can at present recover only a very minor amount of fresh water. Some of the lakes and inland seas, such as the Great Salt Lake, the Caspian Sea, and the Dead Sea, are saline or brackish and therefore do not serve as sources of fresh water. Among surface bodies, this leaves only fresh-water lakes and streams to provide man with water.

Streams

Runoff begins when water reaches the ground as precipitation and then moves overland or into the ground. As water runs off the higher areas of the land toward valleys, most of it eventually flows into definite channels

2-5
Measuring flow of water with current meter and width of stream with tape to obtain discharge. (Photograph by E. F. Patterson, U.S. Geological Survey)

to form streams. The channels normally form a system of streams that includes a large stream and its tributaries collecting water from a region. The region drained by such a system is a *drainage basin.*

In terms of stream flow, there are three general types of streams: ephemeral, intermittent, and perennial. Ephemeral streams carry only surface runoff and flow only during and shortly after precipitation periods. They are likely to be small, with ill-defined courses through low areas, and include many of the uppermost tributaries in a drainage basin. Intermittent streams flow during wet seasons, when the ground water table is high enough to intersect the stream channel and sustain the stream flow between rains. During the dry season, however, the ground water table falls, and the stream goes dry because it receives neither ground water nor surface runoff. Perennial streams flow continuously because the precipitation is great enough and the ground water table stays high enough to provide a continuous supply of water. The main streams and most of the main tributaries of a drainage system are perennial streams.

Stream-water supplies come mostly from perennial streams. The damming of intermittent streams and the collection of water in reservoirs provides water in arid regions. The extent to which a stream can supply water for man's needs is dependent mostly on its *discharge,* the volume of water per unit of time flowing past a point on the stream. Discharge is usually expressed in cubic feet of water per second. Since flow, measured at any given time, may be quite variable, the discharge becomes meaningful only if there is a continuous record of discharge. The characteristics of the flow of a stream can be seen in a graph called a hydrograph, on which discharge is plotted against time. From the study of hydrographs, it can be easily seen that streams vary greatly in their flow characteristics. Total yearly discharges, the distribution of time and seasons during which flow occurs, the range in discharge from day to day, and the occurrence of peak or minimum discharges all have a wide range. Not only do hydrographs vary for different rivers, they also vary for different places along the same river.

Analysis of the runoff in a drainage basin cannot be made solely by use of the hydrograph. Climatic factors may influence the runoff in ways not shown in a hydrograph. Of these climatic factors, precipitation is the most important. Runoff does not take place rapidly if precipitation occurs as snow and the snow lies on the ground a long time. If rain falls heavily and for a long time, the runoff may be increased significantly. The distribution of precipitation over the basin, whether uniformly distributed or widely scattered, may cause individual hydrograph readings to vary considerably. Runoff may be increased if a rain has been shortly preceded by other rains and if the soil is saturated with moisture. Other climatic factors, including temperature, wind, and humidity, affect evaporation and transpiration and consequently the runoff.

The characteristics of the drainage basin also influence runoff. If the soil in the basin is open and permeable, precipitation can soak in more readily than if the soil is impermeable, as clay soil might be. The topography also influences runoff, since water flows more easily down steeper slopes. Flat topography, on the other hand, retards drainage. Runoff is also governed by the number and the length of streams in the basin; it is enhanced if there are a sufficient number of streams that form a well-integrated drainage system.

With so many variables influencing stream flow, it is evident that runoff is highly variable across time and place. Therefore, the analysis of runoff is quite complex. However, knowledge of runoff is important, since it aids in our understanding of and planning for adequate water supplies, flood control, and soil conservation.

Figure 2–6 shows the major drainage areas of the United States. Outstanding, of course, is the huge Mississippi River basin, with its many subbasins such as the Missouri, Upper Mississippi, Ohio, and Arkansas river basins. This great basin system drains half of the United States and contains the heart of the nation's agricultural lands. Much of the industrial complex lies in the Great Lakes-St. Lawrence basin and in other small basins in the northeast. The Columbia and Colorado river basins drain the

2-6
Major drainage areas of the United States. (Source: "Water Resources Development Map of the United States," U. S. Geological Survey, 1969.)

water-poor areas of the western United States. The Great Basin has entirely interior drainage, with evaporation taking a heavy toll of what little precipitation falls there.

The average annual runoff in the United States is shown in Fig. 2–7: One can easily see that the eastern half of the country has greater runoff than the western half, with areas of heavy runoff in the western half limited to mountainous areas, especially along the North Pacific Coast. The eastern half of the United States is also an area of water surplus, that is, the average precipitation exceeds the potential evapotranspiration (evaporation transpiration by plants). In contrast, the western half is an area of water deficiency, with the potential evapotranspiration exceeding the average precipitation. One might conclude from this that the eastern United States is adequately supplied with water; but still, one must remember that the bulk of the population lies in the eastern half of the country. Thus, although runoff is greater there, so are demands for withdrawable water. In California, in contrast, one-tenth of the entire population of the United States lives in an area of water deficiency.

Stream valleys, especially those of major rivers, are the most heavily populated of any areas. This has always been true, even in the earliest days

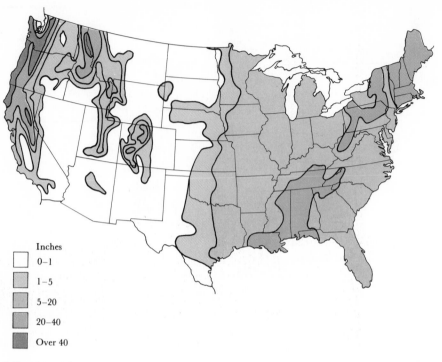

Inches
- 0–1
- 1–5
- 5–20
- 20–40
- Over 40

2-7
Average annual runoff.

of civilization, when important cultures developed along the Tigris-Euphrates and Nile Rivers. Not only is water obtainable from the rivers; it is also readily available from wells in the alluvial (river-deposited) gravel, sand, and silt of the flood plain. The alluvium itself is fertile and useful to agriculture.

In many places, the demand for water as a useful resource for industrial, agricultural, and domestic purposes has run afoul of the other demands made on it as a navigable, aesthetic, waste-carrying power source. Where population pressures are high, all the many uses for stream and lake waters come into conflict. Industrial waste burned on the surface of Lake Erie near Cleveland in 1970; the recreational and navigational potentials of that part of the lake were nil at the time. Agricultural withdrawals of water from the Rio Grande have resulted in complex apportionment of water among Colorado, New Mexico, Texas, and the Republic of Mexico. Much of the water is withdrawn for use along the course of the river. In addition, much is lost

2-8
The rich flood plain of the Missouri River valley in Iowa. (Photograph by R. D. Miller, U.S. Geological Survey)

through evaporation. The Rio Grande is scarcely larger where it enters the sea than it is near its headwaters in Colorado. Its tributaries are also heavily drawn upon for water. The possibilities for multiple usage of the river are negated by nearly total withdrawal and evapotranspiration of the water.

The average runoff in the United States is 1200 billion gallons per day, or about 8½ inches of water per year (out of a United States average of 30 inches of precipitation). As we have already seen, however, the discharge of a stream is a highly variable characteristic. During times of flood or of other high discharge, we simply cannot intercept the complete flow and retain it for later use in time of drought. Nor can all the flow be withdrawn from every stream in the country, the Rio Grande notwithstanding, because the other purposes must be served. J. C. Maxwell estimates that about 800 billion out of 1200 billion gallons per day of the runoff can actually be withdrawn leaving, by necessity, the rest in the streams for those other purposes.

2-9
Salt cedar, a common phreatophyte, Gila River valley near Phoenix, Arizona. (Photograph by E. F. Patterson, U.S. Geological Survey)

Water Supply from Lakes and Reservoirs

If anything, lakes are more geologically and biologically fragile than are streams. The water in streams moves faster, so that waste does not accumulate in one place. Any withdrawal of the water supply from lakes, ponds, and reservoirs must take into account the other purposes for which the waters are used. Removal of water from reservoirs is more rapidly offset by inflow than is the replenishment of lake waters, which have lesser inflow. Hence, withdrawal of lake water should be carefully regulated.

Evapotranspiration takes a heavy toll of water stored in lakes and reservoirs. In areas of high aridity, such as along the Colorado River, vast amounts of water evaporate from the river and especially from reservoirs such as Lake Mead. It is estimated that in Texas three times as much water is lost by evaporation from reservoirs as is used for municipal and industrial purposes. To prevent this unwanted loss, attempts have been made to reduce evaporation by spreading the surface of lakes with a "skin" of certain evaporation-retarding chemicals such as hexadecanol. The practicality of doing this is still unknown, although we do know that disturbance of the water surface by storms and boats tends to break up the protective film. Just what the effect of the chemicals is on water quality and biological community needs further investigation, although so far this effect does not seem to be harmful.

The building of large dams and reservoirs has been a subject of controversy. On the one hand, there are obvious benefits, such as water supply, power, and recreation derived from reservoirs. On the other hand, there are such detrimental effects as the short life of a reservoir that fills with sediment and the changes, which extend even to coastal areas and deltas, in the regimen of the stream below the reservoir. One of the most notable examples of change is the effect of the high dam at Aswan, Egypt. The previous geological and biological order has been complexly altered—the seasonal floods, the ground water, the channel, the flood plain and delta erosion, the salinity of the eastern Mediterranean, the fishing and wildlife habitat, even disease, and probably a host of other facets of the Nile River regimen have all been rearranged by this alteration of the environment by man. Not the least of the changes is an alteration in the availability and distribution of the water supply to that thirsty land.

Dams are obstacles to navigation and to the migration of spawning fish. It is also entirely possible that there may be microclimatic changes that occur when large reservoirs are built. Although adequate studies are lacking, some people think that there has been an increased humidity and a greater growth of vegetation in the areas around the Lake of the Ozarks, Table Rock Lake, and other reservoirs in Missouri. We shall examine some of the geological effects of constructing reservoirs in a later chapter.

Obviously, the building of reservoirs cannot go on forever. Along any stream there are only a limited number of sites where geological conditions are suitable for building dams safely and for impounding water without major leakage. Nearly all major streams in the United States have been dammed, and few sites remain for the construction of more dams. Bitter controversies have occurred, and some proposed dams, such as the one at the lower end of the Grand Canyon, have not been built, because of serious questions about their benefits and their deleterious effects, not the least of which is the destruction of precious wildlife habitat and scenery.

GROUND WATER

One of the prime sources of water for nearly all of man's uses is within the ground. Wells have been dug for many years; records indicate that elaborate subterranean water systems and wells existed in Middle Eastern cultures thousands of years ago. In modern times man has bored wells by the hundreds of thousands and has exploited ground water stores as never before, often even ruthlessly. It is now necessary that every aspect of ground water geology be studied very carefully, so that we may receive the maximum benefit from this splendid resource. As is the case with all resources, ground water has its limitations—there is a finite quantity of it, it is sometimes hard to find, it may not be located in places that best suit man's endeavors, and it may be in short supply. Like so many other things in the natural environment, it is fragile and liable to suffer damage when assailed by man's technology. On the other hand, ground water has its advantages. It is usually of reasonably high purity and cool in temperature, useful qualities in most respects, particularly in industrial processes where it is used as a coolant. Ground water is found widely; fairly pure and abundant ground water lies under most areas of the continents, including the continental shelves. In a few limited places the ground water is hot and can be utilized for energy.

Occurrence of Ground Water

As we have already noted, the movement of water into the ground is part of the hydrologic cycle. About 10 percent of the water that falls as precipitation ultimately enters the ground. The body of ground water we are considering here is essentially meteoric water, composed of water that enters the ground after precipitation or from influent lakes or streams. Ground water includes neither water of hydration which is essentially locked into mineral matter nor *connate* water of normally greater depths whose origin is sea water that becomes trapped in marine sediments that later form sedimentary rocks. Perhaps the distinctions are somewhat arbitrary, as connate water in near-surface beds may easily become part of the ground water supply. *Juvenile* water, originating from magma and volcanic eruptions, may also be added to the ground water supply from time to time.

Ground water lies in either loose soil nearest the surface, in more or less broken-up rock beneath the soil, or the consolidated bedrock beneath this. As water moves downward due to the pull of gravity, it sinks as far as the permeability of the rock will allow and fills the reservoir of soil and rock to a point which depends on how much water there is (Fig. 2-10). Two zones are formed: the lower zone, the *zone of saturation,* in which the openings in the reservoir material are filled; and the upper zone, *the zone of aeration,* characterized by water, that is either wetting or permeating the rock particles, and air in the openings. The zone of aeration includes a layer of soil moisture at the top, a relatively dry intermediate layer, and a layer called the capillary fringe just above the zone of saturation. This capillary fringe is an area in which some upward movement of

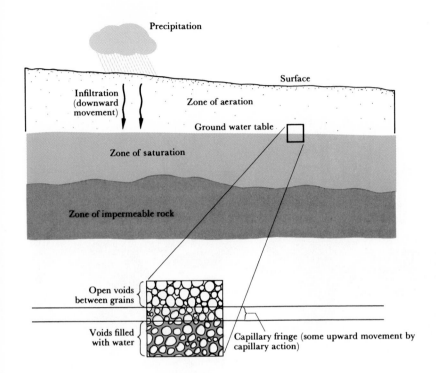

2-10
*Relationship of the zone of aeration, ground water table, and zone of
saturation.*

water occurs at, and slightly above, the ground water table because of
capillary movement of water from the saturated zone. The boundary between the zones is called the *ground water table*. To the geologist studying
ground water, the behavior of the water in the zone of saturation is of the
greatest importance.

Recharge and Infiltration

Recharge is the addition of water to the ground water supply. It may come
from rain or snow or from surface water infiltration. *Infiltration* is the
movement of water through the soil and rock toward the zone of saturation.
Whether water runs off, evaporates, or infiltrates depends on a number of
factors: the nature of the precipitation, the vegetation, and the condition of
the ground. Precipitation may be of varying degrees of intensity or of
greater or lesser duration, and its distribution may be variable or even.

2-11
Excavations made for artificial recharging of ground water at Rohrers Island, Montgomery County, Ohio. (Photograph by S. E. Norris, U.S. Geological Survey)

Heavy vegetation may intercept a large amount of precipitation on its leaves, and plant roots may remove some of the water that enters the ground withdrawing it mostly from the zone of soil moisture. A ground surface without plant cover receives the full brunt of falling rain, and if the rainfall is intense may have its openings plugged with small grains washing into them. Vegetative cover, on the other hand, helps prevent this plugging and allows easier infiltration. The composition of the ground also affects infiltration; it may be either loosely or firmly compacted unconsolidated soil or solid rock. Rock is less porous or permeable than soil and tends to impede infiltration. The slope of the ground also affects movement of water into the ground; steep slopes allow greater runoff, whereas, flat, poorly drained surfaces retain water longer, so that more water eventually can soak in.

Recharge is also seasonal. Seasonal variations in recharge depend on climatic factors such as distribution of precipitation between arid and wet seasons, temperature, and growing seasons. Evapotranspiration may deplete the soil moisture and rob supplies of ground water. However, infiltration increases during rainy seasons and brings in more water than is lost through evapotranspiration. The soil moisture and ground water are then recharged.

Sometimes we can add to the ground water supplies by means of artificial recharge. Such is the case when man adds water to the ground by spreading it over the surface or injects it through wells. In the first case, precipitation which might otherwise run off as surface flow is impounded and then allowed to infiltrate the ground. This can be done directly from the reservoir in which the water is impounded or by spreading the water over land elsewhere by distributing it from the reservoir. A drawback is that evaporation may cause a great amount of the water to be lost, especially in arid regions, before it can infiltrate the ground.

When water is added to the ground water through wells, there is less chance for loss through evaporation. Recharge through wells can be accomplished by using gravity flow or by pumping, sometimes through wells reaching just into the zone of aeration, down to the depth of the water table, or to an artesian layer. Even though there is less evaporation, other problems arise from recharging through wells. Air and sediment present in the water used for recharging may plug the reservoir if an airlock is created or if the pore spaces become filled with sediment. Some expense is then incurred to remove the sediment and prevent aeration.

Porosity

One of the influences on the behavior of ground water is porosity. Porosity is the volume of voids divided by the total volume of material. We will consider voids to include spaces among solid particles which may be occupied by air or other fluids. The total volume, V_t, equals the volume of voids, V_v, plus the volume of solid material V_s. The porosity of a material, expressed as a percentage, is

$$P = \frac{V_v}{V_t} \times 100.$$

Voids in a rock may consist of many different kinds of openings (Fig. 2-12). Since a considerable amount of ground water occurs in soil, loose sediment, or sedimentary rock such as sandstone, *intergranular* voids are common.

High porosity—
rounded
grains, uni-
form size
(good sorting),
e.g., sandstone

Low porosity—
rounded
grains, many
sizes
(poor sorting),
e.g., sandstone,
conglomerate

Medium porosity—
angular grains,
uniform size
(good sorting),
e.g., sandstone

Very low porosity—
angular grains,
many sizes
(poor sorting),
e.g., arkose,
breccia

Solution
porosity—
mild solution
along crystal
boundaries,
e.g., limestone

Porosity
along open
joints or
bedding planes
(many rocks)

Vesicular
porosity—
(may not be
interconnected),
e.g., basalt

2-12
Types of porosity.

Rounded grains of sediment of uniform size provide the highest porosities in these materials—theoretically up to 47.6 percent in a cubic packing of spheres of uniform size. Increasing angularity of grains results in a decrease of porosity because the bumps on some grains more or less fit into indentations on others. Also, there is lower porosity if the grains are not uniform in size and are poorly sorted. In this case, small grains occupy space among larger ones and rob the material of its possible voids. Porosity is lower in well-cemented rocks, where cement such as silica, calcite, clay, or iron oxide occupies space between grains .

Porosity also exists where approximately planar surfaces are developed in the rock. Such surfaces may include fractures, bedding planes, and frost or dessication cracks. Fractures occur in any kind of rock, but their distribution, density, and degree of openness may vary quite a bit from one area to another or even between adjacent rock units.

Solution porosity develops where mineral matter in rock has dissolved and has left open spaces. The most common occurrence of solution porosity is in limestone, where voids may range from pinpoint

size to immense caverns. Porosity in crystalline rocks, especially igneous
and metamorphic rocks, is low. In these rocks, the crystals interlock and
occupy the available space. Cavities formed by solution porosity in limestone or dolomite and fractures in crystalline-sedimentary, igneous, and
metamorphic rocks are likely to be the most important types of porosity
in these rocks. In any kind of rock there may be some voids such as ves-
icles and lava tubes in extrusive rocks, worm borings in shale, and other
sundry holes. Deep in the earth, however, pressures are so great that
open spaces of any kind are obliterated.

Ground Water Flow

Ground water is almost always in motion. It percolates through the ground
in response to differences in energy or heads from one place to another due
to differences in pressure and elevation. The flow of ground water is toward streams, lakes, or oceans. The path followed by the water may be
quite variable, with some water flowing more or less directly along the
water table and some of it flowing deep into the ground to return upward
later. In any case, the water moves down a pressure gradient.

   Darcy's law states that the velocity of ground water equals the hydraulic gradient times the coefficient of permeability. The hydraulic gradient is the loss of energy due to friction per unit of distance traveled. The
coefficient of permeability is an expression of the degree to which an earth
material can transmit the flow of a fluid such as water, oil, or gas. This coefficient is not the same in a given rock for all fluids, since it is dependent
on such things as the viscosity and the density of the fluid. Although here
we are considering only water and not petroleum, there are nevertheless
variations in temperature, pressure, and salinity that cause some differences
even within the characteristics of water.

   The coefficient of permeability, besides being dependent on the fluid,
is also dependent on the medium. Earth materials vary enormously in their
permeability. Permeability depends on pore size, the way the grains are
packed together geometrically, and the interconnection of voids. Permeability is roughly related to porosity. If there are no voids, the rock is not
permeable; but even where voids do exist, they must be connected and
must be large enough to prevent molecular forces from extending all the
way across them, thereby stopping the flow of the fluid through friction.
Loose, or open, packing of grains gives rock or soil a higher porosity.

   The way in which voids are joined also influences permeability. Some
paths through rock may be quite direct; others may be tortuous, causing
considerable loss of energy as the fluid moves along. Coarse, fragmental
rocks such as sandstone and conglomerate consistently have the highest per-

meability because they have a considerable number of voids that are well connected. Limestone, if fractured, may be quite permeable. Rocks that have high permeability are called *aquifers.*

Shale and claystone are rocks consisting mostly of clay. The term *clay* refers to both a group of minerals and particles of very small size. The minerals are, for the most part, hydrous aluminum silicates. The size of these particles is less than .005 mm. Clay minerals generally have an open structure, but the openings are very tiny. Because of the smallness of the openings, permeabilities are very low. Movement of water in clay is capillary; there is almost no movement in well-packed shale. Hence, shales are impervious to water flow unless they are fractured. Such impervious rocks are called *aquicludes.*

Rocks that possess very high permeability due to open fractures which allow movement of water may not behave at all according to Darcy's law. In these rocks there may even be turbulent flow through open channels. Almost any kind of rock could have fractures allowing this kind of movement. In most cases, open channels in rock are found near the surface of the ground, as fractures tend to be closed below a depth of 200–300 feet. Two rocks notable for having openings that allow great flow of water are basalt and limestone. Neither of these has significant intergranular porosity. Basaltic lava flows are nearly always highly fractured. These fractures, called joints, form when the lava sheet cools and shrinks; they provide good passages through which water can flow. In the Thousand Springs area of Idaho, water discharges from the lava flows of the Snake River plain. In limestone, solution resulting from water flowing through the carbonate rock enlarges joints and may produce vast networks of open channels. Continued flow and solution produces *karst* areas with sinkholes and caverns. Missouri, Kentucky, Indiana, and Florida, for example, are noted for having numerous limestone aquifers and springs, some of which discharge enormous amounts of water.

The absolute velocity of ground water is very slow indeed compared with that of water in surface streams. Water may barely move, or it may fairly race along at a few hundred feet per day. Most rates of flow lie in the range of a few feet per year to a few feet per day.

Artesian Water

The term *artesian,* according to one definition (AGI, 1960), refers to ground water that is under a hydrostatic head great enough to force the water to an elevation above the aquifer containing the water. The surface formed by connecting all such elevations to which water will rise is called the *piezometric surface.* Aquifers that contain artesian water are confined by overlying aquicludes. Figure 2–15 shows an aquifer.

2-13
Artesian well, flowing about 2500 gallons per hour, near Roswell, New Mexico. (Photograph by E. F. Patterson, U.S. Geological Survey)

2-14
Large, flowing well near Roswell, New Mexico. (Photograph by E. F. Patterson, U.S. Geological Survey)

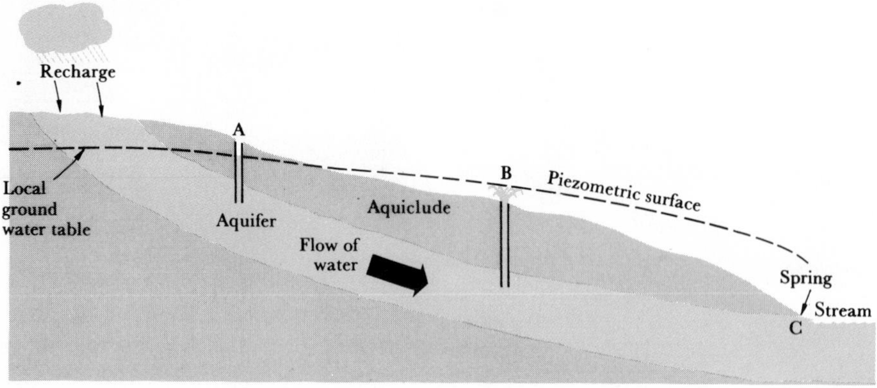

2-15
Artesian conditions.

A local ground water table is present in the aquifer (others not shown in Fig. 2–15 may be present in other beds). The confining bed (aquiclude) may be saturated below the water table but is essentially impermeable, so that movement of ground water is limited to the downward flow in the aquifer. The height of this water table depends on recharge and infiltration in the recharge area. The water table in the aquifer would not be present in the vicinity of points A or B in Fig. 2–14, but the piezometric surface toward which water would rise in wells at points A and B is present. At well A, the water will rise to the piezometric surface, which is below the ground surface in the figure. At well B, the water will flow out of the top of the well because the ground surface is below the piezometric surface. Both wells are considered to be artesian. The piezometric surface slopes downward, away from the ground water table, in the aquifer because of the loss of energy due to friction along the aquifer. At places where the confined aquifer intersects the surface, such as at point C in Fig. 2–15, stream flow may be sustained by artesian water or by artesian springs. Artesian springs may also occur where water reaches the surface through joints.

Withdrawal of Ground Water

Ground water is usually removed by man through wells. In some instances man utilizes water supplies from natural springs. Most wells are "ordinary," that is, the wells penetrate the zone of saturation in unconfined rocks or soil, and the water rises in the well to the ground water table. In artesian

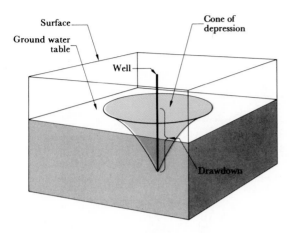

2-16
Removal of water around well by pumping.

wells, on the other hand, the water rises to the piezometric surface, and if this surface is below the ground surface, the well must be pumped. All ordinary wells must be pumped.

The depth of water wells ranges from shallow, to a few feet, to tens of feet, or perhaps even to hundreds of feet. A relatively small number of wells measure thousands of feet deep, generally for reasons such as the high cost of retrieving water from great depths, lack of porosity and permeability of the rocks, and the higher concentrations of dissolved minerals in the waters. The shallower wells are sometimes dug by hand, but most wells are now drilled or bored by either rotary drills or cable (impact) rigs.

Pumping water out of the ground water reservoir tends to lower the water table or the piezometric surface. Such lowering is called *drawdown* (Fig. 2-16). In many areas, e.g., Las Vegas, Nevada; west Texas; and Arizona, to cite only a few, excessive pumping has resulted in drawdown of many tens or hundreds of feet over a wide area, and recharge has been insufficient to compensate for it. The ground water has been depleted so much that it would take years to reestablish its former levels, even if pumping ceased.

A *cone of depression* (Fig. 2-16) is produced in the vicinity of a pumping well. The shape of the cone depends on pumping rates, continuity of pumping, reservoir characteristics, coefficient of permeability, and other factors.

All removal of water from the ground water reservoir causes changes in the environment. This is particularly true with regard to the subsurface

2-17
Drilling for water, Baton Rouge, Louisiana. (Photograph by E. F. Patterson, U.S. Geological Survey)

environment of the ground water, but may also be true of the surface environment. In the subsurface, withdrawal of ground water immediately causes changes in the pressure system, the direction and rates of flow of water, and the position of the water table. The surface may be affected to the point of disturbing vegetation that sends roots to the water table. Also, in extreme cases of heavy withdrawal, the surface may subside. Mexico City is built on lake beds that were formed rather recently, geologically. These beds had contained much water, and withdrawal of water through untold thousands of wells caused subsidence of the ground by a few meters. Some buildings have sunk so low that they must now be entered via the second floor.

A further effect of subsidence is the decrease of porosity as the medium becomes more compact. All changes in the environment related to the withdrawal of ground water can be blamed on two basic things: the nature of the geologic and hydrologic conditions (porosity, permeability, pressures, and so forth) and the nature of the withdrawal of the water (quantity, rate, and so forth).

If water is withdrawn from a ground water reservoir, some water remains behind, wetting the mineral grains of the reservoir. The quantity of water that can be drained out is called the *specific yield*; the quantity of water that remains in the reservoir is called the *specific retention*. For example, a reservoir might have a volume of 1000 billion cubic feet containing 100 billion cubic feet of water. Of this, 20 billion cubic feet can be pumped out, giving a specific yield of 2 percent; the specific retention is 80 billion cubic feet or 8 percent. The specific yield is determined by knowing the rock characteristics, including the porosity, the amount of water pumped, and the volume of rock from which the water is pumped.

Saltwater Intrusion

Encroachment of sea water into wells on islands or along coastlines may also present difficulties in obtaining sufficient ground water supplies in these areas. A boundary (interface) exists between fresh and salt water where an aquifer meets the ocean. For some purposes this boundary can be thought of as sharply delineated, but as there is some mixing across it, the boundary, more properly, is a zone of mixing. As long as there is a head of ground water above sea level and inward from the shore, the boundary will not be horizontal, contrary to what one might expect from the density difference between fresh and salt water. Instead, the boundary slopes inland, with salt water positioned in a wedge under the fresh water (Fig. 2–18). The slope of the boundary is such that at any given point, the height of the fresh water column below sea level (H_2 in Fig. 2–18) is about 40 times the height of the column above sea level (H_1). This can be seen by examining the densities of fresh and salt water:

$$\text{density of fresh water} \times (H_1 + H_2) = \text{density of salt water} \times H_2$$

or

$$1.000 \text{ g/cc } (H_1 + H_2) = 1.025 \text{ g/cc } (H_2).$$

Obviously, if the densities vary, so will the 40-to-1 ratio. This would occur, for example, where brackish lagoonal or estuarine waters form the interface with fresh water. And even so, there are additional factors that may render the complete description of any given boundary quite complex.

Large populations along seacoasts have caused heavy pumping of ground water in these areas. For example, Long Island, New York, has experienced salt water intrusion, as wells there have been heavily pumped. Because lowering the water table brings about a corresponding rise of the salt water, to the extent shown above, many wells have experienced intrusion of salt water. On small islands the ground water may exist as a lens of fresh water floating on salt water; therefore, recovery of ground water may require especially delicate tactics (Fig. 2–19).

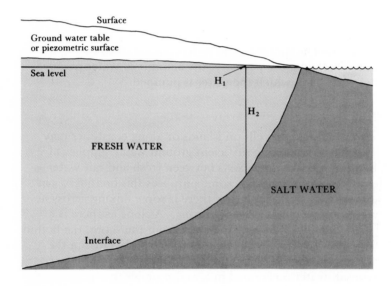

2-18
Salt water-fresh water interface along coastline.

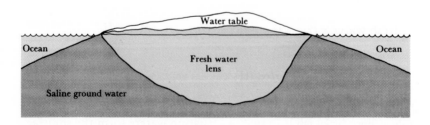

2-19
Lens of fresh ground water on island.

REFERENCES

American Geological Institute. *Glossary of Geology and Related Sciences,* 2d ed., Washington: AGI, 1960.

Bue, C. D. *Principal Lakes of the United States,* U. S. Geological Survey Circular 476, 1963.

Cloud, Preston. "Mineral Resources from the Sea," in *Resources and Man,* National Academy of Sciences, National Research Council, San Francisco: Freeman, 1969.

Davis, S. N. and J. M. DeWiest. *Hydrogeology,* New York: Wiley, 1966.

Flawn, P. T. *Environmental Geology: Conservation, Land-use Planning, and Resource Management,* New York: Harper & Row, 1970.

Havens, J. S. *Recharge Studies on the High Plains in Northern Lea County, New Mexico,* U. S. Geological Survey, Water-Supply Paper 1819-F, 1966.

Horne, R. A. and A. D. Little. *Marine Chemistry—The Structure of Water and the Chemicals of the Hydrosphere,* New York: Wiley, 1960.

Kazmann, R. G. *Modern Hydrology,* New York: Harper & Row, 1965.

MacKichan, K. A. and J. C. Kammerer. *Estimated Use of Water in the United States, 1960,* U. S. Geological Survey Circular 456, 1960.

Mason, Brian. *Principles of Geochemistry,* 3rd ed., New York: Wiley, 1966.

Maxwell, J. C. "Will there be enough water?" *American Scientist* 53, 9 (1965): 97–103.

Pickard, G. L. *Descriptive Physical Oceanography,* Long Island City, New York: Pergamon Press, 1963.

Piper, A. M. *Has the United States Enough Water?,* U. S. Geological Survey, Water-Supply Paper 1797, 1965.

Skinner, B. J. *Earth Resources,* Englewood Cliffs, N. J.: Prentice-Hall, 1969.

Strahler, A. N. *The Earth Sciences,* 2d ed., New York: Harper & Row, 1971.

Todd, D. K. *Ground Water Hydrology,* New York: Wiley, 1959, p. 336.

U. S. Congress, Senate, Select Committee on National Water Resources. *A Preliminary Report on the Supply of and Demand for Water in the United States as Estimated for 1980 and 2000,* by Nathaniel Wollman. 86th Cong., 2d sess., 1960, Comm. Print 32, p. 131.

U. S. Department of the Interior, U. S. Geological Survey, *The National Atlas of the United States of America,* Washington: Government Printing Office, 1970.

Wenk, Edward, Jr. "The physical resources of the ocean," *Scientific American* 221, 3 (Sept. 1969): 166–176.

Wisler, C. O. and E. F. Brater. *Hydrology,* New York: Wiley, 1959, p. 408.

Man will have to devise new ways of obtaining water if he is to have an adequate supply of it in the future. (Photograph by Evelyn Wilde)

WATER 3 PRESENT AND FUTURE

INTRODUCTION

In the previous chapter we explored the geological significance of water. In the present chapter we will explore its human significance. We will briefly look at the current use of water, some projections for future needs, and some methods by which we may supplement or redistribute our water supply in the future. Two factors are largely responsible for a shortage of water at any given place: the water may be present but not of usable quality, or it may be in the wrong place. Both of these problems will be considered.

One problem that looms increasingly larger is that of obtaining an adequate water supply for the future. In the past there seemed to be little or no problem in finding water; it made little difference where or how man obtained water or what he did to it. The abundance and purity of water attracted settlers to the New World. However, people began to run into problems when they pushed westward into the arid and semiarid parts of this country. At that time the country began to fill with people at what, we must now admit, was an extremely high rate. Water usage increased also, not at a corresponding rate, but even faster because of both accelerated industrial development and agricultural practices of irrigation.

Domestic usage is becoming inflated: Americans must be the cleanest people on earth, and their cars must be the cleanest automobiles on earth. Their grass is the best watered, and their buildings are the best air-conditioned. It would be interesting indeed to count the private and motel-owned swimming pools. The movement of rural populations into cities has created large urban concentrations of people; supplying these cities with water of adequate quantity and purity has become a complicated and difficult problem of geology, engineering, and economics. New York and Los Angeles are probably the two largest phreatophytes in the world. Los Angeles has sent its roots out in the form of aqueducts reaching hundreds of miles in many directions.

WATER USE

It is difficult to pinpoint when water problems of the modern age began, but we might start with the year 1949. In that year the city of New York had a severe shortage of water, and city officials went so far as to request people to cut down on water usage. This was shocking, for to the average American, wasting water is regarded virtually as a right and a privilege. As a result, at least in part, of the New York shortage, President Truman appointed a commission mandated to develop a national water policy. Actually, official bodies charged with considering water resources have existed, at one time or another, since President Theodore Roosevelt's time.

The problems continue to grow. In recent years about 50,000 wells that pump ground water from the Ogallala Formation in eastern New Mexico and West Texas have been drilled. Over 5,000,000 acres are irrigated with this water. So much water has been removed from the ground that ground water tables have dropped alarmingly—in some places over 120 feet (Fig. 3-1). The rapid removal of vast quantities of water can never be made up by recharge if present practices continue, because the recharge over the area averages less than an inch per year. Here, we have a situation in which water is an essentially nonrenewable resource; as such it is literally *mined* from the ground. The problems still continue. The water supply is adequate for many purposes and for many regions, but in some cases usage rates so closely approach the limits of the available supply, severe shortages would ensue if drought upset this delicate balance.

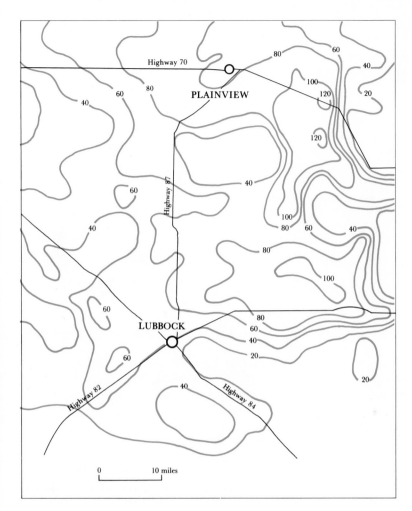

3-1
Lowering of the water table from excessive pumping for irrigation pur-
poses in an area around Lubbock and Plainview, Texas. Contours show
number of feet of lowering. (Source: Adapted from "Hydrologic In-
vestigations Atlas No. HA 330," U.S. Geological Survey, 1969.)

A great many scientists have examined the supply of water for the
future and have been nearly unanimous in saying that we must begin now
(this was also said years ago) to do something about it. John C. Maxwell
has suggested that approximately 60 percent of the runoff might be
available for use. He has calculated that a total water usage of 15,000

3-2
Irrigating with large, self-propelled sprinklers on the high plains of Nebraska. (Photograph courtesy USDA—Soil Conservation Service)

gallons per day per person, taken from a maximum supply of 750 billion gallons per day, will suffice to supply a population of 230 million without much change in our way of life. Since the estimated population of the United States in 1971 was 207 million, Maxwell's limit could be reached by the year 1980.

Table 3-1 Some average water requirements necessary to produce certain products

Water requirements (in gallons)	Product
Home use	
70 (per day)	Per capita home use in U. S.
3–5	Flush toilet
30–40	Bath
5	Per minute of shower
10	Wash dishes
30	Automatic washing machine
300	Water lawn one hour
Food production	
250,000	1 ton of sugar
250,000	1 ton of corn
375,000	1 ton of wheat
115	Enough wheat for one loaf of bread
125,000	1 ton of potatoes
1,000,000	1 ton of rice
2,500,000	1 ton of cotton fiber
16,000	1 gallon of milk
200,000	1 ton of alfalfa
7,500,000	1 ton of beef
300	1 barrel (31.5 gal) beer
Manufacturing	
250–500	1 ton of bricks
82,000	1 ton of kraft paper
150,000	1 ton of nitrate fertilizer
184,000	1 ton of fine book paper
468	Refine 42 gal barrel of crude oil
10	Refine 1 gal gasoline
1,115,000	Refine 100 barrels of synthetic fuel from coal
660,000	1 ton of synthetic rubber
240,000	1 ton of acetate
350,000	1 ton of aluminum
30,000 (cooling water)	1 ton of pig iron
32,000 (average)	1 ton finished steel

(Sources: *Nuclear Energy for Desalting*, AEC, pp.3–4; *Water and Industry*, USGS pamphlet, pp. 3–4; *World Book Encyclopedia*, v. "W," p. 94; *Scientific American*, Sept. 1963, p. 93.)

Water use in the United States is presently growing at a rate of approximately 25,000 gallons per minute. Use of water in the United States can be divided into three broad categories: industrial, irrigation, and domestic and rural. (Table 3-1 shows the amount of water needed for various uses.) Of these three, industry uses by far the greatest amount, accounting for

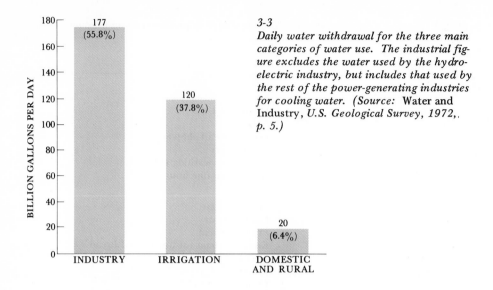

3-3
Daily water withdrawal for the three main categories of water use. The industrial figure excludes the water used by the hydroelectric industry, but includes that used by the rest of the power-generating industries for cooling water. (Source: Water and Industry, *U.S. Geological Survey, 1972,. p. 5.)*

nearly 56 percent of the use. Figure 3-3 shows the relative amounts consumed by these three categories of users. The figure shown for industrial use excludes the water which passes through the turbines of our hydroelectric system. This water will be excluded from our definition of "use" because it is not actually withdrawn, but represents, more precisely, an "in stream" use. In addition, the generation of electric power other than hydroelectric power accounts for the use—mostly for cooling— of some 129 billion gallons of water per day, which is over 70 percent of the total industrial use. As a matter of fact, among the four common categories of industrial use of water (cooling water, process water, boiler feedwater, and sanitary and service water), cooling water represents about 90 percent of all water withdrawn by industry. Actually, industry uses a great deal more water annually than it withdraws, because much industrial water is recirculated and reused. If industry used its water only once, its withdrawal rate would be significantly increased.

A major distinction needs to be made between the use of water for industry and its use for irrigation, aside from the difference in withdrawal. The water "used" in industry is returned, ultimately, to the natural water system. Nearly as much water is returned as was originally taken, with only about 5 percent to 6 percent being "consumed" or lost during processing. Contrast this figure with that of "consumption" through evapotranspiration during irrigation, where as much as 85 percent of the water is

3-4
Patterns formed by large circular irrigators, Holt County, Nebraska. (Photograph courtesy USDA—Soil Conservation Service)

lost. Figure 3-5 shows the geographic distribution of water use for industry and irrigation in the United States. The change in the withdrawal rates for the three main categories between 1950 and 1965 is shown in Fig. 3-6.

Water is used in numerous other ways, e.g., transportation and navigation, recreation, as a habitat for living creatures, and as a carrier and dilutant for sewage and industrial effluent. All these, of course, are in addition to the work water does as a geologic agent of weathering, erosion, transportation, and deposition.

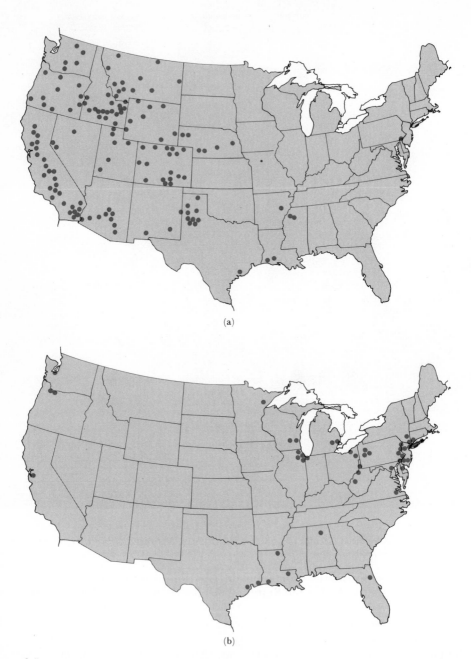

(a)

(b)

3-5
Major uses of water: (a) irrigation; (b) industrial. Each dot represents ½ to 1 cubic kilometer of water per day. (Source: The National Atlas of the United States of America, 1970.)

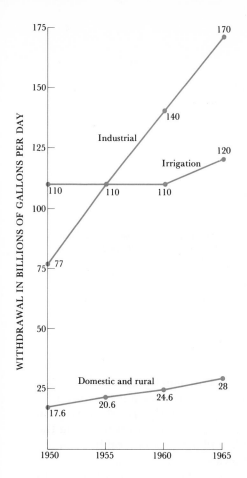

3-6
Change in water withdrawals for the three main categories of water use for the years 1950–1965. (Source: Estimated Use of Water in the United States, 1965, *U. S. Geological Survey Circular 556, 1968, p. 50.)*

FUTURE USE

One of the most comprehensive studies of future water needs was that conducted in 1960 by the United States Senate Select Committee on National Water Resources. One of the publications based on that study was written by Wollman, who projected the supply and demand for water to the years 1980 and 2000. Figure 3–7 shows a comparison of Wollman's projections and the amount of water withdrawn in 1960, according to figures provided by A. M. Piper of the United States Geological Survey. Already, we can see a problem if Maxwell's figure is reasonable and if the projections by Wollman are accurate. If 750 billion gallons per day is the maximum amount we may reasonably expect to withdraw from our runoff, then obviously the demand of 888 billion gallons per day projected for the year 2000 is excessive.

As we have already noted, there is a finite amount of water in the hydrosphere. There is also a fairly constant amount of water at any one point in the hydrologic cycle at any given time. Our supply is finite and

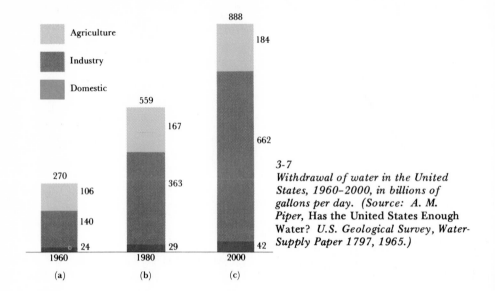

3-7
*Withdrawal of water in the United
States, 1960–2000, in billions of
gallons per day. (Source: A. M.
Piper,* Has the United States Enough
Water? *U.S. Geological Survey, Water-
Supply Paper 1797, 1965.)*

measurable: within certain limits we know how much water we have, and
we know where it is, i.e., in the ground, in surface lakes and streams, in
polar icecaps, and in the oceans. The icecaps presently provide us with no
water, nor will they in the immediate future. Recovery of water from the
oceans is becoming economically feasible and offers some real hope of
joining ground water and surface water as a supplier of water.

It is very difficult to predict the future. The forecasters of the future
of water supplies range from those who refuse to admit that problems exist
to those who attempt to maintain a scientific rationality in their fore-
casting to those prophets of doom whose journalism foretells disaster. As
things come to pass, however, continual re-evaluations must be made of
projections of future populations and future supply and demand of water
and mineral supplies. We are certain of a number of factors that we can use
as a basis for predicting future needs. First, it is inevitable that the population will increase. Second, we definitely will need more water to support
this population. Third, it will not be as simple to obtain adequate supplies
of pure water as it has been in the past. But it is obvious that the questions:
How many people will there be? *How much* water will we need? *How
will we obtain* it? are very difficult to answer.

It is certainly not easy for a person to understand the problems of
water supply if he lives in an area of moderate growth and has never experienced a shortage of water. Perhaps his well has never gone dry, or perhaps he has never had to ration water for washing his car. When a severe
water shortage occurs and the wells go dry, drinking water is rationed, and

3-8
Irrigation of alfalfa field, using ground water. The sophisticated well installation can deliver 1000 gallons per minute into the ditch. (Photograph courtesy USDA–Soil Conservation Service)

domestic supplies must be hauled in tankcars from points miles away, the problem suddenly becomes very clear to him.

The seriousness of water shortages becomes apparent when industries must shut down and farming must be abandoned because of an insufficient water supply or an accumulation of salt in the ground due to continuous irrigation.

It is apparent that obtaining enough water is problematic. Many of these problems are geological. We must know how to inventory our supplies; we must learn more about the occurrence of water, especially of ground water, about the behavior of water and its movement through the hydrologic cycle, about the effects on the environment as we withdraw water from surface and from ground reservoirs in ever-increasing amounts, and about the effects of pollution on water quality. One hundred years ago, in the dusty streets of Tucson, Arizona Territory, vendors sold cups of water from goatskin bags. Where will we buy our water one hundred years from now?

DESALINATION

In general, water containing as much as 1000 parts per million (ppm) of dissolved solids is undesirable for most uses, although water containing 2000 or 3000 ppm can be used for some purposes. Water with as much as 4000 ppm is used only in a very few places for irrigation. Drinking water should contain less than 1000 ppm. Industry usually requires less than 400 or 500 ppm.

It is not entirely necessary that all water used in industrial or even domestic situations have the high degree of purity as drinking water. The purity of most municipal supplies now is actually much higher than necessary for many purposes. Purity requirements are set fairly high, and therefore brackish or somewhat brackish waters could be substituted for water of higher quality in selected cases, e.g., where corrosion is not a factor. This substitution could result in considerable cost-cutting.

The concentration of solids in the ocean averages about 35,000 parts per million, indicating the enormity of the task of desalting this water to a level of usable purity. Sea water yields about 200 pounds of salt for every 1000 gallons processed. A serious problem with using desalinated water today is the cost of transporting the fresh water to where it is needed. Less of a problem exists if the water is to be used near the coast or if inland brines are being desalinated. However, movement of the water to inland areas involves the added costs of pumping uphill, laying pipelines, and building distributive facilities. The amount of agricultural activity located near enough to the coast to have desalinated water available to it is small compared to the total. This, coupled with the fact that agriculture cannot generally afford the price of desalinated water, virtually rules out the sea as a source of agricultural water in the near future.

In the United States, fresh-water supplies for industrial and municipal use cost around 30 to 35 cents per 1000 gallons, delivered to the user. Agricultural supplies are cheaper (they have to be) and cost about 5 cents per 1000 gallons. The cost, not including delivery to the plant, of desalinated water now averages 75 cents to one dollar per 1000 gallons. Thus, at least for most places in the United States, the price of fresh water converted from sea water is still not competitive with the cost of fresh water from the land. Desalinated water, then, will have its major use in industrial concerns in areas near seacoasts. If fresh water supplies reach limits critical enough to warrant sufficient purification of sea water, desalinated water will also be used in municipal supplies—again, in areas near the coast. Thus, the Los

Table 3-2 The five pilot plants authorized by the Office of
Saline Waters in 1958

Location	Process
San Diego, California	Multiple-flash distillation
Freeport, Texas	Long-tube vertical distillation
Roswell, New Mexico	Forced-circulation vapor-compression
Wrightsville Beach, North Carolina	Freezing
Webster, South Dakota	Electrodialysis

Angeles, San Diego, and San Francisco areas on the west coast; the Houston-Beaumont and New Orleans areas on the Gulf Coast; and the cities of the industrial, heavily populated east coast from Boston to Washington are prime possibilities to become future users of water from the ocean.

In 1952 the federal government established the Office of Saline Waters (OSW) to investigate methods of large-scale desalination. In 1958 five pilot plants were authorized (Table 3–2) to test different desalination processes. This program has generated much valuable data to date and has provided physical facilities for experimentation, which has led, among other things, to operations that can provide water almost as cheaply as can some municipal water treatment plants.

Of the methods currently in use or being tested, the processes involving distillation or use of a membrane are considered to have the most potential. Distillation processes include the vertical tube, multiple flash, and solar distillation processes. The membrane process is usually accomplished by either electrodialysis or reverse osmosis.

Distillation Processes

Multistage flash distillation Former Secretary of the Interior Stewart Udall called the multistage flash process "the workhorse of the desalting business." In this process (Fig. 3–9) seawater is routed through several chambers via a series of coils, and steam has been formed in each chamber. The interaction of this steam with the coils carrying the relatively cooler seawater is twofold: first, the steam condenses as pure water on the relatively cooler coils and is collected for later use; second, the seawater in the coils is heated somewhat. As the seawater passes through successive coils, it becomes increasingly hotter. Therefore, before it actually enters the distillation process, it needs to have only a small amount of energy added to bring it up to the desired temperature. It is then injected into a low-pressure chamber, where it boils immediately, or "flashes," to steam.

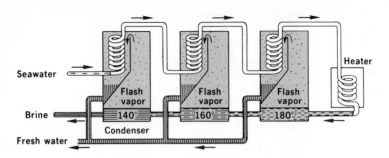

3-9
Multistage flash distillation. (Source: Office of Saline Waters)

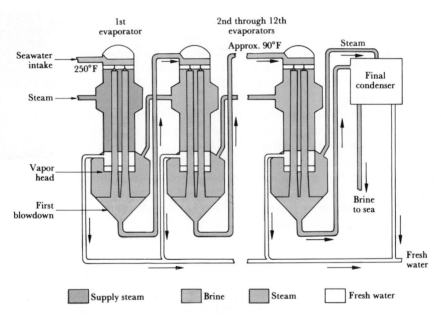

3-10
Long-tube vertical distillation. (Source: Office of Saline Waters)

The salt water is passed through a series of chambers which have progressively higher vacuums and lower temperatures.

When the Cuban government shut off the water supply to the American Marine base at Guantanamo Bay, Cuba, in 1964, a unit using this process in San Diego was disassembled and shipped to the base. The construction of a new desalination plant would have taken three years, but the San Diego plant was disassembled, shipped to Cuba, and put into operation in about five months. The plant was producing about 1.4 mil-

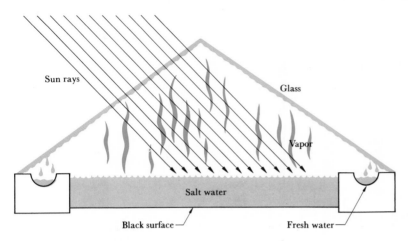

3-11
Solar distillation. (Source: Office of Saline Waters)

lion gallons of water per day at a cost of approximately one dollar per thousand gallons. That plant is still serving the base at Guantanamo.

Long-tube vertical distillation   The principle involved in long-tube vertical (LTV) distillation is essentially the same as that in multistage flash (MSF) distillation (Fig. 3–10). The main difference is that in the LTV process, the brine, pulled by gravity, flows down the inside of long tubes. Steam, supplying heat to the brine inside, flows around the outside of these tubes and condenses the brine to fresh water as it comes into contact with the relatively cooler tubes.

Solar distillation   Two thousand years ago Julius Caesar used solar distillation of salt water to provide fresh water for his troops. For as long as men have been going to sea, sailors have used this same method to supplement their supply of fresh water. Even today, this method is used by island inhabitants to increase their fresh-water supply. In this process salt water is put into a shallow, broad basin which is covered by a transparent dome. Sunlight penetrates the transparent dome and warms the water, causing evaporation (Fig. 3–11). The water vapor formed in this way condenses on the dome and runs down to be caught in a trough. Although inexpensive, in terms of energy needed to run the process, the process is not generally practical for a large-scale operation, since it depends on sunshine, a definite drawback in many areas. In addition, the yield is low. Under ideal conditions something less than 0.2 gallons per square foot of basin per day can be produced. In comparison, Chicago's Central Water Filtration Plant can produce nearly 1.75 billion gallons of water per day. This plant occupies about 61 acres of land area. To produce the same amount of

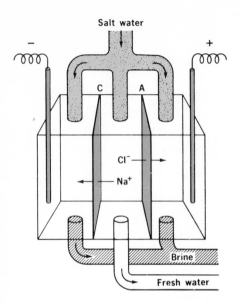

3-12
Schematic of desalination through electro-dialysis. (Source: Office of Saline Waters)

water using solar distillation would require well over 300 square miles of basin area.

Membrane Processes

Electrodialysis Of all the desalination methods in commercial use today, electrodialysis is the only one that removes the salt from the water, as opposed to removing the water and leaving the salt behind. This process (Fig. 3–12) depends on the fact that salts in solution are in the form of ions—positively charged cations and negatively charged anions. In this process the salt water is introduced into a chamber whose sides are ion-permeable membranes. In chambers located on either side of this entry chamber are positive and negative terminals to which the charged ions of dissolved material migrate. The membrane is impermeable to water molecules, and therefore the water in the entry chamber becomes increasingly fresh as the salt migrates to the charged terminals. Several of these terminals may be set up in series to insure high-quality water.

This kind of operation works best in brackish water, that is, water containing more salt than one part per thousand and less than 35 parts per thousand. The process is not efficient on normal marine water and is best used where the water supply is only slightly salty.

Reverse osmosis In the normal osmotic process fluids flow through a semi-permeable membrane in response to a pressure gradient. When salt water

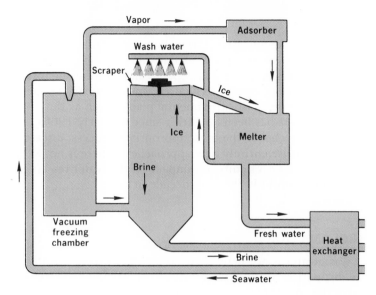

3-13
Schematic showing desalination using freezing process. (Source: Office of Saline Waters)

and fresh water are separated by an appropriate membrane, osmotic pressure will cause the fresh water to flow through the membrane to dilute the salt water. If the salt water is put under pressure (at least 350 pounds per square inch), this natural process is reversed, and the water from the salt side is forced through the membrane to the fresh-water side, leaving the salts behind. This method is very promising. New membranes, which can withstand high pressures and can maintain their efficiency, are being developed, and this promises to reduce the cost of desalination significantly. If the new developments in this process meet expectations, this could become the most desirable process, since it has a very low energy requirement.

Freezing This process (Fig. 3-13) falls under neither of the previous categories, but should, perhaps, be included for the sake of completion. In this process seawater is pumped into a freezing chamber, where a slush of ice crystals and brine is formed. This slush is pumped to a separating chamber where the ice crystals, because of their lower density, float to the top and are removed. A disadvantage of this process is that a film of brine which is difficult to remove adheres to the fresh-water crystals. One major advantage of this process is that it takes less energy to freeze seawater than to evaporate it.

Summary

All processes of desalination have advantages as well as disadvantages, and no one process would be able to serve equally well in all situations. The choice of a process for a conversion plant will depend, among other factors, on geographical features, source of energy to be used, and chemistry of the water to be processed. With a series of experimental plants, such as the ones the Office of Saline Waters has been operating since the early 1960s, we will be able to intelligently choose the most appropriate process, discard unworkable processes, and devise entirely new approaches. There is no doubt that desalination will make a significant impact on the water resources of many areas of the world in the next decade or two.

WASTEWATER RECLAMATION

The reclamation of wastewater is not new. Few communities in the United States can claim the distinction of using water wholly untouched by another public utility facility. The amount of water that has previously passed through an upstream wastewater system depends on its geographic location and on whether the stream is flowing above or below average. For every million gallons of water pumped out of a stream and into a city via its municipal water system, nearly the same amount is returned to the stream as sewage. The sewage may have been processed through a sewage treatment plant, but it is not in the same condition, generally, as when it was taken from the river. Mixing and diluting sewage with the river water and using natural filtration usually brings the water to an acceptable level of purity before it is withdrawn downstream by the next community. Therefore, each town downriver is using drinking water that is purified sewage dumped by all the towns upstream that have used the river for sewage disposal.

This situation will become even more common in the future. Some of the more enlightened cities situated in areas where the water supply is barely sufficient to meet present demands and where a drought could cause massive problems are already considering "closed circuit" water systems. In this scheme the sewage is processed and the reclaimed water is put back into the city's own water system. Thus, the same water can be continually recycled by a community, and only small amounts of "new" water have to be introduced to offset any losses due to consumptive use or minor leaks in the system. Among the cities currently using, building, or planning wastewater reclamation and reuse facilities are San Diego, Pomona, and Lake Tahoe in California; Long Island, New York; and Dallas, Texas.

Some cities have been forced to reuse their wastewater because of uncontrollable circumstances. This happened in Chanute, Kansas, in a five-month period from October 1956 to February 1957 (the tail end of the 1952–1957 drought which hit Kansas and which has been described as perhaps the "worst drought in recorded hydrologic history"). The Neosho River ceased to flow in September 1956, leaving Chanute, a town of 12,000 people, without a water supply. After considering several proposals which would provide an emergency supply of water, the town started a recycling process in early October. The recycled water met the State bacteriological standards for drinking water, but there were problems with froth, color, and odors. Nevertheless, the town did not run out of water and was able to sustain itself until the drought broke and the river started flowing again.

The reclamation and reuse of industrial and domestic wastewater may offer one of the best alternatives to seeking new supplies of water. If reuse becomes routine, even in areas where increasing population density has pushed water supplies to the limit, it may be necessary to bring in only small amounts of "new" water to match the population and industrial growth.

WATER IMPORTATION

A major problem with water is that it is not always where it is needed most. Distribution is such that some areas have considerably more water than is needed, while other areas never have enough. For centuries man has been involved in transferring water, via canals and aqueducts, from places of excess to places of scarcity. Archeological evidence shows that extensive irrigation existed in the Mesopotamian Valley as early as 4000 B. C. The vicinity of Baghdad, a thousand years ago, had as much as 3000 miles of irrigation canals, including one canal which was nearly 300 miles long. Here in the United States we have been involved with irrigation and the transfer of water for many decades.

The current proposals for the transfer of water from one place to another, however, differ in magnitude from all previous projects. In California a project was completed in 1913 to provide some 150,000 acre-feet per year to the eastern side of the Sierra Nevada range. The aqueduct was extended in 1940 to provide 320,000 acre-feet per year and again in 1968 to carry 472,000 acre-feet. Early in this century New York City developed systems in the Hudson Basin to accommodate 986,000 acre-feet per year. The Big Thompson Project in Colorado provides about 230,000 acre-feet per year; the Moffat Tunnel brings about 100,000 acre-feet per year to Denver from the Fraser River on the western side of the continental divide. In contrast, the North American Water and Power Alliance, proposed by the Ralph M. Parsons Company of Los Angeles (Fig. 3–14) is being designed to transfer 110 million acre-feet per year. Built into the design is an ulti-

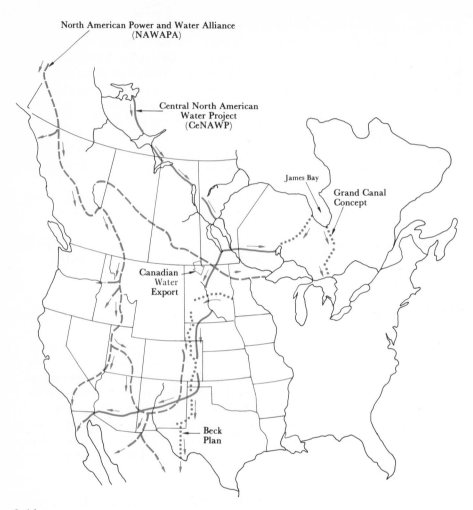

3-14
Schematic map showing five schemes for the regional transfer of water currently under consideration. (C. C. Warnick, "Historical Background and Philosophical Basis of Regional Water Transfer," in Arid Lands in Perspective, *ed. William G. McGinnies and Bram J. Goldman, Tucson: University of Arizona Press, 1969, p. 358. Reprinted by permission.)*

mate capacity of 250 million acre-feet per year. The water would be brought from northwestern Canada to serve seven Canadian provinces, thirty-three states in the United States, and three of the northern states of Mexico. This project would transfer an amount of water equal to roughly half the amounts of water presently withdrawn from all sources in the regions to be served.

Whether or not interbasin transfers on the scale of that mentioned above will become a reality in the next few decades is currently a matter of conjecture. However, smaller-scale, shorter-distance transfers involving international agreement among no more than two nations may certainly help to ease water problems in many parts of our country, as well as in other parts of the world. Some of the currently planned water-transfer projects are listed in Table 3-3.

Table 3-3 Summary of plans proposed for the regional transfer of water

Project name	Approximate date of proposal	River basin(s) for source	Countries involved	States involved
United Western	1950	Columbia River North Pacific coastal streams	United States Mexico	11 Western states
California Water Plan	1957	Northern California rivers	United States	California
Pacific Southwest Water Plan	1963	Northern California streams Colorado River	United States Mexico	California, Ariz., Nev., Utah, N. M.
Snake-Colorado Project	1963	Snake River	United States Mexico	Idaho, Nev., Cal., Ariz.
North American Power and Water Alliance (NAWAPA)	1964	Alaskan and Canadian Rivers with Columbia River	United States Canada Mexico	Western Texas Lake states
Yellowstone-Snake-Green Project	1964	Yellowstone River Snake River	United States	Mont., Idaho, Wyo., Lower Col. states
Pirkey's Plan Western Water Project	1964	Columbia River	United States Mexico	Ore., Wash., Cal., Utah, Ariz., Nev.
Dunn Plan Modified Snake-Colorado Project	1965	Snake and Columbia Rivers	United States Mexico	Idaho, Ore., Wash., Utah, Ariz., Nev., California
Sierra-Cascade Project	1965	Columbia River	United States	Ore., Nev., California
Undersea Aqueduct System	1965	North Coast Pacific Rivers	United States	Oregon, California
Southwest Idaho Development Project	1966	Payette River Weiser River Bruneau River	United States	Idaho

Table 3–3 (cont'd)

Project name	Approximate date of proposal	River basin(s) for source	Countries involved	States involved
Canadian Water Export	1966	Several Canadian Rivers	United States Canada	All Western states
Central Arizona Project	1948 1967	Lower Colorado River basin	United States Mexico	Utah, Nev., Ariz., Cal.
Central North American Water Project (CENAWP)	1967	Canadian rivers	United States Canada Mexico	Great Lakes, Western
Smith Plan	1967	Liard River McKenzie River	United States Canada Mexico	17 Western states
Grand Canal Concept	1965	Great Lakes and St. Lawrence River	United States Canada	Great Lakes states
Beck Plan	1967	Missouri River	United States	S. D., Neb., Kan., Col., Tex., Okla.
West Texas and Eastern New Mexico Import Project	1967 (1972 due)	Mississippi and Texas Rivers	United States	Okla., Tex., New Mex., Louisiana
Pacific-Mead Aqueduct Augmentation by Desalinization	1968	Pacific Ocean	United States Mexico	California, Arizona
Yukon-Taiya Project	1968	Yukon River	United States Canada	Alaska

(C. C. Warnick, "Historical Background and Philosophical Basis of Regional Water Transfer," in *Arid Lands in Perspective,* ed. William G. McGinnies and Bram J. Goldman, Tucson: University of Arizona Press, 1969, p. 348. Reprinted by permission.)

REFERENCES

American Chemical Society. *Saline Water Conversion—II.* Symposia held March 27, 1961 and March 27–8, 1962. Advances in Chemistry Series No. 38, 1963.

Brown, H. A. "Super-chlorination at Ottumwa, Iowa," *Journal of the American Water Works Association,* (July 1940): 1147.

Estimated Use of Water in the United States, U.S. Geological Survey Circular 556, 1968.

Haney, P. D. "Water reuse for public supply," *Journal of the American Water Works Association* (February 1969): 73–78.

Kneese, A. V. and B. T. Bower. *Managing Water Quality: Economics, Technology, Institutions,* Baltimore: Johns Hopkins Press, 1968.

Linsley, R. K. and J. B. Franzini. *Water-Resources Engineering,* New York: McGraw-Hill, 1964.

Maxwell, J. C. "Will there be enough water?" *American Scientist* 53, 9 (1965): 97–103.

Metzler, D. F., *et al.* "Emergency use of reclaimed water for potable supply at Chanute, Kansas," *Journal of the American Water Works Association* (August 1958): 1021.

Piper, A. M. *Has the United States Enough Water?* U.S. Geological Survey, Water-Supply paper 1797, 1965.

Spiegler, K. S., (ed.) *Principles of Desalination,* New York: Academic Press, 1966.

Stephan, D. G. and L. W. Weinberger. "Wastewater reuse—has it arrived?" *Water Pollution Control Federation Journal* 40, 4 (1968): 529–539

Warnick, C. C. "Historical background and philosophical basis of regional water transfer," in *Arid Lands in Perspective,* ed. William G. McGinnies and Bram J. Goldman, Tucson: University of Arizona Press, 1969.

Weinberger, L. W., D. G. Stephan, and F. M. Middleton. "Solving our water problems—water renovation and reuse," *Annals of the New York Academy of Science* 136, (1966): 131–154.

The interaction between lichens and bare rock may be one of the first steps in the formation of soil, upon which man is so dependent. (Photograph by George Sheng)

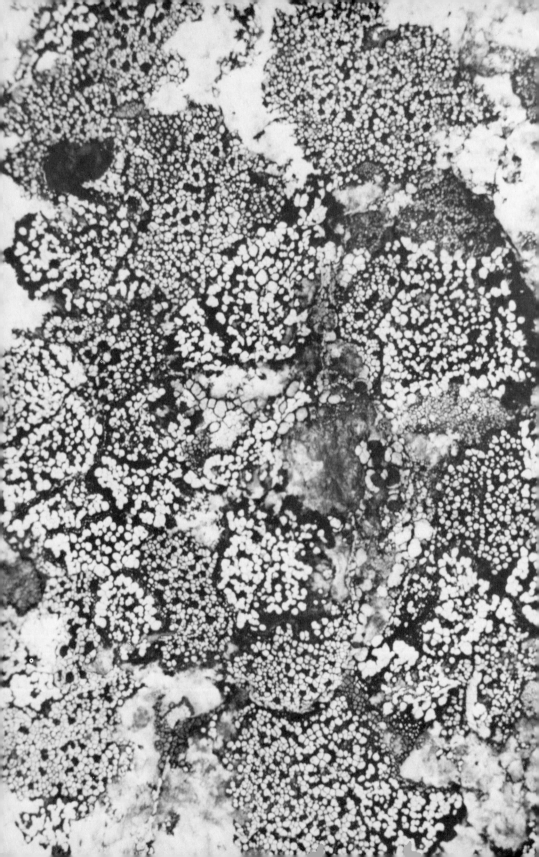

SOILS

INTRODUCTION

On an object as large as the earth, the thin film of soil—usually only
about one to six feet thick—which covers most of the land surface could
hardly be considered significant. Yet soil is of vital importance to man,
for our very existence depends on this "insignificant" layer. From it come
the inorganic nutrients essential for the continued health and growth of
the plants and animals without which man could not survive. Although
these same inorganic substances are present in other forms, such as igneous
rocks, they generally occur in forms that are inaccessible to most organisms. It is only after the soil-making process has begun that these essential
elements are recombined into forms usable by large numbers of diverse organisms. Without soil, man could not exist.

4-1
Glacial boulder split into 26 pieces by frost action, Sequoia National Park, California.
(Photograph by F. E. Malthis, U.S. Geological Survey)

SOIL FORMATION

Soil covers most of the land surface of the earth in a somewhat continuous blanket, usually without sharp, distinct breaks between one soil type and another. The formation of this soil blanket from rock material is simply the response of geologic material to a fundamental principle of geology. This principle, an underlying principle of science, states that materials attempt to reach a state of equilibrium with the conditions which surround them. In a more geological context we may call this the "Law of Stability of Rocks and Minerals." According to this law, rocks and minerals are stable only in the environment of their formation. When removed from that environment, they tend to change (albeit slowly) into new rocks and minerals which are stable in the new environment. If this scientific principle did not exist in nature, we would have only bare rock, exposed at the surface, and we probably would not be here to puzzle over this fact. Only the most primitive living things, such as lichens, are able to exist on bare rock. One change that takes place as a condition of this law is *weathering*, which involves two fundamental processes: *mechanical disintegration* and *chemical decomposition*.

Physical Weathering

Physical weathering, or mechanical disintegration, is a process in which rock material is broken into increasingly smaller pieces by the physical forces of

4-2
The schematic drawing shows the volume-to-surface-area ratio change as material is broken. The photo shows this in nature. (Photograph by Donald W. Levenson)

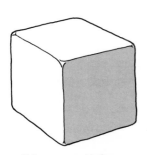

Volume: one cubic foot
Area: 864 square inches

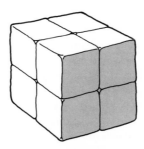

Volume: one cubic foot
Area: 1728 square inches

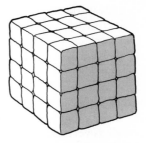

Volume: one cubic foot
Area: 3456 square inches

nature without involving any chemical change. These forces initiate the breakdown of fresh bedrock into pieces which have a relatively higher surface area to volume ratio, thus providing surfaces which can be acted upon by chemical weathering. The surface area to volume ratio increases geometrically as the material is broken into increasingly smaller pieces (Fig. 4–2).

Agents of physical weathering   These agents include frost wedging, plants and animals, unloading, heat expansion and contraction, and others. Frost wedging is the result of the alternate freezing and thawing of water in rock material and soil. In most areas it is undoubtedly the most common, and probably the most effective, of the physical weathering agents, accounting for a high percentage of the physical weathering which takes place. This agent is most noticeable in the higher latitudes and altitudes, where chemical weathering is less effective because of the colder temperatures (which slow down chemical reactions) and because, often, water in these locations is frozen and therefore useless in the process of chemical weathering.

When water freezes, its volume increases by about 9 percent. This expansion is accompanied by fantastic forces. At about $-8°$ F., which is a common temperature in many parts of the world, the force of crystallization (Fig. 4–3) of ice amounts to almost 30,000 pounds per square inch (nearly 2100 tons per square foot). This is about 40 times greater than the force necessary to break granite. This provides a wider crack for a larger amount of water to enter the next time, and this continues until the rock is literally split wide open.

Most mountain peaks are surrounded by a field of boulders that are the products of frost wedging. This is especially true if the mountain is located in the temperate zone, where the repeated alternations between freezing and thawing are much more frequent, and therefore effective, than in the colder climates, where freezing is more permanent and thawing is rare. There are perhaps five times as many freeze-thaw cycles annually in the southern parts of Canada as there are in the Canadian territories north of the Arctic Circle.

An interesting relationship between rock fragments and temperature results in what is often referred to as *patterned ground*, or *stone polygons* (Fig. 4–4). A mass of solid rock conducts heat much better than does a comparable volume of loose, fine-grained soil. Consequently, when a rock fragment is close to the surface, the soil heat is conducted from under the rock fragment. As this process continues, the ever-thickening layer of ice pushes the rock fragment up to the surface. In some areas, such as in arctic or tundra climates, the larger rock fragments may be moved horizontally, as well as vertically, and are often sorted into narrow bands. These bands usually intersect to form a series of polygons outlined by ridges of rock fragments.

Plants and animals are also significant forces in physical weathering. Anyone who has walked down an old, tree-lined sidewalk can attest to the ability of plants to physically break up rock material. In many parts of the country where solid rock crops out, it is common to see places where a tree started to grow in a crack in the rock and wedged the pieces of rock apart

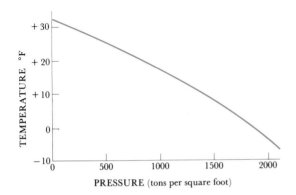

4-3
*The increase in pressure of confined pure water at
temperatures below freezing. In this system, confining pressures of 2100 tons per square foot
are readily reached. (Data from Bridgeman, 1911)*

4-4
*Aerial view of polygonal ground in northern Alaska. Polygons are about 100 feet across
and have ice wedges in cracks between them. (Photograph by R. E. Wallace, U.S. Geological Survey)*

as the tree grew larger. Smaller plants contribute to this process by rooting in small holes and tiny cracks which increases their size. In these cases, weathering immediately attacks any new surface. There seems to be a general consensus that the total amount of rock material broken in this manner is quite large, but measurement of its extent is very difficult.

Animals do their part in the physical weathering process, too. Burrowing animals bring a great deal of fresh or partly weathered rock material to the surface, where it is exposed to more efficient chemical weathering action. Charles Darwin, after making close observations in his own garden, estimated that worms alone brought more than 10 tons of material per acre to the surface annually. In arid regions, where there are few worms, this is done by ants, termites, and very small, burrowing animals.

We often wonder how cracks in the rocks, which allow water to seep in or plants to root, begin. One way (but not the only one) is a result of the release of confining pressure which may surround a rock body. This is called *unloading.* When rocks are formed at some depth in the earth, they are in an environment of high pressure. When the overlying material is removed through erosion, this pressure is reduced, and the rock tends to expand slightly. Tension cracks are a common accompaniment to this expansion.

4-5
Joint in granite widened by growth of tree roots. (Photograph by G. K. Gilbert, U.S. Geological Survey)

4-6
Breaking up of granite by unloading, Sierra Nevada, California. (Photograph by G. K. Gilbert, U. S. Geological Survey)

If the rock was formed as a result of lithification of marine mud at the bottom of the sea, it was formed under the pressure of the overlying water. Each 10 meters of seawater add the equivalent of one additional atmosphere of pressure (15 pounds per square inch) to the system. If the rock is formed as the result of the cooling of a magma deep in the earth, we find that about 3.5 to 4.0 meters of overlying rock equal one atmosphere of pressure. If the magma is deep, the pressure environment is great. During the process of erosion (or of retreat of the sea), the confining pressure is lessened, and the rock mass expands in accordance with these new conditions. This is most apparent in massive rocks such as granite. Granite from Stone Mountain, Georgia, when quarried out in blocks, expands one-tenth of one percent along its length. This amounts to nearly one-eighth of an inch of expansion for ten-foot slabs. In many mine operations, blasting is done at the end of the day so that rock faces which have been newly uncovered by the blasting will have the night to adjust to the new low-pressure environment.

4-7
Development of granite boulders by exfoliation and disintegration, Fresno County, California. (Photograph by G. K. Gilbert, U.S. Geological Survey)

In discussions about factors which bring about physical weathering, many people suggest that physical weathering is accomplished by the *expansion and contraction* of rocks due to solar heating. It seems reasonable that on very hot days the rocks would expand because of the heat and then would contract as the temperature cools. Most individuals have had the typical experience of pouring a hot liquid into a glass or cup, causing it to break, or the converse experience of putting a cold glass into hot water, causing similar results. It is only natural to apply this experience to what happens to rocks. Indeed, this idea of solar heat as a disruptive force has been believed for a long period of time. In desert areas we often find boulders which have been shattered for no apparent reason. The sounds of rocks breaking in the desert have been compared to gunfire. Thermal expansion and contraction have been held responsible for these phenomena.

So far, laboratory tests of this idea have not verified it. Various pieces of rock material have been test-heated in furnaces to determine at what temperature they shatter. Results showed that they begin to rupture only at temperatures much greater than those in any desert. D. L. Griggs, in 1936, carried out a classic experiment along these lines. He heated a three-inch cube of granite to about 140°C for about five minutes, then cooled it with an electric fan to about 30°C for about ten minutes. He repeated this for 89,400 cycles, which equals 244 years of diurnal solar heating. After experimenting for three years he could find no change in the rock, even under close microscopic examination. We are left, as is often the case in geology, with what seems to be a dichotomy. Perhaps our experimental design is failing to include some factor which is present in nature; perhaps the answer lies in the time factor, since desert rocks have been subjected to millions of diurnal cycles, instead of the few tens of thousands in our experiments. On the surface, then, it would appear that solar heating has limited value as an agent of physical weathering. There are a number of other factors which do cause a certain amount of physical weathering in particular regions and under certain circumstances. These include wind, running water, glaciers, and gravity. Perhaps you can add others.

Chemical Weathering

Chemical weathering, or chemical decomposition, is the second major type of weathering. We have seen that physical weathering breaks the rock into increasingly smaller pieces but does not change the rock's chemical composition. Chemical weathering also breaks a rock into smaller pieces but does produce changes in the chemical composition. This is called chemical decomposition, or rotting away of the rock. The result of this process is the formation of new minerals at the expense of the original minerals and their components.

Agents of chemical weathering   These include hydrolysis and solution by water, oxidation, carbonation, plants, and others. *Hydrolysis and solution by water* are actually independent processes of chemical weathering, but since they occur simultaneously in nature, it is more expedient for us to explain both of them here. Hydrolysis, in chemical weathering, refers to the reaction between the H^+ or OH^- ions present in water and the mineral elements present in rocks. This process produces both soluble and insoluble products. The more soluble products are taken into solution by the water and are usually carried out of the system. Of course, water may cause the breakdown of certain minerals. This is demonstrated by the dissolving of such minerals as halite (NaCl) and gypsum ($CaSO_4 \cdot 2H_2O$).

The hydrolysis reaction can be stated in the following generalized form:

$$MeAlSiO_n + H_2O \longrightarrow Me^+ + OH^- + HAlSiO_n + HSiO_n + Al(OH)_3$$

In this formula "Me" indicates metallic ions such as "K," "Na," or "Ca," which are commonly found in the feldspars, amphiboles, and pyroxenes. The "$HAlSiO_n$" is silica, which goes into solution and is the product of the reorganization of the $AlSiO_n$ structure found in the feldspars, and others,

4-8
Solution channels in a bed of limestone. Acids dissolve in chemical weathering. (Photograph by M. R. Mudge, U.S. Geological Survey)

to that structure found in the clay minerals. The $Al(OH)_3$ may or may not be present, depending on the other conditions. The dissolving power of water may be greatly enhanced, for certain rocks and minerals, by dissolving CO_2 gas in the water. As rainwater falls through the atmosphere, it picks up and takes into solution CO_2. As this rainwater soaks through the soil it absorbs even more CO_2 into solution. The result of this is the formation of a weak acid called *carbonic acid*, which greatly increases the dissolving power of water on many minerals. This formation is given in the following reaction:

$$H_2O + CO_2 \longrightarrow H_2CO_3^- \quad \text{(carbonic acid)}$$

The carbonic acid actually occurs in solution as two ions, $H^+ + HCO_3^-$. The bicarbonate ion may be partially broken into $H^+ + CO_3^-$, the latter being the carbonate ion. It should be noted that throughout these reactions there occurs release of positive hydrogen ions which furthers hydrolysis. Carbonic acid's power to dissolve is especially apparent when it works on calcite or limestone.

It is possible, on the basis of knowing these and other chemical reactions, to predict how various rocks and minerals will respond to chemical weathering. Table 4–1 indicates what happens to the common elements found in the common rocks during chemical weathering. It should be particularly noted that the solubility is not the same for all elements.

Table 4–1 Reactions of common elements to the chemical weathering environment

Element	Reaction	Product	Solubility
Fe	oxidizes	hematite	nonsoluble
Fe	oxidizes with hydration	limonite	nonsoluble
Fe	occasionally dissolves in carbonic acid	Fe ions	soluble
Ca	dissolves in carbonic acid	Ca ions	soluble
Mg	dissolves in carbonic acid	Mg ions	soluble
Na	hydrolyzes or is carbonated	Na ions	soluble
K	hydrolyzes or is carbonated, but is fixed or	absorbed by clay	nonsoluble
	a minor amount escapes as $\longrightarrow$	K ions	soluble
$SiAlO_n$	group hydrolyzes	clay minerals and hydrated silica	nonsoluble / soluble
SiO_2	as quartz or chert: for practical purposes should be considered as insoluble. Remains quartz or chert.		

Reprinted by permission of the author and publisher from W. D. Keller, *Principles of Chemical Weathering*, rev. ed., Columbia, Missouri: Lucas Brothers, 1959, p. 54.)

4-9
Sinkhole caused by solution of limestone underlying the field. (Photograph by M. R. Mudge, U.S. Geological Survey)

Mineral composition of the granite	Chemical composition	Products of weathering	
		Soluble	Insoluble
Orthoclase	$KAlSiO_n$	K^+ (minor), soluble silica	Clay with K^+ absorbed on it
Plagioclase	$NaCaAlSiO_n$	Na^+, Ca^{++}, soluble silica	Clay with K^+
Hornblende	$CaFeMgAlSiO_n$	Ca^{++}, Mg^{++} soluble silica	Clay mineral, hematite, and/or limonite
Biotite	$KFeMgAlSiO_n$	K^+ (minor), Mg^{++}, soluble silica	Clay mineral, hematite, and/or limonite
Quartz	SiO_2	None	Quartz grains

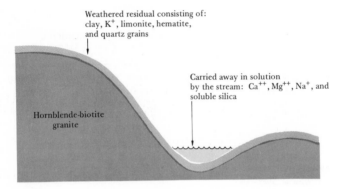

4-10
The table gives the composition of a hypothetical granite and the weathering products resulting from each mineral. The diagram shows the disposition in nature of each of the weathering products. (Reprinted by permission of the author and publisher from W. D. Keller, Chemistry in Introductory Geology, *4th ed., Columbia, Missouri: Lucas Brothers, 1969, p. 52.)*

To understand these effects, you might visualize a hill upon which granite crops out. Include in your mental image a small stream at the base of the hill. What products would be formed during the weathering process? Ignoring the effect of erosion, we might suggest that any soluble products will be transported out of the area in solution by the stream and that any insoluble products will be left behind on the hillside to form a mantle of soil. Referring to Fig. 4-10 we find a hillside covered with reddish, iron-rich clay which is gritty because of the quartz grains left behind. We have a stream which is rich in dissolved material, mostly Ca and Mg ions, making the water "hard."

Oxidation is the loss of an electron (e^-) from one of the atoms in a mineral. Oxidation often involves the combination of elemental oxygen that has become ionic with the mineral; however, this is not always the case.

4-11
Lichens growing on sandstone. This represents an early stage in the weathering and eventual breakup of the rock. (Photograph courtesy USDA–Soil Conservation Service)

Metallic iron can be oxidized to pyrite (FeS_2) by the presence of sulfur without any oxygen present.

Carbonation is the combination of carbonate (CO_3^-) or bicarbonate (HCO_3^-) ions with minerals. This process occurs most commonly on rocks containing calcium, magnesium, or iron. The following equation illustrates this method of chemical weathering. The result is the formation of the soluble bicarbonates of calcium and magnesium and the "dissolving" of the dolomite.

$$CaMg(CO_3)_2 + 2CO_2 + H_2O = Ca(HCO_3)_2 + Mg(HCO_3)_2$$

One would seldom think of living *plants* as agents of chemical weathering, but they are, indeed, very important agents of this process. During the process of photosynthesis, hydrogen ions are released by the plant from the surface of the rootlets. These roots are in search of nutrient ions, such as calcium, potassium, and magnesium. Through an ion exchange process, the hydrogen ions are exchanged by the rootlet for the needed nutrients contained in the surrounding soil or rock material.

Certain plants play an additional role in weathering; these are plants which accumulate elements. Although accumulator plants have been studied mainly in connection with the accumulation of minor elements (such as selenium) which are detrimental to livestock, there are those which accumulate the major elements, such as silicon, aluminum, iron, magnesium, and calcium. These accumulator plants certainly have geologic implications when we consider that the heavy loss of certain major elements determines some soil types.

There is a distinct possibility that silica-accumulators figure significantly in the alteration of igneous material to lateritic soil (soil in which a high percentage of the silica has been removed and elements such as aluminum, magnesium, and iron have been relatively enriched). Many tropical plants, such as those found in zones where laterite forms, contain several percent of silica dry weight. Up to 20 tons dry weight of new growth above ground and several more tons below are annually added per acre in many tropical areas. It has been estimated that an overgrowth of silica-accumulating plants, which contain an average of only 2.5 percent silica and add 16 tons dry weight of new growth per year, would extract approximately 2000 tons of silica per acre in about 5000 years. This is approximately equivalent to the silica contained in one acre-foot of basalt. From this standpoint, silica-accumulators could transform igneous bedrock into lateritic soil in a geologically short time. (For a more detailed discussion of this subject, see the bibliographic reference for Lovering, 1959.)

There are other agents and processes of chemical weathering which you may want to review. These include chelation; ion exchange, such as in the dialysis of clay minerals; and hydration. These processes are important in certain circumstances, but are considerably less important than those covered in more detail in this chapter.

We have demonstrated a number of mechanisms which can alter rock material and form the soils on which we are so dependent. You might keep the first part of this chapter in mind when reading Chapter 12, where we talk about trace elements and their effect on our health. It will be worth remembering that the processes we have discussed are often responsible for releasing those trace elements into the natural system.

Rates of Weathering

Although several studies have been done in past years on the rate of weathering, so far few generalizations can be made. However, we can consider a few items which should give you a better understanding of this subject.

Variables involved   There are many variables encountered in this problem. They include: composition of the rock, size of the rock fragments being weathered, presence or absence of macro- and micro-fauna and flora, topographic relief, position of the water table relative to the material being weathered, composition of the ground water involved. climate of the area, adjacent geologic material, relative solubility of the minerals involved, and probably many others not readily apparent.

One such "nonapparent" factor is the order of crystallization of igneous minerals, as defined by Bowen's Reaction Series (Fig. A–2). According to this scheme, certain minerals tend to crystallize from a magma during the time the magma is within a certain temperature range. (We are oversimplifying here, but we are interested only in the overall framework, not in the details.) Figure A–2 shows that a combination of olivine-pyroxene-calcium-rich feldspars would be expected to form first and might be expected to form a gabbro. Last, one would expect to find a biotite-sodium-rich feldspar-potassium feldspar-quartz combination formed from the magma and producing a granite. If we look at the law of stability of rocks and minerals, we might expect (if all other variables remain the same) the gabbro to weather more rapidly than the granite, due to the fact that at the earth's surface the minerals making up the gabbro are farther from their environment of formation (in terms of energy) than are those minerals making up the granite. Therefore, we would expect the granite to be the more stable of the two.

Examples of weathering rates   Igneous rocks weather slowly. There are areas of igneous rock worn smooth by the passage of the Pleistocene glaciers, which retreated some 10,000 years ago, that show no appreciable weathering in that length of time. On the other hand, rocks made up mainly of calcite (limestone and marble) weather more rapidly in the temperate climate of the midcontinent. In visiting a number of old cemeteries in Missouri, one finds that most of the marble headstones erected in the late 1800s show a great deal of weathering. Interestingly, there seems to have been a general trend around the turn of the century toward using fewer marble and more granite headstones. Perhaps this was related to the fact that the two materials are relatively durable but marble was found to be incapable of withstanding the onslaught of the chemical weathering conditions found in this region. Other "tombstone studies" have shown that "a good-quality sandstone weathered very little in 200 years, a slate of 90 years age had clear engraving, but a marble stone had partly crumbled into sand." One study done in England near the turn of the century indicated that, depending on the type of limestone, it took 240 to 500 years to produce one inch of weathering.

A study of the weathering of the Great Pyramid of Khufu (Cheops) at Giza, Egypt, which provided some interesting notes on weathering rates, was carried out during 1959. The pyramid was built about 2800 B.C. and was faced with smooth, well-fitted limestone blocks which protected the inner core of the pyramid from weathering. Beginning in the ninth century this outer facing was removed for use in building other structures in the vicinity. Of the four varieties of rock material used (gray, hard, dense limestone; gray, soft limestone; gray, shaly limestone; and yellow, limy, shaly sandstone) the gray, hard, dense limestone has been so little affected by weathering in the past thousand years that marks made by the quarry tools are still visible. The gray, soft limestone shows weathered pits an average of 2 cm deep; the gray, shaly limestone has developed weathered niches as deep as 20 cm; and the yellow, limy, shaly sandstone turns to rubble easily. The annual loss due to weathering is estimated to be only about 0.01 percent of the total volume; at this rate the pyramid should last for another 100,000 years, providing man does not interfere and accelerate this rate. These thoughts on weathering rates should be modified when one remembers that the desert environment brings about slower than normal rates of weathering.

No discussion of weathering rates would be complete without including the fate of the two Egyptian obelisks, both now called "Cleopatra's Needle." These granite structures (both made from the same granite), each deeply engraved with many hieroglyphics, were erected around 1600 B.C. in Egypt. After some 3500 years in the desert climate they showed only very slight weathering. One was taken to London and the other was sent to New York (circa 1880), where it was set up in Central Park. The combined effects of physical and chemical weathering have obliterated much of the story depicted by the hieroglyphics. Despite attempts made in the 1920s and 1930s to preserve the structure by applying shellaclike preservatives, the combined effects of frost wedging, alternate wetting and drying, and the corrosive effects of the CO_2-rich New York air have erased part of the picture story. The obelisk which was moved to London apparently shows appreciable weathering but has not been as affected as the one in Central Park (Fig. 4–12).

When looking at the insoluble residue analysis of limestones in the midcontinent United States, one can estimate that it would be necessary to weather approximately 30 feet of limestone to form one foot of soil. It has been estimated that, on the average, it takes approximately 5000 years to dissolve one foot of limestone in the Mississippi Valley; this implies that 150,000 years of weathering are necessary to form a foot of soil in the midcontinent region of the United States.

(a)

(b)

4-12
*Obelisk of Thothmes III ("Cleopatra's
Needle") in Egypt (a), and in Central
Park (1918), New York (b), showing
differences in weathering. (The Metropolitan Museum of Art, photograph (b)
taken in 1918, photographs courtesy of
the Metropolitan Museum of Art and
New York City Department of Parks)*

Variable Factors in Soil Formation

Soils (Figs. 4–13 and 4–14) are the result of the physical and chemical reactions conditioned by the variables involved in soil formation. Among
these variables are: parent rock material, climate, topography, the biosphere, and time. A closer look at each of these factors will give you a
better appreciation of the entire process.

Parent material, whether residual or transported, is the basis of all
soils. Of the five variables involved in soil development, parent material,
time, and topography are passive in the process. Intuitively, one might
suspect that parent material is the strongest factor in determining the ultimate soil that will be developed. However, this is true only of soils which
have been recently developed from fresh rock material, (that is, young
soils). In the development of soil, the time factor and the influence of the
parent material are inversely related to each other, while the influence of
climate and other factors are directly related to the length of time. The
soil, though formed from a similar granite found in Missouri, Montana, and
New Mexico, will be different in all three cases, *even if the granite is essentially identical in all three cases.*

Soils formed from rock material in which magnesium is either absent
or only minimally present will contain, of course, little or no magnesium.
In many cases, lack of a trace element such as boron makes an entire area
unfit for growing certain crops. This lack is a direct result of the absence of
boron in the basic chemistry of the parent material. The converse, unfortunately, is not necessarily true. That is, just because the parent rock is rich
in inorganic nutrients does not mean that the resulting soil will be rich in
those nutrients. Time and climate may join to strip the rock and deprive
the area of these nutrients, leaving a very poor, nutrient-deficient soil.

Time is a necessary, though passive, variable in soil formation, because
the development of the mature soil profile is a very slow process. Soil
scientists feel that the development of acid soils in regions having a forest
cover and blanketed by sandy, well-drained material may take no more
than 100 to 200 years. The estimates vary considerably, but an average
of several thousands of years seems to be acceptable. Tropical soils, once
formed, are very resistant to change and may maintain their stability for
five million years or more.

Topography, another passive factor, also exerts its influence in soil formation. In general, it is magnitude and orientation of slope which are the
prime variables of topography. Magnitude of slope has a dual effect on the
local soil picture. First is the problem of erosion: the steeper the slope,
the greater the erosion (generally) and the thinner the soil profile, for soil
is carried away as rapidly as it is formed. Second, "slope" can cause any
constituents dissolved out of the soil to be carried away, or at least trans-

4-13
*A soil profile, showing a silty loam (top 28 inches)
overlying a silty clay loam. (Photograph courtesy
USDA—Soil Conservation Service)*

4-14
Generalized soil profile showing the various horizons.

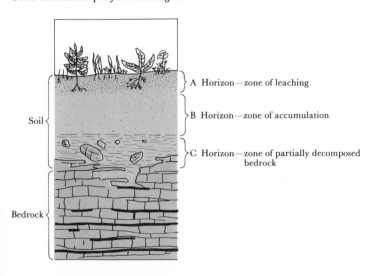

Soil

A Horizon—zone of leaching

B Horizon—zone of accumulation

C Horizon—zone of partially decomposed
bedrock

Bedrock

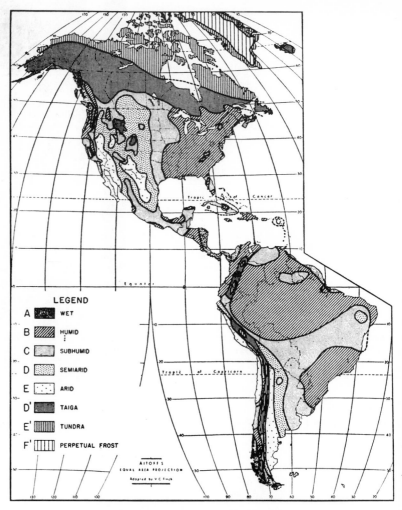

4-15
Generalized map of the world distribution of the principal climates.
(Source: U.S. Department of Agriculture Yearbook, 1941)

ferred from soil horizon to soil horizon. In the absence of slope, that is, in poorly drained flatlands, not only will the soil horizons tend to be thicker, but the character of the soil will be totally different from that in areas of steeper slope.

Slope orientation relative to compass direction has an influence on soil formation. In the Northern Hemisphere, south-facing slopes develop soils of quite a different character from those developed on north-facing slopes. Because south-facing slopes are more in line with the direct rays of

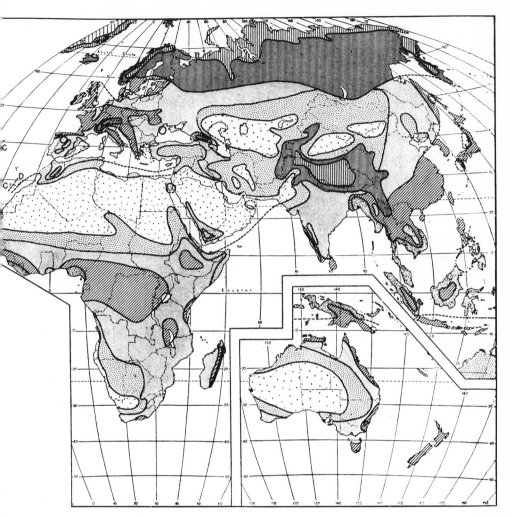

4-15
(cont'd)

the sun, their soils tend to be warmer and drier and support vegetation
quite different from that on north-facing slopes.

Of all the processes involved in soil formation, climate is of primary
importance. While the parent rock and plants contribute the materials
from which soils are made, climate determines the process of soil development. Climate and soil type are so integrally related that a world map of
soils closely resembles the map of world distribution of climate types.
Compare Fig. 4–15 (world climates) with Fig. 4–17 (zonal soil groups).

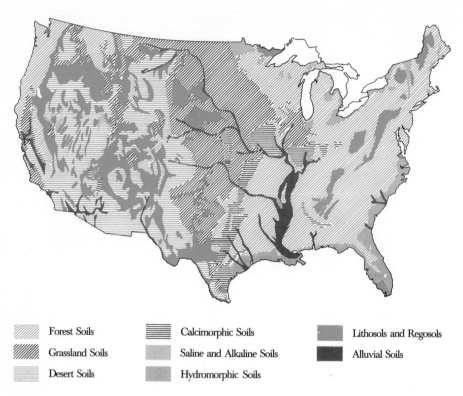

	Forest Soils		Calcimorphic Soils		Lithosols and Regosols
	Grassland Soils		Saline and Alkaline Soils		Alluvial Soils
	Desert Soils		Hydromorphic Soils		

4-16
General pattern of eight great soil groups.

Of all the factors which constitute "climate," temperature and precipitation exert a strong, perhaps the strongest, influence. The broad pattern of soil arrangement reflects these influences; in North America we can see their influence at work (Fig. 4-16). In the eastern part of North America we see that the soils are zoned primarily according to the influence of temperature. As we proceed from the humid east coast of the United States to the much drier interior, soil zoning becomes increasingly controlled by rainfall (Fig. 4-15). The separate influences of temperature and moisture can often be identified and measured; however, since the two generally work together, their separate influence is not easily discernable.

In any given area, the soil climate may differ greatly from the atmospheric climate. We have been looking at climate as it affects soil development directly; climate also affects soil development indirectly. Climate is an important influence on chemical reactions. Many chemical reactions are much more active at high temperatures than at low temperatures. In areas of relatively high temperatures and heavy rainfall, leaching of the soluble components from the upper parts of the soil profile is both common and rapid. In these regions decay of organic material is considerably accelerated and releases substances into the ground water system which greatly enhance the water's power to dissolve. Often, this rapid decay of organic material is the direct result of the presence of more abundant and varied animal life which attacks and devours the rotting organic matter. The converse is generally true, also. In cold climates chemical reactions are retarded, decomposition of organic matter is slow, and the fauna is neither as abundant nor as varied. It is very hard for any type of change to take place in frozen ground. However, leaching is often pronounced under these conditions, thereby giving the soils a distinct character.

One must keep in mind that for biological or chemical reactions to occur, water must be present. In very dry climates (hot or cold), leaching and decay are retarded, and normally soluble minerals (calcium carbonate, salts, and alkalies) become heavily concentrated in the soils.

Biological activity serves two purposes in soil development: chemically, it changes the local soil conditions, and physically, it adds bulk to and churns the soil. Chemically, the decomposition of vegetation is more rapid because of the presence of living organisms. Soils developed under different types of vegetation vary also. Soils developed in forest regions generally have more horizons, and the "A" horizon is more highly leached; grassland soils have a very rich, thick (a foot or more) layer of organic matter and usually have a poorly developed "B" horizon.

Grasses tend to maintain soil fertility by bringing calcium and magnesium up from some depth in the soil and transferring them into the stems and leaves of the grass. When the grass dies, these inorganic nutrients are deposited on the surface of the soil and thus help to maintain a circulation of inorganic nutrients. Humus (partly decomposed organic matter) is important in soil development in two ways. In the process of humus formation, humic acids are formed and released in the soil, increasing leaching and generally changing the chemistry of the soil.

Humus also provides organic colloids—tiny bits of matter ranging in size from approximately 10 angstroms to 1000 angstroms which cannot be seen through ordinary optical microscopes. Colloids are smaller than the wave length of the visible light they must reflect in order to be seen. (The shortest wave length of visible light is approximately 4000 angstroms.) Colloids have an extremely high surface to volume ratio and hold a surface charge which can attract and hold ions. Plants require many of these ions for their growth processes and can obtain them with different degrees of ease from colloidal particles through a process known as ion exchange. This process is also known as base exchange, since these ions are often ions of calcium, sodium, and potassium, all of which are bases.

Colloids contribute to the character of the soil in other ways. These particles hold layers of water molecules to their surface; this water is not completely lost under ordinary evaporating conditions and therefore helps to keep the soil soft and moist. Colloids may fill the interstices in a soil and impede the drainage of water, which may increase the stickiness or leave relatively large pore spaces through which water and air can pass freely, depending on whether the colloids are in the disperse (individual) or flocculated (large numbers held together) state. Some bacteria have the function of taking nitrogen from the air, which cannot be used by plants, and converting it to nitrate, which can be used by the plants. Thus, the presence or absence of these organisms changes the chemistry of the soil.

Biological activity also affects the physical conditions of soil formation. The stems, leaves, and roots of plants help to keep soil loose while decay is taking place. Soils containing large amounts of vegetable matter do not become hard and crusty. Amimals affect the soil largely through their physical action: digging and burrowing disturbs, agitates, and generally reworks the soil. Generally, profile development is poorer in an area inhabited by numerous animals because of their reworking action. Ants, termites, gophers, moles, field mice, and other digging animals continually bring material from lower horizons to the surface, causing a constant recycling of soil. As burrows collapse, the surface is lowered, and the material again slowly sinks to some depth. Earthworms create interconnected openings which give good aeration to the soil; but perhaps even more important, they change the texture and chemistry of the soil as it is passed through their digestive tracts.

SOIL CLASSIFICATION

For over four centuries man has attempted to classify the soils which cover much of the earth's land surface. Indications are that the earliest recorded attempt was made for tax purposes; presumably, the tax assessment was based on the productivity of the soil. In Russia, efforts to relate soil productivity to taxes, about a century ago, laid the foundation for the scientific study of soil, which in turn led to our current classifications.

It is not our object, nor is it within the scope of this text, to go into the complexities of soil classification, but you should be aware of the current status of this subject. Currently, there are two main systems of classification in use in the United States. Table 4–2 summarizes the major features of a classification system first published in 1938 and modified to its present form in 1949. This system identifies several great soil groups, and it is these groups which are usually represented on world soil maps, such as the one shown in Fig. 4–17.

Table 4–2 Classification of soils into great soil groups

Order	Suborder	Great Soil Groups
Zonal soils	Soils of the cold zone	Tundra
	Desert soils	Desert soils Sierozems Brown soils Reddish-Brown soils
	Semi-arid, subhumid, and humid grassland soils	Chestnut soils Reddish Chestnut soils Chernozems Prairie soils
	Forest soils	Podzols Gray podzolic soils Brown podzolic soils Red-Yellow Podzolic soils
	Lateritic soils	Bauxite Laterite soils
Intrazonal soils	Saline and alkali soils	Soils of imperfectly drained arid regions
	Hydromorphic soils	Bog soils Low-Humic Gley soils
	Calcimorphic soils	Brown Forest soils Rendzina soils
Azonal soils	Regosols	Lithosols Regosols Alluvial soils

(Source: Soil Survey Staff, U.S. Department of Agriculture, 1960)

Cold Zone Soils

These are the soils of the tundra, which in the summer commonly show a few inches of acidic, brown, loamy soil, with permafrost underneath at a depth of about one foot. Utilization of this soil is hampered severely by low temperature.

Desert Soils

These soils are often quite coarse, since weathering is mostly restricted to physical weathering. Since there is little water for leaching, the soils are often limy and/or salty. In low swampy places these soils tend to accumulate salts. Often, there is a concentration of clay in lower horizons. Lack of water presents a problem in any attempts to utilize these soils. Salt also complicates this problem, for even if there is not enough salt present to

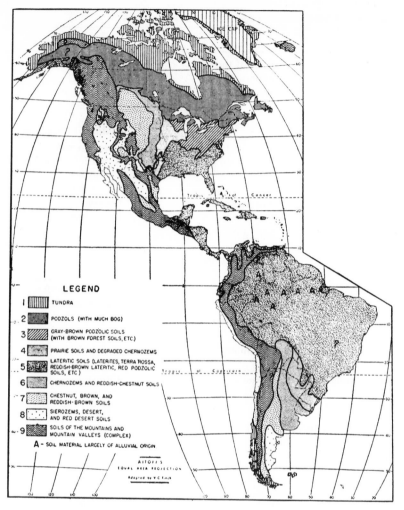

4–17

*World distribution of the principal zonal soil groups. (Source: U.S.
Department of Agriculture Yearbook, 1941)*

make the soil sterile, the amount is usually sufficient to upset the optimum
balance of inorganic material needed for plant nutrition (see Fig. 4–18).

Grassland Soils

These soils are represented by Chernozem, Chestnut, and Prairie soils, the
latter two being transitional between the Chernozems and Podzolic soils

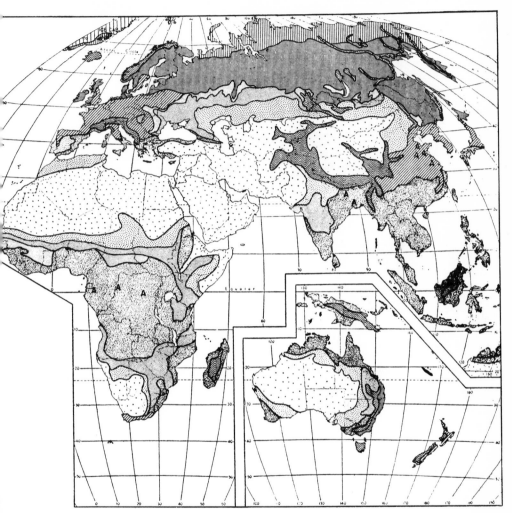

4–17
(cont'd)

of wetter regions and the desert soils of drier areas. This group of soils is typically black and high in both humus and ions of calcium, magnesium, and potassium. In terms of agriculture, these soils are the most useful to man.

Temperate Forest Soils

These are the Podzol and Podzolic soils, which show strong to moderate leaching, are composed of layers of clay or dense accumulations of colloids,

4-18
Relative hazard of plant nutrient losses by leaching: I,
slight hazard; II, moderate hazard; III, severe hazard; IV,
very severe hazard. (Reprinted by permission of the
publisher from L. B. Nelson and R. E. Uhland, "Factors
that Influence Loss of Fall Applied Fertilizers," Proceedings of the Soil Society of America 19, *1955, p. 492.)*

and tend to be of limited agricultural value. However, with the application of lime and fertilizers, these soils can become almost as useful in
agriculture as are the grassland soils.

Latosols

These soils are formed under conditions of high temperatures and heavy
rainfall. These conditions cause significant chemical weathering of the parent rock and removal of most of the silica through leaching. This leaves high
concentrations of the oxides and hydroxides of aluminum, manganese, and
iron in the soil. The lack of humus and the excessive leaching make this soil
almost useless for agricultural purposes. Although Latosol soils support native rainforest vegetation, they become bricklike when exposed to the atmosphere through the removal of this vegetation. In fact, in some areas of
southeast Asia these soils are used as a building material.

Other Soil Types

The Intrazonal soils are usually limited to peculiar local environments and are represented by the saline and alkaline soils of arid climates, the bog, marsh and meadow soils, and the calcimorphic soils in which the parent material (limestone) overcomes any conditions imposed by the climate. The Azonal soils, often grouped together as "mountain" soils, consist of either transported soils or very coarse soils which are the result primarily of the physical weathering of the parent material.

The United States Department of Agriculture has developed a new soil classification which, it claims, is based entirely on the morphological qualities of the soils. The Department feels that this will allow the classification of a given soil profile to remain unchanged regardless of the outcome of some of the current controversies over soil genesis. This classification has been altered from time to time in a series of "approximations." The current classification, adopted in 1965 as the official classification system of the United States Department of Agriculture, is the Seventh Approximation. Table 4-3 compares the terminology of this classification with the approximate equivalents in the Great Soil Groups. Since this classification does not create new groups, the previous discussion of the Great Groups is still pertinent.

Table 4-3 Seventh Approximation soils orders, their meanings, and approximate Great Soil Groups equivalents

Order	Meaning	Approximate equivalents in Great Soil Groups
1. Entisols	Recent soils	Azonal soils
2. Vertisols	Inverted soils	Swelling clays (Grumusols)
3. Inceptisols	Young soils	Brown Forest and Gley soils
4. Aridisols	Arid soils	Desert soils
5. Mollisols	Soft soils	Chestnut, Chernozem, and Prairie soils
6. Spodosol	Ashy soils	Podzol and Podzolic soils
7. Alfisols	Al-Fe soils (Pedalfers)	Gray-Brown Podzolic and degraded Chernozems
8. Ultisols	Ultimate soils (leached)	Red-Yellow Podzolic and Lateritic soils of the U. S.
9. Oxisols	Oxide soils	Latosols
10. Histosols	Organic (tissue) soils	Bog soils

(Source: Soil Survey Staff, 1960)

NUTRIENT MOBILITY

The fact that soils are the basis for our existence and have profound influences on our health and well-being earns them a definite place in a text of this kind. If the soils are good and contain a well-balanced collection of mineral nutrients (both bulk elements and trace elements), we live better and are healthier. Conversely, if the soils we depend on are poor and deficient in either bulk or trace elements, then we, our plants, and our livestock suffer accordingly. It seems worthwhile, therefore, to discuss briefly the problem of nutrient mobility in soil and to indicate that normally this mobility results in losses. Figure 4–18 depicts nutrient loss due to leaching in various parts of the United States. The problem of nutrient loss is further complicated by the fact that some elements are more mobile than others, and the mobility of elements differs with various environmental conditions.

The inorganic nutrients presently known to be essential for plant growth are listed in Table 4–4. It should be noted that the list of trace elements essential for plants is expanding as research uncovers previously unknown trace-element requirements. This list will undoubtedly have other elements added in the near future.

Since the ultimate source of all inorganic nutrients (nitrogen may be a possible exception) is igneous rock, we should assess the amount of these elements in this rock type. However, igneous rocks are not abundantly exposed at the surface. Sedimentary rocks cover a large portion of the land surface (estimates range from 60 percent to 75 percent), and shales represent about 80 percent of the exposed sedimentary rocks. It follows, then, that plants derive their natural inorganic nutrients from igneous rocks through shales. Table 4–5 lists some average weight percentages of five of the major nutrients found in these two rock types.

The original amount of phosphorus in the parent rock is of great importance, since little of this element comes from other sources. Phosphorus is essential to plant growth; therefore, the amount of phosphorus in the parent material may, in effect, put an upper limit on the growth potential of the plants which grow in this soil.

To give you an appreciation of the full range of nutrient loss in soils, we shall compare soils which show a large net loss during formation. The first group is best represented by the Red-Yellow Podzolic and Lateritic soils, or Ultisols, of the United States; the Lateritic soils are well represented by the Chestnut, Chernozem, and Prairie soils, or Mollisols.

Ultisols

This soil category includes the soils in the southeastern portions of both the United States and Asia, the central and eastern portion of South Amer-

Table 4-4 The elements known to be essential for plant nutrition and growth

Major nutrients essential for plant growth

Nitrogen	(N)	Sulfur	(S)
Phosphorus	(P)	Magnesium	(Mg)
Potassium	(K)	Calcium	(Ca)

Trace elements essential for plant growth

Iron	(Fe)	Iodine	(I)
Manganese	(Mn)	Zinc	(Zn)
Boron	(B)	Copper	(Cu)
Chlorine	(Cl)	Molybdenum	(Mo)

Table 4-5 Average percentages by weight of five elements in igneous rocks and shales

Element	Igneous rocks	Shales
Calcium	3.59	2.22
Magnesium	1.96	1.34
Potassium	2.59	2.69
Phosphorus	0.13	0.074
Nitrogen	0.005	0.051

Reprinted by permission from R. W. Simonson, "Loss of Nutrient Elements During Soil Formation," *Nutrient Mobility in Soils*, ed. O. P. Engelstad, Soil Science Society of America, Special Publication No. 4, 1970, p. 23.

Table 4-6 Tabulation of the range of percentage by weight of five of the major nutrients from four soils and their source material

	Ca	Mg	K	P	N
1.	0.02–0.10	0.03–0.12	0.62–1.04	0.01–0.04	0.02–0.26
2.	0.99	0.26	4.43	0.17	---
3.	0.07–0.26	0.08–0.49	0.58–1.05	0.0007–0.024	0.02–0.16
4.	18.82	9.86	1.25	0.014	---
5.	0.54–0.70	0.53–0.82	1.64–1.70	0.039–0.079	0.070–0.240
6.	0.66	0.82	1.66	0.044	0.050
7.	0.93–1.72	0.52–1.25	1.82–2.23	0.052–0.092	0.04–0.17
8.	1.07	1.12	2.27	0.044	0.03

1. Ultisol (Red-Yellow Podzolic soil), Georgia
2. Fresh Granite, Georgia
3. Ultisol (Red-Yellow Podzolic soil), Virginia
4. Fresh Limestone, Virginia
5. Mollisol (Brunizem), Iowa
6. Underlying Loess, Iowa
7. Mollisol (Chernozem), Nebraska
8. Underlying Loess, Nebraska

(Reprinted by permission from R. W. Simonson, "Loss of Nutrient Elements During Soil Formation," *Nutrient Mobility*, ed. O. P. Engelstad, Soil Society of America, Special Publication No. 4, 1970, pp. 24, 25, 32, 33.

ica, the central portion of Africa, and the northeastern part of Australia (Fig. 4-17). Of the major nutrients, the amounts of exchangeable calcium, magnesium, and potassium are quite low (Table 4-6). This low level of exchangeable bases is characteristic of Ultisols and reflects a long history of weathering and leaching. The phosphorus in these soils, while having properties similar to those of the other three nutrients, is actually less mobile and remains in the soil in larger quantities, especially if the soils contain high levels of aluminum and iron.

Mollisols

This group of soils includes those which exhibit only modest nutrient losses during formation (Table 4-6). These soils are very fertile and are prevalent in the midcontinent of the United States and prairie provinces of southern Canada. They are also the major soils in a wide, east-west belt which lies just north of the Black, Caspian, and Aral Seas and extends from eastern Europe into the USSR. The Argentinian pampas, also, generally consists of these soils.

The levels of exchangeable calcium, magnesium, and potassium may be extremely high in the Mollisols in contrast to their low levels in the Ultisols. However, although losses of phosphorus are not often readily apparent in these soils, indications are that this element is mobile, suggesting at least some loss of phosphorus during soil formation.

We are all aware that there is a definite loss of nutrients during erosion of soil after it has been cultivated; however, too often, we fail to realize that as a result of the soil formation process the soils may already be deficient in inorganic nutrients even before initial cultivation.

SOILS AND MAN

We must consider how soils relate to both man and the environment. To do this we must examine usable soils in terms of their locations and of man's present use of them.

Geography of Soils

The map in Fig. 4-17 shows, in general terms, the worldwide distribution of broad soil groups. From this map we can readily see that the usable soils (i.e., podzolic and chernozemic) of the world are concentrated in certain areas and are not evenly distributed. The soils common to most of the land surface area are: mountain, lateritic, and desert soils.

Mountain soils are characterized primarily by the presence of large amounts of unweathered or poorly weathered parent-rock material and by their shallowness (1-6 inches thick). Usually, the topography of the areas in which these soils prevail is so steep that the processes of soil formation simply cannot compensate for erosion and mass wasting.

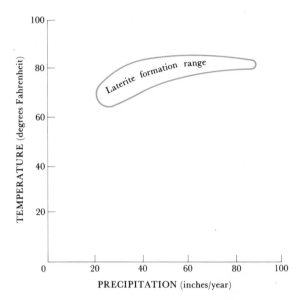

4-19
Optimum range of temperature and precipitation con-
ditions for the formation of laterite. (Reprinted by
permission from B. S. Persons, Laterite Genesis,
Location, Use, *New York: Plenum, 1970, p.6.)*

Lateritic soils are characterized by iron enrichment in certain layers of the soil profile. This type of soil development is particularly prevalent in regions with well-defined wet and dry seasons (Fig. 4-19) and is enhanced in flat areas in which iron leached from upper layers moves downward instead of being carried away. Laterite is formed *in situ* from the weathering of rocks. *Latosols* is a more appropriate name, since it includes those soil types in which iron, aluminum, or silica is concentrated. These soils form residual deposits in tropical and subtropical latitudes (Fig. 4-20).

Desert soils are estimated to represent about 17 percent of the earth's land surface, but the extent of desert area varies from continent to continent. For instance, Australia is about 44 percent desert, Eurasia 15 percent, and Africa 37 percent. Special weathering conditions are present in desert regions. Because of the lack of water, any weathering which takes place at or above the surface is physical weathering, but some chemical weathering does occur in the subsurface. Although the weathered material is usually well distributed by wind and sheet wash, the regolith (the surface mantle of material) is often quite coarse. The presence of salts, which accumulate during evaporation from low areas, is widespread in desert soils.

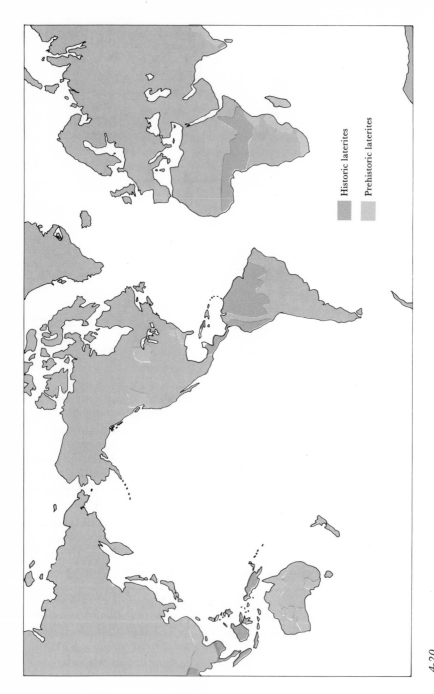

4-20
World distribution map of past and present laterite deposits. (Reprinted by permission from B. S. Persons, Laterite Genesis, Location, Use, New York: Plenum, 1970, p. 6.)

4-21
Obtaining soil samples with auger. (Photograph by E. F. Patterson, U.S. Geological Survey)

The available water is usually insufficient to induce downward leaching of materials from desert soils, but upward movement of mineral matter does occur commonly. The high rate of evaporation at the surface often draws up subsurface moisture, and any mineral material that is dissolved in this moisture is left at the surface as a crust. It is common to find crusts ("Hardpan" or caliche) of calcium carbonate, calcium sulphate, and iron oxides on the surface of desert soils.

Since plant growth is sparse in desert areas, supplies of organic matter are limited. This, coupled with strong oxidizing conditions at the surface and winds sufficient to remove dead vegetation, explains why little humus is found in desert soils. However, the most severe limitation mitigating against the use of desert soils is the lack of water. Many, if not most, desert soils contain a sufficient amount of usable inorganic nutrients, and in most cases the presence of salts at the surface is not prohibitive. Yet, the lack of water essentially makes the deserts of the world useless to us.

4-22
Measuring runoff from sloping plots of loam soil. (Photograph courtesy USDA—Soil Conservation Service)

The foregoing discussion shows that a significant portion of the land surface is currently unsuited for man's productive agricultural use. The soil either consists of unweathered or poorly weathered material on steep slopes (mountain soils), is water deficient and often salty (desert soils), or has been deprived of essential nutrients because of nutrient mobility, therefore turning bricklike shortly after being put into cultivation (lateritic soils). In these days of increasing population pressure and search for resources, we frequently hear that there still remains much land that is not being used, the implication being that we should cultivate it. The fact is, however, that man is using essentially all the land that can be used. While it is true that present utilization is not maximized and that man will ultimately be forced to cultivate land which is presently marginal, the really good soil is already in use. Indeed, some people feel it is being dangerously overused.

Erosion of Soils

Complicating the problem of fixed limits of usable soil available for production to serve man's vital needs is erosion, which steadily and increasingly is wearing away the remaining good soils. Erosion has been continuous throughout geologic time, the rate varying depending on prevailing conditions. Yet, there is little we can do to stop this natural disintegration of

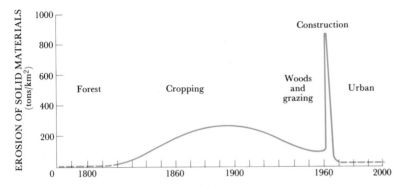

4-23

The sequence of erosional events in an area near Washington, D. C., as the dominant feature of the area changes from a forest through cultivation, back to woods, to a construction phase, and finally to that of an urban area. (Reprinted by permission from Sheldon Judson, "Erosion of the Land, or What's Happening to Our Continents?" American Scientist, Journal of the Society of the Sigma Xi *56, 4, 1968, p. 366.)*

the continents. But the crucial question is: What effect have man and his activities had on this rate?

Several attempts have been made to assess man's influence on the rate of erosion. One such study focused on a lake near Rome. Archaeological evidence indicated that significant human activity in that area dated from approximately the second century B. C. Examination of cores of sediment taken from the lake bed showed a sudden increase in sedimentation at that time, and the rate of sedimentation, though variable, has remained high since then. Extrapolation of these data indicates that erosion was lowering the surrounding land surface at about 2 to 3 cm per 1000 years prior to the invasion of man, and at an average rate of some 20 cm per 1000 years since.

Additional studies have indicated that when land undergoes intensive cultivation, the rate of erosion is three or more times greater than that when the land is under forest cover. One such study describes an area near Washington, D. C. (Fig. 4-23). Under original forest conditions the erosion was about 0.2 cm per 1000 years. In the early nineteenth century a rapid conversion of the forestland to farmland brought a rate of erosion of about 10 cm per 1000 years. With the partial reversion to grazing and forest land in the middle 1900s, the rate dropped to perhaps 5 cm per 1000 years. Construction in the area during the 1960s caused an extremely high rate of some 10 meters per 1000 years. Fortunately, urbanization has reduced the

4-24
Soil erosion in a wheat field, an example of increased erosion when soil is laid bare.
(Photograph by K. N. Phillips, U.S. Geological Survey)

rate in recent years to approximately 1 cm per 1000 years. With this evidence there can be no doubt that man's occupancy of an area significantly increases the erosion rate.

We might ask how these rates compare with those in other parts of the world. We can arrive at an approximate comparison by analyzing both the dissolved and solid material which some of the major rivers of the world carry. Since the source of these materials is the land area in the drainage basin which feeds the rivers, the analysis should give us a fairly reasonable estimate. In Table 4–7 these data are tabulated for three major areas which represent approximately 10 percent of the earth's land surface. In this table, The Great Basin, St. Lawrence, and Hudson Bay drainage areas are not considered in the United States' figures. To make the figure comparable, the United States' data has been adjusted to account for presumed increase of erosion rates brought about by man's cultivation of the land, since neither the Amazon nor the Congo basins has been significantly affected by this factor.

These kinds of studies have led to the estimation that before man came on the scene, the rivers of the world annually carried 9.3×10^9 metric tons

4-25
Erosion in the Ducktown district, Tennessee, caused after smelter fumes killed vegetation. (Photograph by A. Keith, U.S. Geological Survey)

of material from the land to the oceans. After man intervened by engaging in extensive cultivation, this figure rose sharply to 24×10^9 metric tons, which is about two and one-half times greater than the original rate.

Although these data are based on assumptions which are at times incomplete and imprecise, by analyzing them you can acquire an approximate, but generally correct, understanding of the complexity and magnitude of the problem of erosion.

Table 4–7 Rates of erosion for three major areas of the world

| Drainage region | Load, tons $\times 10^6$/yr | | | Erosion cm/1000 yr |
	Dissolved	Solid	Total	
Amazon River Basin	232	548	780	4.7
United States	292	248	540	3.0
Congo River Basin	99	34	133	2.0

(Reprinted by permission from Sheldon Judson, "Erosion of the Land, or What's Happening to Our Continents?" *American Scientist, Journal of the Society of Sigma Xi* 56, 4, 1968, p. 367.)

REFERENCES

Bloom, A. L. *The Surface of the Earth*, Englewood Cliffs, N. J.: Prentice-Hall, 1969.

Blumenstock, D. I. and C. W. Thornthwaite, "Climate and the world pattern," in *Climate and Man*, Yearbook of Agriculture, U. S. Department of Agriculture, 1941, pp. 98–127.

Bridgeman, P. W. "Water in the liquid and five solid states under pressure," *Proceedings of the American Academy of Arts and Sciences* 47 (1911–1912): 441–558.

Bridges, E. M. *World Soils*, London: Cambridge University Press, 1970.

Carroll, D. *Rock Weathering*, New York: Plenum Press, 1970.

Cline, M. G. "Basic principles of soil classification," in *Selected Papers in Soil Formation and Classification*, ed. J. V. Drew, *et al.*, SSSA Special Pub. 1 (1967): 381–392.

Crocker, R. I. "The plant factor in soil formation," in *Selected Papers in Soil Formation and Classification*, pp. 179–190.

Donahue, Roy L. *Soils—An Introduction to Soils and Plant Growth*, 2d. ed., Englewood Cliffs, N. J.: Prentice-Hall, 1965.

Emery, K. O. "Weathering of the Great Pyramid," *Journal of Sedimentary Petrology* 30, 1 (1960): 140–143.

Gibson, J. S. and J. W. Batten. *Soils—Their Nature, Classes, Distribution, Uses and Care*, University: University of Alabama Press, 1970.

Gilluly, J., A. C. Waters, and A. O. Woodford. *Principles of Geology*, 3rd ed., San Francisco: Freeman, 1968.

Goodchild, J. G. "Notes on some observed rates of weathering of limestones," *Geological Magazine* 27 (1890): 436–466.

Griggs, D. T. "The factor of fatigue in rock exfoliation," *Journal of Geology* 44 (1936): 783–796.

Judson, S. "Erosion of the land, or what's happening to our continents," *American Scientist* 56, 4 (1968): 356–374.

Keller, W. D. *Principles of Chemical Weathering*, rev. ed., Columbia, Missouri: Lucas Brothers, 1956, 1968.

————. *Chemistry in Introductory Geology*, 4th ed., Columbia, Missouri: Lucas Brothers, 1969.

Kellogg, C. E. "Soil," *Scientific American Offprint*, 821, 1950.

Livingstone, D. A. *Chemical Composition of Rivers and Lakes*, U. S. Geological Survey Professional Paper 440-G, 1963.

Lovering, T. S. "Significance of Accumulator plants in rock weathering," *Bulletin of the Geological Society of America* 70 (1959): 463–466.

McNeil, M. "Lateritic soils," *Scientific American Offprint*, 870, 1964.

Nelson, L. B. and R. E. Uhland. "Factors that influence loss of fall applied fertilizers and their probable importance in different sections of the U. S.," *Proceedings of the Soil Society of America* 19(1955): 492–496.

Ollier, C. D. *Weathering*, Edinburgh: Oliver and Boyd, 1969.

Persons, B. S. *Laterite—Genesis, Location, Use*, New York: Plenum Press, 1970.

Simonson, R. W. "What soils are," in *Soil*, United States Department of Agriculture Yearbook (1956): 17–30.

_____ . "Outline of a generalized theory of soil genesis," in *Selected Papers in Soil Formation and Classification, op. cit.,* pp. 191–208.

_____ . "Soil classification in the United States," *ibid.,* pp. 415–428.

_____ . "Loss of nutrient elements during soil formation," in *Nutrient Mobility in Soils: Accumulation and Losses,* ed. O. P. Engelstad, SSSA Publication 4 (1970): 21–46.

Soil Survey Division. "Soils of the United States," in *Soils and Men,* Yearbook of Agriculture, U.S. Department of Agriculture, 1938, pp. 1019–1161.

Soil Survey Staff. *Soil Classification—A Comprehensive System—7th Approximation,* Soil Conservation Service, United States Department of Agriculture, 1960.

Thomas, G. W. "Soil and climatic factors which affect nutrient mobility," in *Nutrient Mobility in Soils; Accumulation and Losses, op. cit.,* pp. 1–20.

Thorp, J. "Effects of certain animals that live in soils," in *Selected Papers in Soil Formations and Classification, op. cit.,* pp. 191–200.

Young, Keith. "Gonwanaland and the developing nations," Environmental Geology lecture notes, Milwaukee, November 9–10, 1970.

_____ . "Man in the geobiocoencse," *ibid.*

Salt, which has always been of great importance to man, is a nonmetallic mineral resource. (Photograph by George Sheng)

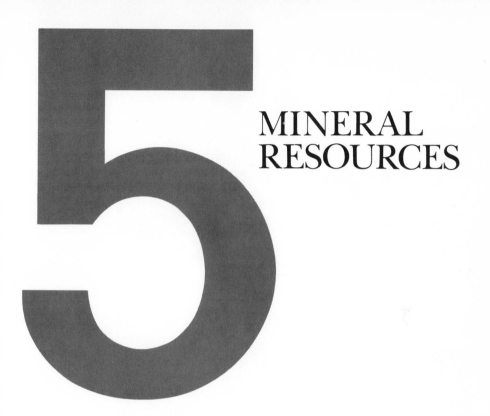

MINERAL RESOURCES

MAN'S USE OF MINERAL RESOURCES

The world's people depend on materials from the earth to sustain life and to provide housing, transportation, and the "extras" of life. Almost every manufactured product is made of materials that have been obtained from the minerals of the earth. The extensive use of minerals, many of which are in short supply, by the developed countries has produced a sense of uneasiness among many scientists, economists, and businessmen throughout the world. This feeling of uneasiness, even urgency, resulting from this drain on our mineral resources often does not affect the layman who does not understand how elements originate. In discussions about the use of our mineral resources, it is not at all unusual to hear statements that all we need to do is reduce our demands to give the earth time to replenish our supply of lead, zinc, or other minerals.

5-1
*Panning for gold, Seward Peninsula, Alaska, 1900. (Photograph by
A. H. Brooks, U.S. Geological Survey)*

ORIGIN OF ELEMENTS

All matter is composed of elements. An element is a substance having *one
atomic number,* which is the number of protons in the nucleus of the atom.
An atom cannot be broken into another substance except through a nuclear
process. Here, we are interested both in compounds made up of several elements and substances made of only one element.

Although the details of the process of creating elements are not fully
known, a model of the general process, consistent with observed data, has
been developed. In 1937 V. M. Goldschmidt compiled a table showing
cosmic abundances of the elements. Later work has altered some of the
details, but none of the major features, of this table (Table 5-1). Any
theory concerning the origin of elements must therefore be consistent
with these abundance data.

Table 5-1 Cosmic abundance of the chemical elements

Element	Atomic number	Atomic weight	Relative percent of atoms
Hydrogen	1	1	86.64
Helium	2	4	13.16
Oxygen	8	16	0.09
Neon	10	20	0.05
Carbon	6	12	0.03
Nitrogen	7	14	0.008
Silicon	14	28	0.003
Magnesium	12	24	0.003
Sulfur	16	32	0.001
Argon	18	40	0.0005
Iron	26	56	0.0005
Aluminum	13	27	0.0003
Calcium	20	40	0.0002
Sodium	11	23	0.0002
Nickel	28	59	0.00009
Phosphorus	15	31	0.00003
Chromium	24	52	0.00003
Manganese	25	55	0.00002
Potassium	19	39	0.00001
Chlorine	17	35	0.000009
Cobalt	27	59	0.000006
Titanium	22	48	0.000006
Fluorine	9	19	0.000006
Vanadium	23	51	0.0000007
Copper	29	64	0.0000007
Zinc	30	65	0.0000007
Lithium	3	7	0.0000003
Strontium	38	88	0.0000002
Krypton	36	84	0.0000001
Scandium	21	45	0.0000001
Germanium	32	73	0.0000001
Beryllium	4	9	7.0 ⎫
Boron	5	11	7.0 ⎬ $\times 10^{-8}$
Selenium	34	79	7.0 ⎪
Lead	82	207	7.0 ⎭
Zirconium	40	91	48.0 ⎫
Gallium	31	70	31.0 ⎪
Yttrium	39	89	31.0 ⎪
Rubidium	37	85	21.0 ⎬ $\times 10^{-9}$
Bromine	35	80	14.0 ⎪
Barium	56	137	14.0 ⎪
Tellurium	52	128	10.0 ⎭
Xenon	54	131	10.0 $\times 10^{-9}$

Table 5-1 (cont'd.)

Element	Atomic number	Atomic weight	Relative percent of atoms	
Arsenic	33	75	69.0	
Molybdenum	42	96	69.0	
Tin	50	119	34.0	
Platinum	78	195	34.0	
Ruthenium	44	102	31.0	
Cadmium	48	112	31.0	
Neodymium	60	144	31.0	
Niobium	41	93	28.0	
Palladium	46	107	24.0	
Dysprosium	66	162	24.0	
Iodine	53	127	21.0	
Cerium	58	140	21.0	$\times 10^{-10}$
Erbium	68	167	21.0	
Osmium	76	190	21.0	
Cesium	55	133	17.0	
Lanthanum	57	139	17.0	
Gadolinium	64	157	17.0	
Iridium	77	193	17.0	
Ytterbium	70	173	14.0	
Mercury	80	201	14.0	
Silver	47	108	11.0	
Thallium	81	204	11.0	
Bismuth	83	209	11.0	
Rhodium	45	103	69.0	
Antimony	51	122	69.0	
Praseodymium	59	141	69.0	
Samarium	62	150	69.0	
Holmium	67	165	69.0	
Indium	49	115	34.0	
Europium	63	152	34.0	
Hafnium	72	179	34.0	$\times 10^{-11}$
Tungsten	74	184	34.0	
Gold	79	197	34.0	
Terbium	65	159	31.0	
Thulium	69	169	31.0	
Rhenium	75	186	17.0	
Lutetium	71	175	14.0	
Thorium	90	232	11.0	
Tantalum	73	181	69.0	$\times 10^{-12}$
Uranium	92	238	28.0	

(Data on relative percent of atoms adapted by permission from R. J. Foster, *General Geology*, New York: Charles E. Merrill, 1969, p. 466.)

Prior to the mid-1950s the prevailing concept used to explain the origin of the elements was one offered by George Gammow and others. This theory suggested that at some time in the past, some 9 to 12 billion years ago, all matter in the universe was concentrated at a central point. Because of this concentration of matter, tremendous temperature and pressure were generated within this mass and were responsible for forming the elements. Using hydrogen as a basic working material, the great temperature and pressure forced hydrogen nuclei to "fuse," forming heavier atoms. The heavier atoms in turn were forced together to form progressively heavier atoms, until most of the elements had been formed. The result of this concentration of matter was an explosion of gigantic proportions, giving this concept the name of the "Big Bang" theory. The explosion flung the newly formed matter outward in all directions from this central point, and it was this material which formed all the galaxies and solar systems.

According to this theory, the initial cosmic abundances which originated in the "Big Bang" remained essentially fixed. The stars formed from this material underwent only very minor changes, such as an increase of the helium abundance with a concurrent decrease of the hydrogen, and a destruction of small amounts of lithium, boron, and beryllium. These were not considered to be significant changes, however, and the initial ratios remained basically constant.

Later work on abundance data showed a system of such complexities that no single process could account for all of them. In 1957 Burbridge and others examined the problem of element creation by using hydrogen "burning" to produce helium, then by using a second process of helium "burning" to produce carbon–23, oxygen–16, neon–20, and perhaps magnesium–24. A third process, in which alpha-particles were successively added to oxygen–16 and neon–20, produced magnesium–24, silicon–28, sulfur–32, argon–36, and calcium–40. In addition, five other proton and neutron processes accounted for the creation of the other elements. The abundances derived theoretically from these processes were consistent with known data, and the processes themselves were consistent with observed facts of stellar evolution.

The recent discovery of technetium–99 in the spectra of stars tends to confirm Burbridge's work. This discovery supports the suggestion that elements are continually being formed in the universe. If all matter had been formed by the "Big Bang" several billion years ago, as suggested by Gamow, then technetium, which has a half-life of some 100,000 years, could not possibly still be detectable in stars. The primary "life-force" is the conversion of hydrogen to helium, with the associated expulsion of huge quantities of energy. However, the process does not stop there, for as the star passes through the various stages of its development, different elements are formed. Hence, the formation of elements occurs exclusively under conditions of "nuclear burning" in the interiors of stars. The lighter elements are produced throughout this normal evolution of a star; but it is only in the final stage of the evolutionary process, the explosion of the star (supernova), that the remaining, heavier elements are created.

Although details of this process are either missing or as yet poorly understood, we do know enough to state that the elements available to us on earth had an extraterrestrial origin and that the processes necessary to produce new elements do not occur on earth. This excludes, of course, those minute amounts of elements formed as products of fission during the natural breakdown of unstable radioisotopes. For several reasons, this process cannot be regarded as a source by which to replenish our natural resources. First, the amount of material produced during fission is insignificant. Second, the elements produced in this breakdown are not those which are commonly used or needed. Third, in many cases, the half-life (the time it takes for half of the parent material to break down to its end-product, is of such duration that the ultimate production of the end-product is completely inconsequential. A clear example of this is the production of lead–206 through the disintegration of uranium–238, which has a half-life of about 4.5 billion years. Fourth, the radioisotopes which are the endproduct of this breakdown account for only a fraction of a percent of the mass of the earth. Thus, it should be clear that we must be capable of living with what we have, without hoping to increase the available resources in the forseeable future.

THE MAKE-UP OF MINERAL RESOURCES

Elements

About 90 elements occur naturally; man has been able to discover others in the laboratory, bringing the known number to 105. Although there are 105 elements, 99 percent of the earth's crust, that thin outer layer which extends from the surface to an average depth of ten miles, is composed, by weight, of only eight elements. These eight, shown in Table 5–2, constitute nearly all the visible rock and soil at, and below, the surface of the earth.

Cosmic Abundance

The chemical elements and present estimates of their abundance in the universe are listed in Table 5–1 and depicted graphically in Fig. 5–2. Two things are evident from Tables 5–1 and 5–2. The first is that about 99 percent of all matter in the universe is composed of either hydrogen or helium. All other elements combined represent only about 1 percent of the total matter in the universe. Since our solar system was made from cosmic mate-

Table 5-2 The most abundant elements in the earth's crust

Element	Weight (percent)	Volume (percent)
Oxygen (O)	46.60	91.97
Silicon (Si)	27.72	0.80
Aluminum (Al)	8.13	0.77
Iron (Fe)	5.00	0.68
Calcium (Ca)	3.63	1.48
Sodium (Na)	2.83	1.60
Potassium (K)	2.59	2.14
Magnesium (Mg)	2.09	0.56
	98.59	100.00

(Adapted by permission from Brian Mason, *Principles of Geochemistry*, 3d ed., New York: Wiley, 1966, pp. 45–46.)

rial, we would expect the overall composition of newly formed systems to be approximately proportional to ours (as shown in Table 5–1), at least during the early history of the system. The second point is that when Tables 5–1 and 5–3 are compared, the major difference between the crustal and cosmic abundances of the elements is the relative lack of hydrogen and helium in the crust. Another minor difference is the absence in the crust of the noble gases which do not combine with other elements. This evidence supports the idea that our solar system was formed from cosmic material and also indicates that the earth, in its nearly five-billion-year history, has not altered the abundances by adding to certain elements. If the earth were capable of manufacturing elements, we could not expect as small a divergence between these two tables as we have.

Compounds and Minerals

When elements are chemically combined, they form *compounds*. These compounds may be either organic or inorganic. If these compounds satisfy certain conditions, we call them minerals. There are many definitions for the word mineral, and perhaps no single definition exists. According to one definition, however, a *mineral* is a compound that fulfills the following conditions:

1. It is a *naturally* occurring *inorganic* substance.
2. Its composition can be represented by a chemical formula.
3. It is crystalline, i.e., has a definite internal structure.
4. It has physical properties that are fixed and controlled by its composition and structure.

This definition excludes some substances that other definitions would include, but in most cases it is applicable.

Abundance relative to silicon = 10^6

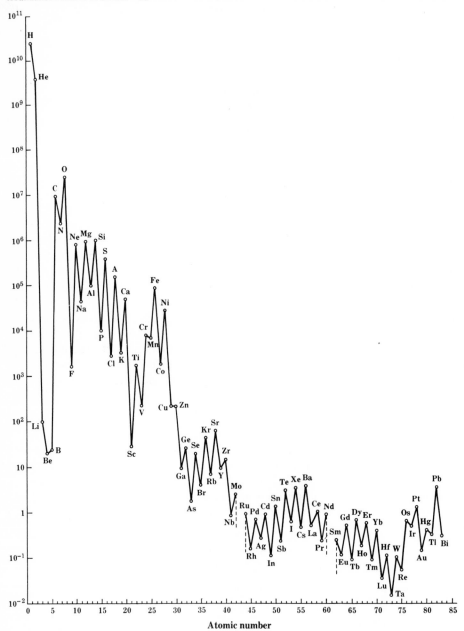

Atomic number

◀ 5-2
Cosmic abundance of elements showing generally decreasing abundance with higher atomic number. The even-numbered elements are more abundant than the odd-numbered elements on either side of them. The vertical scale is logarithmic. (From Distribution of the Elements in Our Planet *by L. H. Ahrens. Copyright 1965 McGraw-Hill. Used with permission of McGraw-Hill Book Company.)*

Table 5–3 Average crustal abundance of the elements (excluding the rare gases and short-lived radioactive elements) in parts per million (ppm). To convert numbers to percent composition, move decimal four places to the left (i.e., Sr=0.0375%).

Element	ppm	Element	ppm	Element	ppm
O	466,000	Y	33	Ge	1.5
Si	277,200	La	30	Mo	1.5
Al	81,300	Nd	28	W	1.5
Fe	50,000	Co	25	Eu	1.2
Ca	36,300	Sc	22	Ho	1.2
Na	28,300	Li	20	Tb	0.9
K	25,900	N	20	I	0.5
Mg	20,900	Nb	20	Tm	0.5
Ti	4,400	Ga	15	Lu	0.5
H	1,400	Pb	13	Tl	0.5
P	1,050	B	10	Cd	0.2
Mn	950	Pr	8.2	Sb	0.2
F	625	Th	7.2	Bi	0.2
Ba	425	Sm	6.0	In	0.1
Sr	375	Gd	5.4	Hg	0.08
S	260	Yb	3.4	Ag	0.07
C	200	Cs	3.0	Se	0.05
Zr	165	Dy	3.0	Ru	0.01
V	135	Hf	3.0	Pd	0.01
Cl	130	Be	2.8	Te	0.01
Cr	100	Er	2.8	Pt	0.01
Rb	90	Br	2.5	Rh	0.005
Ni	75	Sn	2.0	Os	0.005
Zn	70	Ta	2.0	Au	0.004
Ce	60	As	1.8	Re	0.001
Cu	55	Y	1.8	Ir	0.001

(Adapted by permission from Brian Mason, *Principles of Geochemistry*, 3rd ed., New York: Wiley, 1966, pp. 45–46.)

Table 5-4 Several minerals containing metallic elements, their formulas, and the weight percent of the principal metallic element. Ore minerals are in italics.

Metallic element	Mineral	Formula	Weight (percent metal)
Chromium (Cr)	*Chromite*	$FeO \cdot Cr_2O_3$	46%
	Crocoite	$PbCrO_4$	10%
Cobalt (Co)	*Smaltite*	$CoAs_2$	28%
	Cobaltite	$CoAsS$	16%
	Bierberite	$CoSO_4 \cdot 7H_2O$	9%
Iron (Fe)	*Hematite*	Fe_2O_3	70%
	Magnetite	$FeO \cdot Fe_2O_3$	72%
	Chromite	$FeO \cdot Cr_2O_3$	25%
	Siderite	$FeCO_3$	48%
	Limonite	$2Fe_2O_3 \cdot 3H_2O$	60%
	Acmite	$NaFe(SiO_3)_2$	24%
Manganese (Mn)	*Pyrolusite*	MnO_2	53%
	Psilomelane	$(Ba,H_2O)_2Mn_5O_{10}$	30%
	Manganite	$Mn_2O_3 \cdot H_2O$	62%
	Braunite	$3Mn_2O_3 \cdot MnSiO_3$	23%
	Carpholite	$H_4MnAl_2Si_2O_{10}$	17%
Molybdenum (Mo)	*Molybdenite*	MoS_2	60%
	Wulfenite	$PbMoO_4$	26%
	Chillagite	$3PbWo_4 \cdot PbMoO_4$	5%
Nickel (Ni)	*Pentlandite*	$(Fe,Ni)S$	40%
	Garnierite	$H_2(Ni,Mg)SiO_4 \cdot H_2O$	30%
	Millerite	NiS	65%
	Connarite	$H_4Ni_2Si_3O_{10}$	32%
Tungsten (W)	*Wolframite*	$(Fe,Mn)WO_4$	51%
	Scheelite	$CaWO_4$	64%
	Chillagite	$3PbWO_4 \cdot PbMoO_4$	11%
Vanadium (V)	*Patronite*	VS_4	28%
	Vanadinite	$Pb_4((PbCl)VO_4)_3$	11%
	Descloizite	$(Pb,Zn)_2(OH)VO_4$	7%
	Rossite	$CaO \cdot V_2O_5 \cdot 4H_2O$	33%
Aluminum (Al)	*Bauxite*	$Al_2O_3 \cdot 2H_2O$	42%
	Corundum	Al_2O_3	53%
	Spinel	$MgO \cdot Al_2O_3$	38%
Cadmium (Cd)	*Greenockite*	CdS	78%
	Otavite	$CdCO_3$	65%
	Cadmium Oxide	CdO	88%
Copper (Cu)	*Native Copper*	Cu	100%
	Bornite	Cu_5FeS_4	63%
	Chalcocite	Cu_2S	80%
	Cuprite	Cu_2O	89%
	Mitscherlichite	$2KCl \cdot CuCl_2 \cdot 2H_2O$	20%
	Chalcopyrite	$CuFeS_2$	

Table 5-4 (cont'd)

Metallic element	Mineral	Formula	Weight (percent metal)
Lead (Pb)	*Galena*	PbS	87%
	Cerussite	$PbCO_3$	78%
	Anglesite	$PbSO_4$	68%
	Dundasite	$Pb(AlO)_2(CO_3)_2$	50%
Mercury (Hg)	*Cinnabar*	HgS	86%
	Livingstonite	$HgS \cdot 2Sb_2S_3$	35%
Silver (Ag)	*Native Silver*	Ag	100%
	Argentite	Ag_2S	87%
	Cerargyrite	$AgCl$	75%
	Matildite	$Ag_2S \cdot Bi_2S_3$	28%
Tin (Sn)	*Stannite*	$Cu_2S \cdot FeS \cdot SnS_2$	28%
	Cassiterite	SnO_2	79%
	Teallite	$PbSnS_2$	30%
Titanium (Ti)	*Ilmenite*	$FeTiO_3$	32%
	Rutile	TiO_2	60%
	Titanite	$CaTiSiO_5$	24%
	Oliveiraite	$3ZrO_2 \cdot 2TiO_2 \cdot 2H_2O$	16%
Zinc (Zn)	*Sphalerite*	ZnS	67%
	Smithsonite	$ZnCo_3$	52%
	Hemimorphite	$HZnSiO_5$	37%
	Zincite	ZnO	80%
	Larsenite	$PbZnSiO_4$	18%

There are about 2500 known minerals, not including varieties. Of these only about 250 are considered to be abundant, and of these only about 25 are common rock-forming minerals. Thus, the minerals, like the elements, are relatively numerous; similarly, most are measurably insignificant.

You will notice that most of the minerals listed in Appendix Table A-1 are silicates (that is, those minerals containing the elements silicon and oxygen). Indeed, about 95 percent of the crust of the earth is composed of silicates. Table 5-2 shows why this is the case: the two elements silicon and oxygen, by weight, make up almost 75 percent of the crust. Therefore, the silicates should be the most abundant minerals.

Concentrations in Deposits

Ore is an aggregate of minerals contained in rock, in which at least one of the minerals becomes profitable when extracted. Thus, all mineral deposits are not ore bodies, but what is considered simply a deposit today may be ore tomorrow. Current economic situations determine what is and is not ore.

To be considered of some value, then, an element must generally be concentrated above the level of its crustal abundance (Table 5-3). Table 5-4

CONCENTRATION ABOVE CRUSTAL ABUNDANCE NECESSARY FOR ECONOMIC
RECOVERY

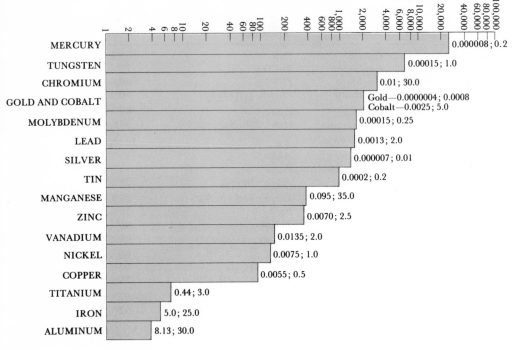

5-3

*In general, the lower the crustal abundance of an element, the greater its concentration
must be before economical recovery is possible. The first number is the crustal abundance (taken from Table 5-3); the second number is the minimum metal content an ore
must have in order to be mined economically.*

lists the primary ore minerals, along with some non-ore minerals, which
comprise most of the metallic elements commonly in use today. In each
case the percent, by weight, of the element has been calculated to illustrate
the range of concentrations commonly occurring in minerals.

A second level of concentration which must be considered is the amount
of the desired mineral present, relative to the enclosing non-ore rock (that
is, country rock). A highly enriched mineral, e.g., native copper—100 percent, might be dispersed so sparsely through the country rock that several
tons of rock must be processed to extract a few ounces of copper. This procedure may be less desirable, perhaps, than mining rock containing a mineral such as bornite, which has only 63 percent copper, but which might be
heavily concentrated in the rock.

Concentrations within ore bodies are always above the level of average crustal abundance for the element being mined. The crustal abundances of metallic elements are inversely related to the concentrations necessary to make their mining profitable. The metal content required for economical production also depends greatly on the metallurgical capacity to refine the element from its ore. Figure 5-3 illustrates the concentration of the metallic element relative to its ore body which is necessary for the mineral element to be mined profitably. Some production below these levels takes place, but only when these metals can be produced as by-products in the processing of other metals. One should keep in mind, however, that the total volume of all economic mineral deposits is very insignificant in comparison to the total volume of the crust.

Geographic Distribution

The vagaries of time and geologic processes have resulted in an extremely irregular distribution of mineral resources on this planet. As Table 5-5 shows, the known reserves of most minerals are mostly present in a very few countries; and each country lacks reserves of at least some useful minerals.

Much of the present distribution of resources is related to the mountain-building processes (Figs. 5-5 and 5-6) and to continental drift. For example, it has been suggested that the tin belts of the world were originally contiguous (Fig. 5-7). Initial formation by mountain-building processes and subsequent separation by continental drift may account for the present irregular pattern. The concept that at one time the continents were joined into one supercontinent and have drifted apart over the last 200 million years is one that has become widely accepted in geology. We will discuss it further in Chapter 10.

When we consider that many of our significant ore deposits were formed by very specialized processes operating under unique conditions in limited areas, it becomes obvious why the present deposits are localized. For example, *pegmatites* are coarse-grained igneous rocks that usually occur as localized masses and often support concentrations of ore minerals along the contact between the igneous body and the intruded rock. *Hydrothermal* processes involve ore minerals which are deposited from fluids in cracks and fractures of existing rock. These fractures may have been formed as the result of the emplacement of an igneous body, and the hot fluids involved in hydrothermal change may have been associated with molten rock which was intruded. The ore bodies are localized because they are the result of events which were themselves very localized.

Table 5-5 Location of known world reserves of selected mineral elements and percent
of the total known reserve held by the country. Only those places holding the
largest reserves are shown.

Mineral element and location	Percent	Mineral element and location	Percent
FERROUS		Cadmium	
Chromium		Europe	37.0
Repub. of So. Africa	74.2	North America (excluding USA)	30.0
Rhodesia	22.6	Australia, Africa, Asia, and	
USSR	1.9	South America	33.0
Cobalt		Copper	
Zaire (Congo)	31.2	USA	27.8
New Caledonia	18.3	Chile	19.3
Zambia	15.9	USSR	12.5
Cuba	15.5	Zambia	9.7
USSR	9.4	Peru	8.0
Canada	8.0	Zaire	6.5
Iron		Lead	
USSR	32.0	USA	37.0
South America	18.9	Canada	12.6
Canada	12.1	Eastern Europe	12.6
Middle East, Asia, and		Australia	10.5
Far East	11.5	Western Europe	8.4
Australia, New Zealand,		South America	5.2
New Caledonia	10.6	Asia	5.2
Manganese		Mercury	
Republic of So. Africa	37.7	Spain	31.2
USSR	25.1	Italy	21.9
Gabon	12.0	People's Republic of China	12.5
USA	8.5	Yugoslavia	12.5
Molybdenum		USSR	9.4
USA	58.2	Platinum Group	
USSR	18.5	Repub. of So. Africa	47.2
Chile	16.2	USSR	47.2
Nickel		Selenium	
Cuba	24.5	USA	26.5
New Caledonia	22.4	Chile	15.9
USSR	13.6	USSR	12.2
Canada	13.6	Zambia	9.5
Indonesia	10.8	Peru	7.4
Tungsten		Zaire	6.4
People's Republic of China	74.4	Canada	3.2
USA	6.7	Silver	
South Korea	3.6	Communist countries, except	
Bolivia	3.1	Yugoslavia	36.0

Table 5–5 (cont'd.)

Mineral element and location	Percent	Mineral element and location	Percent
Vanadium		USA	24.0
USSR	59.3	Mexico	13.3
Repub. of So. Africa	19.8	Canada	11.6
Australia	14.8	Peru	9.6
NONFERROUS		**Tin**	
		Thailand	32.4
Aluminum		Malaysia	13.8
Oceania	34.3	Indonesia	12.7
Africa	27.1	People's Republic of China	11.5
Jamaica	10.3	Bolivia	
USSR	5.1	**Titanium**	
Yugoslavia	3.4	Norway	20.4
Surinam	3.4	USA	17.2
Hungary	2.6	Canada	17.2
People's Republic of China	2.6	USSR	17.1
		India	10.2
Zinc		United Arab Republic	6.8
USA	27.2		
Canada	20.2		
Eastern Europe	11.3		
Western Europe	11.3		
Asia	8.1		
Australia	7.3		
NONMETALLIC			
Boron			
USA	50.0		
USSR	20.8		
Turkey	16.7		
Phosphorus			
Morocco	42.2		
USA	31.2		
USSR	11.9		
Sulphur			
Near East and S. Asia	44.1		
Eastern Europe	15.8		
USA	12.4		
Central and South America	9.3		
Canada	6.3		

(Data from U. S. Bureau of Mines, *Bulletin* 650)

5-4
Beryl crystals in pegmatite. The hammer is about one foot long. (Photograph by G. C. Eddy, U.S. Geological Survey)

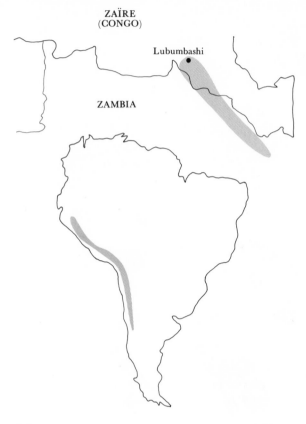

5-6
*Alignment of the zones in which porphyry copper deposits
occur along the western part of South America and in
which the copper deposits known as the Zambian Copper
Belt occur in the central part of southern Africa. The
Zambian belt, which may have been deposited as copper-
rich sediments along a shoreline during the Precambrian,
presently produces 12%–15% of the world's copper.*

◀ 5-5
*Alignment of the zones in which porphyry copper deposits (in color) and disseminated
molybdenum deposits (gray) in the western United States. The molybdenum deposits occur near the eastern margin of the North American Cordillera.*

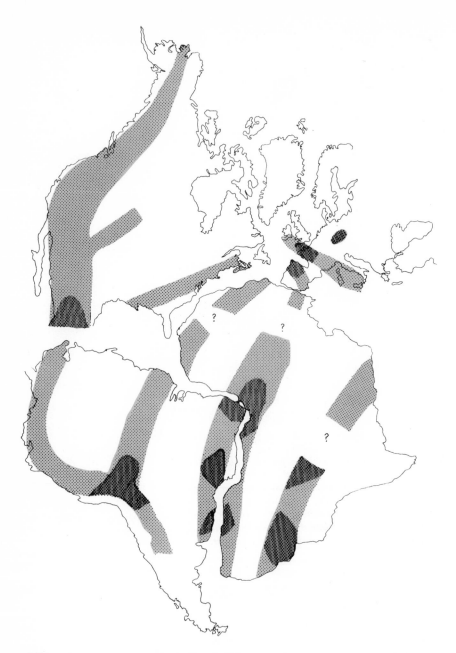

5-7
*Possible relation of the major tin belts around the Atlantic Ocean. (Reprinted by permis-
sion from R. D. Schuiling, "Tin Belts on the Continents Around the Atlantic Ocean,"*
Economic Geology **62**, *1967, p. 548.)*

Table 5-6 Classes of mineral products

Value	GROUP		
	Rock	Single minerals	Elemental
Large volume, low priced	Sand, gravel, limestone for cement	Salt (halite)	Iron
		Clays and feldspar	Lead
↑	Potash and phosphates		Tin
		Graphite	
		Mica	Silver
	Building stone (limestone sandstone granite)	Quartz crystals Industrial diamonds	Uranium
↓			Radium
Small volume, high- priced	Ornamental stone (granite marble)	Gem diamonds	

The Formation and Classification of Mineral Deposits

Mineral deposits are formed through various processes, many of which are highly complex. Aggregations of mineral matter constitute *rocks*. In general, rocks form in three ways: (1) by the solidifying of molten earth material (magma), (2) through the accumulation of cemented sediments or chemically precipitated materials, or (3) by the alteration, effected by heat and pressure, of already-existing rocks. These three processes produce igneous , sedimentary, and metamorphic rocks, respectively (Appendix A). Each of the three great rock groups can be further subdivided, and because there are many variations in their origins and mineral composition, numerous rock classifications are possible. However, the nature of a given mass of rock can generally be classified in terms of two factors: (1) the elements available to form its minerals and (2) the particular conditions of temperature, pressure, and chemical activity in the area.

There are many ways in which products from the earth can be grouped, depending on whether we wish to categorize them by economic value, origin, usage, or by some other way. To simplify classification we can group them as shown in Table 5-6, which also gives examples of possible categories. From the table we can see that rocks can be used as bulk products, or that separate minerals can be selectively extracted from them, or that individual elements can be recovered from ore minerals in a rock.

MINERALS FROM THE LAND

Until now man has obtained nearly all of his mineral supplies from deposits on the land, especially those deposits that were most accessible, concentrated in easily worked masses, and easily processed. In ancient times, for example, man used rock to build the pyramids of Egypt, and mined copper from the Sinai peninsula or gold from the Andes, where these materials could be recovered with minimal technological knowledge. Later, the invention of drilling, blasting, and other mining techniques enabled man to mine below the surface. Now, man's advanced technology helps him produce materials from both progressively more inaccessible places and increasingly lower-grade deposits.

There are many important minerals. Some, like platinum, are quantitatively minor but extremely useful and vital. In terms of widest usage and greatest quantity, however, there are a few minerals which can be identified as "most important." Let us look at some of these—both metals and nonmetals.

METALLIC RESOURCES

The Ferrous Metals

This group consists of iron and those metals which are added to iron to make a variety of iron and steel products. In addition to iron, this group includes manganese, silicon, nickel, cobalt, chromium, rhenium, vanadium, molybdenum, niobium, tungsten, and tantalum.

Of all the iron ore processed, approximately 96 percent is processed in blast furnaces to make pig iron. Of the remaining 4 percent, about 3 percent is sent directly to steel furnaces, and the remaining 1 percent is devoted to various miscellaneous uses. The pig iron produced in the blast furnaces is the major resource needed by the steel industry, which consumes about 90 percent of this pig iron. The bulk of the remaining 10 percent is remelted in coke-burning furnaces to form cast iron.

In addition to iron, cast iron contains carbon, silicon, and smaller amounts of phosphorus, sulfur, and manganese. Cast iron is very hard, brittle, and heavy and cannot be worked at any temperature. Instead, it is cast in preshaped molds to form a variety of products. The small amount of pig iron not supplied to either the steel industry or the cast iron works is used to make wrought iron. This alloy has a lower carbon content than cast iron, but is tough, soft, and easily worked.

The pig iron used by the steel producers is alloyed with the other mineral elements of this group to form a variety of steel types suitable for many purposes. Table 5-7 lists the properties imparted to steel through the addition of these particular elements.

Two elements in the ferrous group, rhenium and tantalum, are not presently important as constituents in alloy steel. They do, however, have very definite uses. Commercially, rhenium is being used primarily in thermo-couples and high-temperature electrical control systems. Small quantities are combined with tungsten to form structural alloys, and some rhenium is used in flash bulbs and in highly specialized parts for variable-wing aircraft.

Tantalum, once used in ferro-alloys, is presently used primarily in electronics, in the production of optical glass, in special alloys such as tantalum-carbide used in making cutting tools, and in synthetic fiber production. Indications are that tantalum alloys will be increasingly used both in the aerospace industry and in nuclear applications. The use of this element in 1969 was 78 percent greater than in the previous year.

The Nonferrous Minerals

The nonferrous group of minerals, so named because they are seldom used in iron alloys, consists of some 47 elements. The most common of these are copper, aluminum, mercury, lead, tin, zinc, beryllium, magnesium, lithium, and titanium. Among the nonferrous minerals, copper and aluminum alone account for most of the quantity used and for nearly two-thirds of the dollar value of the entire group of 47 minerals. The projected demands for these two mineral elements through the end of the century are also among the highest in this group, largely due to the projected increase in the use of electricity. Currently, the generation, distribution, and utilization of electrical power accounts for nearly half the copper and 13 percent of the aluminum consumed in this country. However, present research into the superconducting properties of supercooled metals and other techniques for transmitting electrical current may, if fruitful, reduce the projected demands for these mineral elements. Future demands for some of the special-property elements in this group are contingent on developments in such fields as electronic communications, copying methods, nuclear power, and the plastic and glass industries.

Some Important Metals

Iron   The most important metal of the industrial world is iron. The second most common metal in the earth's crust, iron is also one of the most readily obtained and useful metals. There is no particular foreseeable problem in obtaining supplies that will last long into the future, although many of the most obvious, most accessible, and richest deposits, e.g., those in northern Minnesota, are rapidly being depleted.

Table 5-7 Common alloy elements in the ferrous group are listed, showing the kind of steel they produce and other uses to which the element may be put

Element	Property of steel	Use of the steel
Chromium	Improves hardness, prevents rust, gives high tensil strength.	Dies, cutting edges for high-speed tools, ball bearings, safes, armor plate for ships and tanks.
	Alloys with more than 10% chromium are called stainless steel.	Flatware, kitchen, hospital, and dairy equipment, transportation items.
Nickel	Increases toughness, resistance to heat and acids, and ductility; highly elastic.	Armor plate, structural steel for bridges, machine parts.
Manganese	Increases strength and resistance to wear.	Rock crushers, large digging equipment and curved railroad rails.
Molybdenum	Increases strength, resistance to heat and corrosion, and makes it more shock resistant.	Auto, aircraft and missile parts, tools, machinery.
Tungsten	Retains hardness at high temperatures.	Tool steel for metal cutting devices, e.g., lathes, saws and drills, permanent magnets.
Vanadium	Increases strength, springiness, and resistance to fatigue.	Auto frames, shafts, axles, locomotive parts, springs, nuclear reactors.
Cobalt	Gives strength at high temperatures.	Permanent-magnet steel, jet engines and gas turbines.
Columbium	Increases heat resistance, shock resistance, strength, and toughness.	Rotary rock drill bits, heavy machinery parts, large anti-friction bearings, bridges, towers and high-rise buildings.

Iron combines readily with oxygen, sulfur, sulfate, carbonate, silicate, and other chemical cations to form a variety of minerals. The oxides, especially hematite, Fe_2O_3, and magnetite, Fe_3O_4, are the richest in iron and the most easily worked to free the iron. Hematite occurs primarily in layers of sedimentary origin, often diluted with SiO_2 in the form of jasper or

Table 5–7 (cont'd.)

Uses other than for ferrous alloys

Leather tanning, paint pigment, dye fixers, anodization of aluminum, refractories. Metallurgical usage accounts for 60%–70% of the total chromium in ore.

Electroplating, nickel-cadmium batterys, nickel-copper alloys, ceramics, chemical uses, dyes and pigments, nickel alloys for gas turbines, turbo superchargers, jet engines, and ship parts exposed to salt water.

Dry-cell batteries, chemical processes, glass making. Over 90% of manganese used is in steel making.

Electronic semiconductors, in chemical reagents, as lubrication, pigments and dyes. Over 50% is used in steel making.

Combined with carbon to form tungsten carbide, an extremely hard substance used as tips for high-speed cutting tools, e.g., bits for mining and petroleum drills, electronics, electric light filaments, fluorescent lamps. About 45% used for carbides, 25% for alloy steels, and 29% for nonsteel alloys.

Chemicals, ceramics, nonferrous alloys. Of the vanadium used, about 84% is used in alloy steel and 10% in nonferrous alloys.

Combined with chromium and tungsten to form stellite, an alloy comparable in hardness to tungsten steel and used for making drill bits and other durable cutting tools. When combined with aluminum and nickel, it forms alnico, an alloy which makes powerful permanent magnets for communication and electronic devices (magnets capable of lifting up 60 times their own weight). Used in paints, dyes, ceramics, and chemical processes.

Electronic alloys and electrical resistance alloys. About 90% of all columbium used is for steel alloys. Its use in 1969 was nearly double its 1968 use.

chert. Several of these deposits, including those in Minnesota and Canada, are very old, many having been formed during the Precambrian age. Their origin is obscure, since the available evidence suggests that they were deposited in ocean waters, but in waters unlike those of modern oceans. We shall not delve into the complex chemical ramifications of sedimentary iron ore deposits, but instead will simply note that they are important, reasonably abundant, and not difficult to locate. Magnetite, on the other hand, appears to be formed mainly from cooling magma and, therefore, is of igneous origin. Such is the case with the great magnetite deposit at Kiruna, Sweden, which is an igneous dike.

Aluminum   Years ago, aluminum was only a laboratory curiosity; now it is abundant and inexpensive. The reason for its past lack of use is that it was difficult to extract from its ores, requiring much energy to break the chemical bonds with which it combines with oxygen in oxides and silicates. In 1886 Charles Martin Hall, only 22 years old, discovered the electrolytic process of extracting aluminum from its ore. This process consists of melting the aluminum in an electric furnace in the presence of another chemical (called a flux) which serves to lower the melting point.

Bauxite is the ore of aluminum. It consists of a number of aluminum oxides and hydroxides and can be considered as either rock or soil material. It forms in wet, tropical climates when silica is leached away from the surface rock material. Oxides of iron and aluminum, being relatively insoluble, are left behind as a residual substance called *laterite*. Bauxite is therefore an aluminum laterite. Although aluminum is prevalent in the earth's crust, especially in silicates, bauxite is the only ore of that metal; nevertheless, supplies are plentiful. Most of the bauxite production occurs in Jamaica, Guiana, Venezuela, Surinam, France, Hungary, and the USSR.

Lead and zinc   Two metals that occur under similar mineralogical conditions and are very often mined together are lead and zinc. They form many minerals, the chief ores of which are galena (lead sulfide) and sphalerite (zinc sulfide). These minerals are considered to be hydrothermal minerals, having been precipitated from aqueous solutions associated with magmatic activity. As magma cools at depth to form igneous rock within the earth, it forms rock material, e.g. granite, composed mostly of silicate minerals. However, small amounts of more volatile substances remain fluid to the end; these substances work their way into, and are finally deposited in, fissures in previously formed minerals. These hydrothermal deposits include important ores such as lead, zinc, copper, and other sulfides, gold, silver, and other minerals.

In the central United States, however, rich deposits of galena and sphalerite are found disseminated in limestone. Igneous bodies are not found near these deposits, and thus the origin of these deposits has been the subject of much debate. However, it is likely that these deposits, too, were formed from hydrothermal solutions.

Major producers of lead are the United States, Australia, Mexico, and Canada; while the major producers of zinc are the United States, Canada, the USSR, Australia, and Mexico.

Copper Most of the important copper deposits are sulfide and oxide minerals which had a hydrothermal origin. In the United States, most copper is produced from igneous rocks that have formed in intrusive masses called *stocks*. The rock is mostly monzonite porphyry, which is very much like granite. Only 1 percent or less copper is present in these rocks. The richer deposits have been depleted, but through technological advances, the capacity to mine increasingly poorer deposits has been created. Enormous amounts of rock are now processed as mining of lower-grade ore goes deeper into the earth. In the United States, Utah, Nevada, Montana, Arizona, and New Mexico produce most of the nation's copper. Chile, Zambia, and the USSR also produce much copper.

Other metals A great many metallic elements are mined and put to use in our technological society. We have briefly examined a few of the important ones. Tin, tungsten, nickel, chromium, gold, silver, mercury, platinum, beryllium, vanadium, and many others could be mentioned—indeed, as could most of the elements, including some of the so-called rare earth elements.

The Outlook

Our society demands that more and more materials be extracted from the earth. Metallurgical, mining, and engineering techniques have thus far been able to fulfill these needs. We have not yet run out of any useful product, but on the other hand, we have seen (Chapter 1) that the usage of metals is increasing exponentially even faster than is the world's population. It is therefore necessary to emphasize that our supplies of earth materials are *finite*. A very few are so abundant that we do not have to worry about their future availability. Many, however, are rare: they occur in geologically limited environments, concentrated only in those places having suitable environmental conditions. Once these supplies have been depleted, the usage situation will become critical. It is a common fallacy to believe that we can then mine increasingly lower-grade ores, since such ores do not exist for many metals. Those that do may be quantitatively insignificant compared to the few concentrated deposits. Discovery of new supplies will become more and more difficult and will require the development of highly sophisticated exploratory techniques requiring highly skilled personnel. The day of the old prospector, his gold pan, and his burro are gone forever.

NONMETALLIC MINERAL RESOURCES

Nonmetallic mineral resources yield nonmetallic elements or are exploited primarily for reasons other than the production of metals. Examples of the exploitation of nonmetallic resources include the mining of native sulfur, a nonmetal; the mining of pyrite, FeS_2, in order to extract its sulfur to make sulfuric acid; and the mining of various gems, mica, talc, and various other rock products. Some nonmetallic products such as diamond or quartz crystals have a relatively high value per unit weight, but most are inexpensive per unit weight. Such materials as sand and gravel, limestone for cement, building stone, and brick clay have a relatively low value per ton. The use of these products must be local, as transportation costs preclude shipping over all but short distances.

Many of the high-bulk, low-price per volume materials are in plentiful supply throughout the world. Nevertheless, there are inherent geological problems connected with their development. Generally, these materials are extracted from surface mines, pits, and quarries. This type of mining involves problems of land-use planning, pollution, and reclamation that may not be present in underground mining. We shall explore these problems in later chapters. The fact that a mineral product is reasonably abundant does not imply that new supplies must not be continually sought and developed or that conservation need not be practiced.

Sand and Gravel

Production of these materials comprises by far the greatest volume and dollar value of all nonmetallic mineral resources. They are used mainly in the manufacture of concrete for construction and paving. These materials are also needed in glass-making (in which silica sandstones of high purity are used), for loose filler material on roads, in sand-blast cleaning techniques, and for many minor purposes. Geologically, sand and gravel are deposited by glaciers, streams, wind, and wave and current action. Dunes and well-worked beach deposits usually contain material fairly consistent in size, whereas glacial deposits are poorly sorted (that is, consisting of different sizes). Stream deposits may contain particles of many differing sizes in many different places. Most sand and gravel is composed of quartz, although other minerals may be present in minor quantities, especially in sediments that have not been thoroughly weathered.

Limestone

Limestone has a great number of uses. Besides being a useful building stone in many areas, it is also used as ballast and road-surfacing material

5-8

Mining potash, Carlsbad, New Mexico. (Photograph by E. F. Patterson, U.S. Geological Survey)

in its crushed form, and as an agricultural soil additive in its pulverized form. It is used as a flux in smelting iron, for the extraction of lime in the manufacture of cement, and for a great variety of industrial processes. Limestone (see Appendix A) is a sedimentary rock composed mainly of calcite ($CaCO_3$), but it often contains magnesium as well.

Evaporites

In the geologic past, shallow, restricted seas sometimes dried up under arid conditions, leaving deposits of chemically precipitated sedimentary rocks composed of gypsum, halite, and potassium salts. Salt has always been of great importance to civilization; indeed, it has even been used as money. The mineral halite (NaCl) is the only significant constituent of rock salt; most beds which are mined for salt are 97 percent or more pure halite. Some salt is also obtained from sea water and other natural brines. Gypsum ($CaSO_4 \cdot 2H_2O$) is used mostly in wall board and plaster. As seen in their tombs, even the ancient Egyptians had discovered the use of gypsum plaster. Potash is an evaporite rock composed of a number of chlorides and sulfates of potassium. Potassium salts, which are highly soluble, represent the ultimate formation occurring under arid conditions. Only when the last puddles of highly concentrated brines dry up will potassium salts be deposited from the sea. The most important use of potash is in fertilizer.

Table 5-8 Strategic and critical metals and minerals (simplified), as compiled by the Office of Emergency Preparedness, February 18, 1972

Aluminum	Diamond	Quartz crystals
Aluminum oxide	Fluorspar	Rutile
Antimony	Graphite	Sapphire and ruby
Asbestos	Iodine	Silver
Bauxite	Lead	Talc
Beryl	Manganese	Tantalum
Bismuth	Manganese ores	Tantalum minerals
Cadmium	Mercury	Thorium oxide
Chromite	Mica, muscovite	Tin
Cobalt	Mica, phlogopite	Tungsten
Columbium	Molybdenum	Tungsten ores
Copper	Nickel	Vanadium
Chromium	Platinum group of metals	Zinc

Sulfur

Another of the most essential nonmetallic minerals is sulfur. Although it occurs in many minerals and as the major native element around volcanoes, its most important commercial source is from the massive salt beds in Texas and Louisiana. The process of sulfur removal involves pumping superheated water down drilled wells to melt the sulfur and then pumping out the sulfur which solidifies at the surface. Most sulfur is used to make sulfuric acid, one of the most important chemicals used in modern technology (over 200 pounds per person of sulfuric acid are used each year in the United States).

We have looked briefly at some of the most widely used nonmetallic mineral products. In addition to these, there are a large number of other minerals useful to man. Some are inexpensive and abundant; others are expensive and scarce. Neither the United States nor any other nation is self-sufficient in each of its mineral resources (whether nonmetallic or metallic). That this is significant is suggested by the fact that the United States maintains a list of strategic and critical materials, many of which are mineral and rock products. (Table 5-8).

REFERENCES

Ahrens, L. H. *Distribution of the Elements in Our Planet*, New York: McGraw-Hill, 1965.

Bateman, A. M. *Economic Mineral Deposits*, 2d. ed., New York: Wiley, 1965.

Burbridge, E. M., G. R. Burbridge, W. Fowler, and F. Hoyle. "Synthesis of the elements in stars," *Reviews Modern Physics* 29 (1957): 547-650.

Burk, C. A. "Global tectonics and world resources," *The American Association of Petroleum Geologists Bulletin* 56, 2 (1972): 196–202.

Clark, K. F. "Structural controls in the Red River District, New Mexico," *Economic Geology* 63 (1968): 553–566.

Emery, K. O. *Some Potential Mineral Resources of the Atlantic Continental Margin,* U. S. Geological Survey Professional Paper 525–C, 1965.

Hoyle, F. *Galaxies, Nuclei and Quasars,* New York: Harper & Row, 1965.

Landsberg, H. H. *Natural Resources for U. S. Growth,* Baltimore: Johns Hopkins Press, 1967.

Lasky, S. G. "Mineral resources appraisal by the U. S. Geological Survey," *Colorado School of Mines Quarterly* 45, 1A (1950): 1–27.

Lovering, T. S. "Non-fuel mineral resources in the next century," *Texas Quarterly* 11 (1968): 127–147.

————. "Mineral resources from the land," in *Resources and Man,* National Academy of Sciences, National Research Council, San Francisco: Freeman, 1969.

Manheim, F. T., R. M. Pratt, and P. F. McFarlin. "Geochemistry of Manganese and Phosphate Deposits on the Blake Plateau," Program for the meeting of the Geological Society of America, New Orleans, 1967, p. 139.

Mason, B. *Principles of Geochemistry,* 3rd ed., New York: Wiley, 1966.

Mason, B. and L. G. Berry. *Elements of Mineralogy,* San Francisco: Freeman, 1968.

Meadows, D. H., D. L. Meadows, J. Randers, and W. W. Behrens III. *The Limits to Growth,* New York: Universe Books, 1972.

McKelvey, V. E. and F. F. H. Wang. *Preliminary Maps—World Subsea Mineral Resources,* U.S. Geological Survey Miscellaneous Geological Investigations Map I–632, 1970.

McKelvey, V. E. "Mineral resource estimates and public policy," *American Scientist* 60 (1972): 32–40.

Park, Charles F., Jr. *Affluence in Jeopardy—Minerals and the Political Economy,* San Francisco: Freeman, 1968.

Petersen, N. S., F. V. Carrillo, and R. C. Vars. *Materials Substitution Study - General Methodology and Review of U. S. Zinc Die-casting Markets,* United States Bureau of Mines Information Circular 8505, 1971.

Population Reference Bureau. "Population and resources: the coming collision," *Population Bulletin* XXVI, 2 (1970): 1–36.

Riley, C. M. *Our Mineral Resources,* New York: Wiley, 1956.

Schuiling, R. D. "Tin belts on the continents around the Atlantic Ocean," *Economic Geology* 62, 4 (1967): 540–550.

Struve, O. "Element formation in stars," in *The Evolution of Stars, How They Form, Age and Die,* ed. T. Page and L. W. Page, New York: Macmillan, 1968.

U.N., Department of Economic and Social Affairs. *Mineral Resources of the Sea,* 1970.

"Vacuuming the Atlantic floor," *Science News* 98 (1970): 134–135.

Manganese nodules, found in some areas of the ocean floor, will supply needed metals for man in the future. (Photograph courtesy John L. Mero)

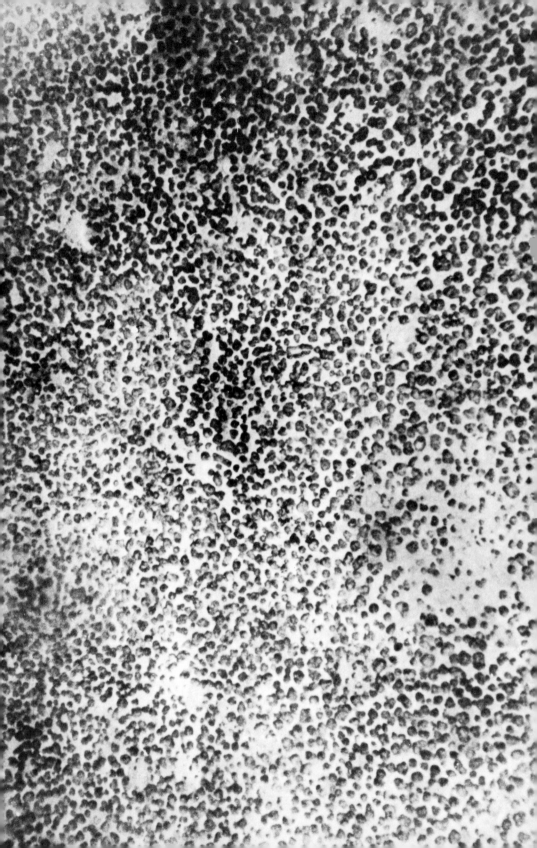

MINERALS FOR THE FUTURE

THE DEMAND FOR MINERALS

Having examined some aspects of mineral resources—what they are, how they form, and where our supplies are located—let us now turn to the problem of providing mineral resources for our future needs. Discovering and developing new supplies are of prime concern in the science of geology.

World demand for the commonly used metals is expected to increase by a factor of four by the year 2000. Several factors will contribute to this projected increase, among them: an increased world population, a higher standard of living in the developed countries, and some degree of development in the presently underdeveloped countries. Although it is difficult to assess the effect that an increased standard of living in the developed countries would have on our mineral resources, it is somewhat easier to gauge possible repercussions from increases in population and the development of underdeveloped countries.

Increased Population

If the world population doubles in the next 35 years, it is obvious that
it would be necessary to approximately double our use of mineral resources
in order to maintain the status quo. Even if it were possible to recycle 100
percent of our metals, we would still experience a net loss in our mineral
reserves, since new stocks would have to be produced merely to accommodate the increased number of people.

Development of Underdeveloped Countries

One characteristic of the world which has remained nearly constant
throughout time is the wide gap between the rich and the poor countries.
An ideal of leaders in many fields has been to bring about a more equita-
ble distribution of the world's wealth with a concomitant improvement in
the overall standard of living. An example of this present inequality
is the fact that the United States, with only 6 percent of the world's population, uses over half the total resources used by the world annually.
Furthermore, 20 percent of the world's population enjoys about 80 percent of the total world income, and this entire 20 percent is concentrated
in the area north of approximately 25 degrees north latitude.

     Let us consider the consequences of equalizing the standard of living
throughout the world. If nothing less than strict equality is acceptable,
then either the standard of living in the United States will have to be
lowered or that of most of the rest of the world will have to be raised to
the level of that in the United States. To Americans, the first alternative
is unacceptable, leaving the alternative of improving the quality of life
in the rest of the world until it corresponds to that in the United States.

     If consumption of materials is an index of the wealth of a nation,
then use of mineral resources should correlate with standard of living.
The per capita use in the United States of several leading metals is tabulated
in Table 6–1. Using per capita consumption for 1969 as the base, we can determine how much of each of these metals the world would have had to
provide in 1969 to bring the standard of living of the rest of the world up
to that existing in the United States. We see that the world would have to
have produced nearly eight times the actual production to equal United
States' consumption. Nearly five times as much copper would have been
required, while iron production would have had to increase by a factor of
three, lead by a factor of six and a half, zinc by a factor of five and a half,
and tin by a factor of eight. For consumption of these commonly used
metals in the rest of the world to have equaled that in the United States in
1969, world production would have had to have averaged approximately
five times the quantity it actually did. Some of the more exotic elements,
which are used almost exclusively by the developed countries, would have

required even larger production increases. Beryllium production, for example, would have to be increased nearly 20 times. If the world population is to double by the end of this century and the standard of living is to be maintained at the level projected above, then in the year 2000, it will be necessary to produce nearly 16 times as much aluminum as was produced in 1969, ten times as much copper, seven times as much iron ore, thirteen times as much lead, over twelve times as much tin, and eight and a half times as much zinc. This tremendous increase would be necessary just to maintain a population which has doubled at a level commensurate with the 1969 standard of living. If the United States' level of affluence were to rise above the 1969 level (as it already has), the standard of living in the rest of the world would have to rise in proportion to this, resulting in even greater production demands.

Table 6-1 Per capita usage of selected metals in the United States for the years 1949 and 1969

Metal	Per capita usage, 1949 (in pounds)	Per capita usage, 1969 (in pounds)
Aluminum	11.86	45.42
Copper	19.24	20.46
Iron ore	1179.15	1366.47
Lead	12.66	13.55
Tin	1.07	0.88
Zinc	9.41	13.55

(Data from *Bureau of Mines Minerals Yearbook, 1969*, I–II and *Minerais Yearbook, 1952*, I)

Mineral Reserves

If the situation discussed above were to become reality, it would surely impose a significant strain on our mineral resources. What reserves do we have that will supply demands such as those described above? Knowing that the rest of the world does not consume as much of the earth's mineral resources as does the United States and feeling reasonably assured that most of the world will not be able to approach our standard of living in the near future, we can best deal with these questions by both looking at the current situation and projecting current trends into the near future. This has been done for a selected group of elements in Table 6–2. The reserve and consumption levels are those estimated by the United States Bureau of Mines for 1969. Some of the estimates of known world reserves are more reliable than others, but each represents our best present knowledge. "Primary" consumption refers to minerals mined for the first time and excludes consumption of recycled scrap. The static index (Column 6) is the number

Table 6–2 Selected mineral elements, their presently known reserves, consumption for 1969, growth rate, and duration if present rate of use continues

(1)	(2)	(3)	(4)	(5)
Mineral element	Known world reserves (millions)[d]	World primary consumption 1969 (millions)[e]	U. S. primary consumption 1969 (millions)[f]	U. S. consumption as percent of world consumption 1969 %
FERROUS				
Chromium	775.0 sh tons	1.9	0.45	24
Cobalt	4,800.0 lbs	44.0	14.00	32
Iron	96,720.0 sh tons	428.0	84.00	20
Manganese	796.0 sh tons	8.2	1.10	13
Molybdenum	10,827.0 lbs	138.0	55.80	40
Nickel	147,000.0 lbs	934.0	319.00	34
Tungsten	2,824.0 lbs	72.7	15.70	22
Vanadium	10.1 sh tons	0.11	0.006	5
NONFERROUS				
Aluminum	1,168.0 sh tons	11.1	4.70	42
Cadmium	1,420.0 lbs	31.3	13.30	42
Copper	307.9 sh tons	7.3	1.54	21
Lead	95.3 sh tons	3.5	1.39	40
Mercury	3.2 flasks[a]	00.3	0.06	23
Platinum gp.	424.0 tr oz	3.3	1.00	30
Selenium	189.0 lbs	2.6	1.07	42
Silver	5,500.0 tr oz	340.0	90.00	26
Tin	4.3 lg tons	0.2	0.06	24
Titanium[b]	146.8 sh tons	1.4	0.45	32
Zinc	123.7 sh tons	5.4	1.40	26
NONMETALLIC				
Boron	72.0 sh tons	0.2	0.09	36
Graphite	100.6 sh tons	0.5	0.06	23
Phosphorus[c]	21,800.0 sh tons	11.4	3.50	31
Sulfur	2,470.0 lg tons	34.4	9.10	26

a. 76-pound flasks at $200/flask
b. Given as titanium equivalents
c. Given as phosphorus content
d. From *Minerals Facts and Problems*, 1970 edition
e. From *Minerals Yearbook*, 1969, Vols. I–II
f. From *Minerals Yearbook*, 1969, Vols. I–II

of years the known reserves would last if used at the 1969 rate. This figure has little meaning, however, because the reserves are not used at the 1969 or at any other constant rate. Rather, the rate of consumption continues to increase. This is what is meant by exponential growth—every year we use more of everything than we did during the preceding year.

Table 6–2 (cont'd.)

(6)	(7)	(8)	(9)	(10)
Static index (years)g	Average growth rate (%)h	Exponential index (E.I.) (years)i	E. I. if world reserves were 5 times as large (years)	E. I. if world reserves were 10 times as large (years)
408	2.6	94	153	180
109	1.5	65	148	190
226	1.8	90	170	207
97	2.9	46	94	116
78	4.5	34	65	80
157	3.4	54	98	118
39	2.5	27	71	95
92	4.4	37	69	85
105	6.4	32	55	66
45	2.7	30	72	96
42	4.6	23	52	66
27	2.4	21	60	84
13	2.6	10	37	56
128	3.8	46	85	103
74	1.3	52	135	182
16	2.7	13	43	62
18	1.1	16	62	98
105	4.2	40	75	91
23	2.9	18	50	70
301	4.4	60	96	111
201	2.0	81	152	186
1912	4.7	96	130	145
72	4.5	32	63	78

g. This is the period of time our reserves would last if only the amount used in 1969 were used in each succeeding year.
h. From *Mineral Facts and Problems*, 1970 edition.
i. The number of years known reserves will last with usage growing exponentially at the average rate of growth given. Calculated by the formula: Exponential Index=$1n(r \cdot s)$, where r equals rate of growth and s equals static index. Formula from *The Limits to Growth*, D. H. Meadows, *et al.*, p. 60.

In projecting the probable growth of mineral use, the United States Bureau of Mines takes into consideration many factors which may potentially affect the use of that mineral (see *Mineral Facts and Problems*, 1970 edition, pages 9–11). This procedure sets both minimum and maximum values for usage, giving a range of growth rates possible for each mineral.

Column 7 lists the averages of these values. Column 8 gives the projected number of years the presently known reserve of a mineral element will last at the current rate of growth. Unlike the static index of Column 6, the rate of growth of usage of the mineral element *has* been taken into account and is reflected by the significantly shortened intervals until depletion. Finally, the last two columns project the number of years the reserve of each mineral element would last, taking into account the growth of the usage rate if our known reserves were five and ten times, respectively, what they are now known to be.

Perhaps the most significant features of Table 6-2 are the last three columns. Column 8 projects possible results in terms of only presently known reserves, if current usage rates continue. The figures in the last two columns hypothesize the likelihood of finding new reserves. However, even supposing that we may discover new reserves does not imply indefinite availability of necessary mineral resources, since it fails to account for exponential growth. Thus, the assumption that if our reserves are projected to last 50 years, then ten times that reserve would last 500 years is false when dealing with exponential growth. In fact, with growth rates of the magnitude of those with which we are dealing, such increased reserves would last only about three, not ten, times as long.

As the world's population grows larger and as presently underdeveloped areas become more developed, it is not impossible that the rates of growth of usage of these materials will also increase. If this occurs, we are destined to see even shorter periods until depletion.

Alternatives Open to Us

The growing affluence of the world will put tremendous strain on all its resources, particularly its mineral resources, in the next few decades. Therefore, we must consider possible alternatives.

Discovery of new reserves   There is little doubt that our presently known reserves will be supplemented in the future by new discoveries. The question is whether or not these additional reserves will be sufficient to satisfy our demands. As one can see from the last two colums in Table 6-2, the discoveries would have to be substantial to make a significant impact on our supplies. What, then, are our prospects for finding new, substantial ore deposits?

Most of the ore we have mined and are now mining became known to us because the ore body cropped out at the surface. This ore constituted the shallow ore which was "easy" to find. It is a fair assumption that most of the available "easy" ore has been discovered in the developed countries. New discoveries can be expected to originate primarily in very remote areas or in underdeveloped countries. Exploration and production in such areas will require the ample financial resources of large companies, but perhaps more importantly will also require wide-ranging governmental support and cooperation from the principals involved. Because there are still many remote areas and underdeveloped countries in the world, one can be optimistic about the possibility of discovering new mineral reserves. However, our optimism should be tempered by the thought that airborne equipment has allowed us to take a broad look at many remote areas of the world without necessitating our entry into them. This has helped us to sort out those areas which look promising from those which show little sign of yielding new reserves with present technology. Furthermore, we are beginning to get a detailed picture of the geologic events which cause the formation of ore deposits, and many areas of the world simply have not undergone these geologic events. So, even though there are extensive territories which could be classified as underdeveloped, there are not many which are totally virgin and unexplored.

An important type of ore body, with which we have not had much experience, is that which does not crop out at the surface. These "blind" ore deposits constitute largely unknown resources. Our techniques for searching out these deposits are primitive and must be refined. The expenses involved in the reconnaissance and preliminary exploration of "blind" ore deposits are much greater than those associated with finding deposits which appear at the surface. We will have to greatly advance our technology before blind deposits can be counted among our reserves.

Substitutions   Substitutes are being sought for many of the current uses of several primary metals. For example, in today's automobiles, plastics are being increasingly substituted for copper, lead, zinc, and tin.

In the last 1960s and early 1970s attempts have been made to replace certain alloy steels, for which the alloying element is in short supply, with alternative types of steel. The shortage of nickel, especially, has encouraged efforts to produce steel substitutes which could replace stainless steel, which has a high nickel content. Some steel makers have converted to a completely tin-free steel with a thin chromium coating for use in making cans which will replace the "tin can."

In the early 1960s zinc-base alloys were used extensively by the automotive industry in such items as assemblies for fuel pumps, carburetors, radiator grilles, horns, heaters, and many others. By the late 1960s usage of these alloys had decreased by some 100,000 tons annually, despite the increase in the number of automobiles produced. Plastics accounted largely for this decrease.

The United States Bureau of Mines instituted a study, as part of a larger program, to "develop and describe a methodology for approaching the problem of substitution between mineral commodities and between mineral commodities and other materials." A report, published in 1971, concerning substitution in the zinc die-casting industry, raised the point that several variables were involved in materials-substitution for any consumer-based product. Among these variables are technology, relative prices and materials, total costs, consumer preferences, time, and legislative-institutional influences. If the material has a functional use in which the physical and/or engineering properties are important, then the factors just listed are relegated to a secondary status, while the engineering specifications become most significant. Factors influencing substitution would be completely different for products with decorative function and those with a functional use as part of a space flight.

One important fact implicit in any discussion of substitution is that since substitutes are also made from materials, there is no way to get something for nothing; the substitute may be limited also. For example, in an effort to ease the lumber shortage in Japan, an artificial wood which looked like wood, could be worked like wood, and had all the other characteristics of wood (except for that good, fresh-lumber smell) was fabricated. Unfortunately, it was being fabricated from petroleum!

Using lower-grade ore   Changes in economic conditions will undoubtedly enable us to use some reserves considered to be submarginal ores in today's economy. What the economic conditions must be and what grades of ore can be used vary with each mineral.

A mistaken concept about the grade of ore and its abundance has been perpetuated by some individuals working outside their area of competence. This erroneous concept originated in the context of the work done more than 20 years ago by a geologist named Lasky. In studying certain types of copper deposits, Lasky formulated a general principle which stated that "in many mineral deposits in which there is a gradation from relatively rich to relatively lean material, the tonnage increases at a constant geometric rate as the grade decreases." Figure 6–1 shows this relationship: the changes in grade are represented by equally spaced vertical lines while the ore tonnages are spaced on a logarithmic scale. This concept is known as the arithmetic-geometric (A/G) ratio.

Although Lasky was careful to state that this principle was true for only certain types of deposits and that the curve represented tonnage of ore, not tonnage of the metal involved, these and other points of the principle have been either misunderstood or conveniently overlooked by some. In fact, many deposits do not show this gradation of ore at all. It is not unusual to find an ore body containing a metallic element in concentrations which may be several thousand times greater than its crustal abundance adjacent to rock containing only the average crustal abundance of that ele-

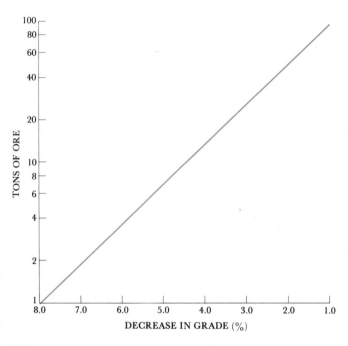

DECREASE IN GRADE (%)

6-1
Typical plot of tonnage of ore plotted against decrease in ore grade as described by Lasky's A/G ratio. This type of plot holds for only certain types of deposits and through only certain ranges.

ment. T. S. Lovering, in *Resources and Man,* gives the following example of an increase in the tonnage, but not in the metal:

> *Another instructive example of the limitations of the A/G ratio involves mercury. In the course of intensive exploration by the U. S. Bureau of Mines and the U. S. Geological Survey during World War II, more than 330 mercury occurrences were inspected and 43 deposits were explored in detail. Commercial ore was developed on 38 of the 43, altogether comprising 370,000 tons of ore averaging 0.8 per cent mercury (16.2 pounds per ton). These same deposits also included 1,220,000 tons of noncommercial mineralized material averaging 0.125 per cent mercury (2.5 pounds per ton) and 285,000 tons averaging 0.08 per cent mercury (1.6 pounds per ton). This is equivalent to 2,960 tons of mercury in the 0.8 per cent ore and 1,525 tons of mercury in*

*the 0.125 per cent "ore," but only 228 tons of mercury in the
0.08 per cent "ore." Thus an arithmetic decrease in grade from
0.8 per cent to 0.125 per cent, a sixfold change, would result in
a threefold geometric increase in tonnage of mercury-bearing
rock--but the total mercury in the larger tonnage would be only
one-half that in the smaller high-grade tonnage. A further decrease
of 0.04 per cent in grade would add a tonnage less than one-
fourth that of the 0.125 per cent ore at a grade only one-tenth
that of the somewhat larger tonnage of high-grade ore!*[*]

In other words, vast additional quantities of low-grade ore add very little
to our supplies.

Some optimists escape from reality by imagining that the availability
of cheap energy will be the key to using submarginal-grade ore. According
to this doctrine, the development of a cheap source of energy ("nuclear
energy" seems to be falsely synonymous with "cheap energy" in this
doctrine) will allow us to profitably mine the poorest of ores. Indeed,
some even state that we will soon be mining average crustal rock to fulfill
our needs. These suppositions are based on a gross misunderstanding of the
economics of mining and processing rock material. By far the largest portion of the cost of a mining operation is devoted to capital and labor. Although a cheap energy source would somewhat reduce the cost, it would
not be sufficient to decrease mining costs to some of the levels which have
been suggested. According to John D. Harper, Chairman of Alcoa (Aluminum Company of America), the cost of nuclear power may be prohibitive
for the aluminum industry.

Recycling In order to alleviate, or at least ease, the consequences of the
shortage of certain metals, it is imperative that we begin to recycle the
metals already in use. When the issue of metal shortages is raised, almost
without exception "recycling" is one of the first remedies suggested, and
it is potentially useful. It has been estimated that there are in the United
States today over 20 million junk automobiles which consist of reusable
metal worth over one billion dollars. There are also untold numbers of
discarded beer cans, refrigerators, washing machines, dryers, and television
sets which would increase the total metal value significantly. Then why is
recycling almost insignificant in today's industrial picture?

First, let us look at scrap steel. To gather the steel products scattered
in junkyards throughout the country, transport them to a processing plant,
and then extract the steel in a form usable by steel mills may cost many

---

[*]*Resources and Man*, Publication 1703, Committee on Resources and Man, National
Academy of Sciences – National Research Council, W. H. Freeman and Company,
San Francisco, 1969, p. 115. Reprinted by permission.

times more than the metal is worth in the present economy. In addition, there has been a substantial reduction in the demand for this kind of scrap by the steel-making industry in the past few years. At least two major trends in the steel industry have contributed to this reduction. First, during the past 15 years American steel makers have exchanged the open-hearth furnace for the basic oxygen furnace, which uses very little scrap. The older, open-hearth process, which preceded the basic oxygen process but now has generally fallen into disuse, had a greater capacity to use scrap; while our present steel-making technology precludes the use of large amounts of scrap. If we are to utilize recycled scrap steel, we will have to change our way of making steel.

The second trend which has contributed significantly to a reduction in the use of scrap steel is the demand for higher-quality steel products. The result of this is a greater variety of "specialty steels" which contain elements such as copper, aluminum, nickel, cobalt, and so forth. This trend has had a twofold effect: it has required the production of alloys whose composition can tolerate only the very slightest deviation, and it has necessitated the addition to steel of elements which are difficult and costly to remove. In earlier, less complex days when steel was nearly all of one quality, recycling scrap was a simpler matter.

This latter problem is not unique to the steel industry. Indeed, it affects all the metal-producing industries, e.g., the aluminum industry. One commercial aluminum alloy calls for 5.5 percent zinc, while many other aluminum alloy specifications call for a maximum of 0.3 percent zinc. Other alloys need up to 0.5 bismuth, while most call for a maximum of 0.05 percent for any one impurity (which usually includes bismuth). Some call for as much as 5.5 percent copper, and others need a maximum of 0.4 percent copper. When scrap aluminum is collected and melted, the composition of the resulting melt is useful for only limited purposes. The cost of attempting to purify the melt by removing the unwanted elements with present technology is prohibitive.

These examples demonstrate that recycling, though imperative, is beset by both economic and technological problems. New methods of making steel and separating elements are needed, as are inexpensive methods of gathering the scrap. A standardization of alloys for particular uses is called for and would help to alleviate the problem. Presently, the only product that meets this qualification is the aluminum beer can, which can be remelted and reused without any great difficulties.

Lowering our level of affluence   One alternative, often overlooked, is the voluntary reduction of our own consumption of minerals. By lowering our standard of living, we would give people in the developing countries an opportunity to increase their standard of living to a level approaching ours. Concurrently, this would extend the lifetime of our reserves, and might even give us the additional time necessary to find and develop new reserves.

Maintenance of the status quo   If it is impossible for the rest of the world to match our high level of consumption, which is presently the case, and if we refuse to lower our materialistic standard of living, we can always attempt to maintain the status quo. This could, undoubtedly, be maintained only by force and through armed conflict, but it is one alternative.

MINERALS FROM THE SEA

The world contains some 326 million cubic miles of water, of which about 97 percent, or 316 million cubic miles, is seawater. This volume of water contains some 50,000 trillion tons of dissolved material, representing more than two-thirds of the 92 naturally occurring elements. Thus, it is obvious why the sea has been called the storehouse of all materials needed by man, since it is able to fulfill his needs for both food and minerals despite a rising population and increased wealth reflected in increased consumption. Those who have unlimited faith in the sea and its treasures claim that we are about to retrieve untold wealth from the sea and that we should not be too concerned with the problem of dwindling mineral resources.

The mineral wealth of the sea includes elements in solution, solid mineral matter on and immediately below the ocean floor, and oil and gas trapped beneath the ocean floor. Postponing our discussion of oil and gas to the chapter on energy, we will restrict our present discussion to the first two types of resources.

Minerals from Seawater

The elements in seawater are listed in Table 6-3 in the order of their concentrations. It is immediately obvious that most of the elements present occur only in very dilute quantities. Indeed, nearly 99 percent of the dissolved solids in seawater is accounted for by the first 16 elements on the list. Concentrations of the others are extremely low, and below the subsequent 35 or so elements on the list, the concentration has decreased by another five orders of magnitude. In general, the most abundant elements in seawater are not the most vital.

Almost everyone has heard that there are great quantities of gold dissolved in seawater and that a fabulous fortune awaits the one who can recover it from the sea. This story is pregnant with both good and bad news. The good news is that it is true that the sea does contain great

Table 6–3 Concentration of elements in seawater and the dollar value of their combined forms

Element	Concentration $(lb/10^6$ gal)	Price/lb (dollars)	Dollar value/10^6 gal — As	$ Value
*Chlorine	166,000.0	0.0143	NaCl	3,389.00
*Sodium	92,000.0			
Magnesium	11,800.0	0.3800	Mg	4,283.00
Sulfur	7,750.0	0.0705	S	521.00
*Calcium	3,500.0	0.0275	$CaCl_2$	481.00
Potassium	3,300.0	0.1250	K_2O (equiv.)	397.00
Bromine	570.0	0.4900	Br_2	266.00
Carbon	250.0	0.3200	Graphite	75.00
Strontium	70.0	0.1300	$SrCO_3$	8.68
Boron	40.0	0.126	H_3BO_3	4.84
Silicon	26.0	0.125	SiO_2	3.13
Fluorine	11.0		CaF_2	
Argon	5.0			
Nitrogen	4.0	0.0245	NH_4NO_3	0.10
Lithium	1.5	0.525	Li_2CO_3	0.75
Rubidium	1.0		Rb	
Phosphorus	0.6	0.121	$CaHPO_4$	0.07
Iodine	0.5	4.00	I_2	2.00
Barium	0.3	0.225	$BaSO_4$	0.06
Indium	0.2		In	
Zinc	0.09	0.1800	Zn	0.015
Iron	0.09	0.2075	Fe_2O_3	0.174
Aluminum	0.09	0.1365	Al	0.011
Molybdenum	0.09	3.280	Mo	0.276
Selenium	0.04	10.50	Se	0.346
Tin	0.03	1.7725	Sn	0.044
Copper	0.03	0.5075	Cu	0.013
Arsenic	0.03	0.047	As_2O_3	0.001
Uranium	0.03		U_3O_8	
Nickel	0.02	1.33	Ni	0.022
Vanadium	0.02	2.21	V_2O_5	0.037
Manganese	0.02	0.3375	Mn	0.006
Titanium	0.009	0.275	TiO_2	0.002
Antimony	0.004	0.57	Sb	0.002
Cobalt	0.004	2.45	Co	0.010
Cesium	0.004		Cs	
Cerium	0.004	1.58	CeO_2	0.005
Yttrium	0.003		YCl_3	
Silver	0.003	26.17	Ag	0.131
Lanthanum	0.003			
Krypton	0.003			
Neon	0.0009			
Cadmium	0.0009	2.60	Cd	0.002
Tungsten	0.0009	5.55	W	0.005
Xenon	0.0009			

Table 6–3 (cont'd.)

| Element | Concentration $(lb/10^6 \text{ gal})$ | Price/lb (dollars) | Dollar value/10^6 gal. | |
			As	$ Value
Germanium	0.0006		GeO_2	
Chromium	0.0004	1.835	Cr_2O_3	0.0008
Thorium	0.0004		ThO_2	
Scandium	0.0004		Sc_2O_3	
Lead	0.003	0.155	Pb	0.00004
Mercury	0.0003	3.466	Hg	0.00087
Gallium	0.0003			
Bismuth	0.0002	4.00	Bi	0.00068
Niobium	0.00009		Nb_2O_5	
Thallium	0.00009	10.00	Tl	0.00080
Helium	0.00004			
†Gold	0.00004	615.57	Au	0.02041

*These elements would combine so as to form approximately 230,000 pounds of NaCl and 17,500 pounds of $CaCl_2$. The dollar value was calculated on this basis.

†Based on $42.22/tr oz

(Concentrations after Goldberg, 1963; prices from the *Chemical Marketing Newspaper,* formerly *Oil, Paint and Drug Reporter,* December 27, 1971, pp. 22–33.)

quantities of gold in solution. The bad news is that an economical recovery is almost impossible.

Gold occurs in a concentration of about 0.0005 troy ounces per million gallons of water. This means that there are about 550 troy ounces (37.5 pounds), or about $23,221.00 at $42.22 per ounce (approximate current price), per cubic mile of water. For the entire ocean, the total is $7338 billion, a formidable sum, indeed!

At the end of World War I a German Nobel Prize-winning chemist, Dr. Fritz Haber, believed that Germany could repay her war debt by extracting gold from seawater. He hired a team and acquired a ship to sail the oceans in search of the highest concentrations of gold. After wasting ten years in which no gold was recovered, Haber abandoned his efforts. It is reported that Dow Chemical Company experimented with extracting various elements, including gold, from marine water in connection with their routine extraction of bromine in their North Carolina plant. In processing some 15 tons of seawater they extracted about 0.09 mg of gold, worth only $0.0001 at that time. This represents almost the total amount of gold extracted from seawater to date.

There are two reasons for our inability to recover gold from seawater economically. First, gold is so dilute in seawater that many millions of gallons of water must be processed in order to recover any significant amount of gold. Large expenditures of energy are required to process these quantities of water, and energy is expensive. Currently, the price of energy involved in gold extraction exceeds the value of gold on the market. The second reason is that for the most part, there are currently no techniques for the selective recovery of a specific element from seawater. Therefore, to extract a single element one must also take large quantities of unwanted solids, mostly salt, out of solution. These twin problems of dilution and nonselective recovery affect nearly all the elements we wish to recover from seawater. Although they may be solved in the future, there are few expectations that they will be solved soon.

Magnesium, bromine, and salt are the only mineral solids now recovered commercially and extensively from the sea. In 1968 a total of 61 percent of the world's magnesium metal, 6 percent of its magnesium products, 70 percent of its bromine, and 29 percent of its salt came from seawater. It seems probable that such elements as iodine, potassium, and sulfur may eventually be recovered commercially, but it is not economically feasible at present. There is a possibility that strontium and boron may be economically retrievable in the distant future, but quantitative recovery of any element lower in concentration than boron seems highly improbable at the present.

There are, however, local conditions that may allow recovery of elements which are too dilute in normal seawater. These alternatives occur in places such as the Dead Sea, which is about ten times as concentrated as seawater. The Red Sea contains cadmium in concentrations that are four orders of magnitude greater than the concentrations in average seawater. (For a "one order of magnitude" increase, multiply by ten; for a "two orders of magnitude" increase, multiply by 100; for a "three orders of magnitude" increase multiply by 1000, and so forth.) Water in some areas of the Red Sea contain concentrations of elements high enough to make them economically attractive; these areas are located in fracture zones beneath the floor of the sea. It is very likely that this situation occurs along other parts of rifted oceanic rises. It may be possible to recover elements from these areas that are not available to us in ordinary seawater because of dilution. Areas like the Great Salt Lake, which is many times saltier than the oceans, also offer possibilities of selective recovery of elements from brine solutions.

Placer Deposits

Placer deposits are bodies of debris which have been deposited by some agent, generally rivers, of erosion. Concentrations of valuable minerals, mixed with sand, gravel, and other rock debris are often found in these deposits. These valuable minerals include diamonds, platinum, gold, tin, and many heavy minerals such as rutile and ilmenite, both of which are sources of titanium. Because of their higher specific gravity, the metallic elements tend to concentrate in low places, separated by gravity from the lighter, less valuable minerals. Many of these placer deposits have been found along the borders of continents in areas as deep as 100 meters below the present sea level. Presumably, many of these were formed during the Pleistocene Epoch, when water was on the land in the form of ice and the sea level was lower than its present level by about this same depth of l00 meters (see Chapter 9). Flooding of the coastlines by the subsequent melting of the glaciers submerged these deposits. Currently, only gold, diamonds, and tin have been recovered economically from these submerged placers, but the value of the material already recovered is several million dollars.

If placer deposits on land can be used as an indication, it is possible that the amount of metals recoverable from these submerged placers could be substantial. Although recovery of minerals from these sources will present special problems of location and exploitation, they offer the best prospects for economical recovery of metals from the sea.

Submarine Sediments

The sedimentary rocks and the sediments of the continental shelf adjacent to the continents also offer some hope for mineral recovery. Of course, the largest share of the mineral wealth of the continental shelf will be petroleum and natural gas. Oil and gas currently account for between 80 and 90 percent of the total dollar value of minerals recovered from the shelf areas adjacent to the United States. However, in addition to the oil and gas deposits, manganese and phosphate deposits have also been discovered on the sea floor in recent years.

Phosphate   The ocean's phosphate deposits (Fig. 6–2), first discovered by the *Challenger* expedition of 1872–1876, occur as phosphatic sands and muds on the sea floor, as consolidated phosphatic beds of Tertiary age, but most commonly as phosphorite nodules. Phosphorite is a complex calcium-phosphate rock. These deposits have been discovered off both coasts of the United States; the west coast of Central and South America, adjacent to Argentina; Japan; South Africa; and most recently, off the central and northern coasts of New South Wales. This latter discovery has added impetus to plans being developed for offshore mining along the coasts of Australia. Although these blankets are thin and poorly defined, some have been found to be of a grade nearly comparable to that of land deposits. These offshore deposits range up to a grade of 29 percent P_2O_5, which is just a few percent below the present cutoff grade for ore that can be recovered on land.

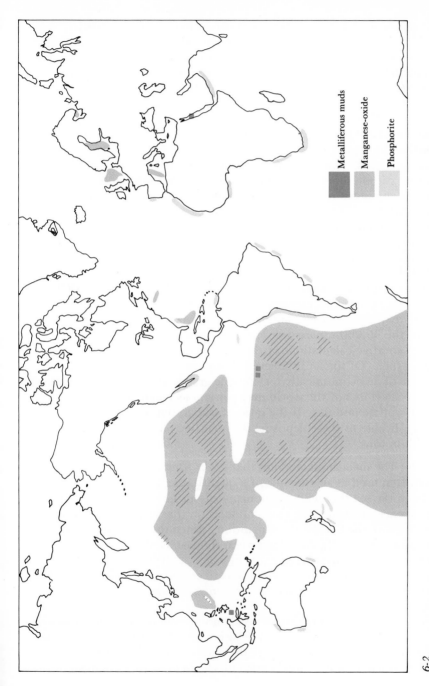

Metalliferous muds

Manganese-oxide

Phosphorite

6-2
Phosphorous, manganese, and metalliferous muds. Nodules are absent or only sparse in the blank areas. In the shaded areas they may cover as much as 20% of the bottom; in the ruled areas, as much as 50%. (Source: U. S. Geological Survey, Miscellaneous Geological Investigations Map I-632, 1970.)

Phosphorite usually occurs in waters rich in inorganic nutrients, especially dissolved phosphate, flowing up from the ocean depths. This situation commonly occurs in coastal waters where the current flows more or less parallel to the coast. The motion of the current causes water to be pulled away from the coast, creating a very subtle "trough" along the coast which is slightly lower than mean sea level. Since water is continuously being pulled away from the shore, the only water free to flow into this trough is that which wells upward from below. The resulting circulation continually replaces the water along the shore with new water rich in inorganic nutrients. One result of bringing water up from depth is the loss of dissolved carbon dioxide. Carbon dioxide may be released from solution by reducing the pressure, increasing the temperature, or agitating the solution. All three of these factors occur in the cirulation pattern just described. The subsequent loss of dissolved carbon dioxide increases the acidity (pH) of the water, which in turn causes any phosphate in solution to precipitate out and form phosphorite.

Phosphorite nodules have been recovered from the ocean floor off the coast of California in depths ranging from about 190 feet to over 8000 feet. Investigations into the occurrence of these nodules off California's coast span more than three decades and have led to an estimate that about 10 percent of the 36,000 square mile shelf off California is covered with nodules. Tentative calculations of the amounts of nodules are placed at one billion tons. Of this, it is believed that no more than five to ten percent will be of an economically recoverable grade. It has been estimated that the ocean floors of the world may contain as much as 300 billion tons of phosphate nodules. If the ratio of five to ten percent recoverable grade is applicable to these deposits, the quantity of nodules available is 15–30 billion tons.

Manganese The manganese nodules that cover some areas of the sea floor are often used as an example of a quantitatively huge, untapped resource from the sea that will supply many of our needed metals in the future. The occurrence of manganese oxide nodules on the deep-sea floor was also first reported by the *Challenger* expedition. Since that time several hundred locations (Fig. 6–2), covering several thousands of square miles, have been recorded. Although occurrences in water as shallow as 400 meters have been recorded, most of the deposits are confined to depths of 3500 to 4500 meters and to the seamounts (submerged mountains) associated with these depths.

The average size of manganese oxide nodules is about 5 cm, although very large ones have been recovered occasionally. Deposits of manganese oxide nodules also occur as crusts on the bedrock of seamounts and average about 2 cm thick, although thicknesses of 10 cm or more have been re-

Table 6-4 Analysis by weight percent, of 54 samples
of manganese nodules taken from the Pacific Ocean.
Maximum, minimum, and average is given for each.

Element	Maximum	Average	Minimum
B	0.06	0.029	0.007
Na	4.7	2.6	1.5
Mg	2.4	1.7	1.0
Al	6.9	2.9	0.8
Si	20.1	9.4	1.3
K	3.1	0.8	0.3
Ca	4.4	1.9	0.8
Sc	0.003	0.001	0.001
Ti	1.7	0.67	0.11
V	0.11	0.054	0.021
Cr	0.007	0.001	0.001
Mn	41.1	24.2	8.2
Fe	26.6	14.0	2.4
Co	2.3	0.35	0.014
Ni	2.0	0.99	0.16
Cu	1.6	0.53	0.028
Zn	0.08	0.047	0.04
Ga	0.003	0.001	0.0002
Sr	0.16	0.081	0.024
Y	0.045	0.033	0.016
Zr	0.12	0.063	0.009
Mo	0.15	0.052	0.01
Ag	0.0006	0.003	
Ba	0.64	0.18	0.08
La	0.024	0.016	0.009
Yb	0.0066	0.0031	0.0013
Pb	0.36	0.09	0.02

(Reproduced by permission of the author and publisher
from J. L. Mero, *The Mineral Resources of the Sea*, New
York: American Elsevier, 1965, Table XXVIII, p. 180.)

corded. Small, sand-sized grains of manganese oxide, about 0.5 mm in
diameter, are commonly contained in some deep-ocean red clays and
organic oozes.

The composition of these nodules is a complex of several manganese
oxide minerals. It appears that elements such as Na, Ca, Sr, Cu, Cd, Co,
Ni, and Mo can be readily substituted for the Mn or Fe in the structures
of these minerals. In addition to these manganese minerals, the nodules
commonly contain geothite (a hydrous iron oxide), opal, rutile, anatase,
barite, nontronite, and a variety of clay minerals. Table 6-4 gives the
analysis of 54 samples of nodules from the Pacific Ocean. This table shows

that the maximum amount of Mn present in these nodules is approximately 41 percent less than that present in deposits of manganese ore on land. However, Ti occurs in amounts averaging 0.7 percent, Fe averaging 14.0 percent, Co averaging 0.4 percent, Ni averaging 1.0 percent, and Cu averaging 0.5 percent. These elements, especially copper, nickel, and cobalt, are presently of considerably more interest than is manganese.

The United States, France, West Germany, the Soviet Union, and several eastern European countries are currently preparing to select mining sites. A United Nations study predicts that the first market to be affected by ocean-floor production will be the cobalt market. It is possible that selected rich beds, like those west of Hawaii, could supply up to half the projected world demand for cobalt by 1980. One American company has begun preliminary testing of equipment, using the Blake Plateau (Fig. 6-3) area off the east coast as a field site. This area was determined to be suitable for testing, as the nodules here are of poor quality. The equipment, which utilizes a vacuum and suction technique, successfully lifted up tons of material from the floor of the plateau, approximately 1000 meters below, and unloaded them in the ore separator on a ship at the surface. As the suction device is moved slowly over the ocean floor, tons of mud, muck, and nodules are dumped into the ship. If tests are successful, it may be possible to build equipment capable of operating at depths of 7000 meters and of recovering and processing up to 25,000 tons of nodules per week. The nodules of the Pacific Ocean may soon be accessible. Estimates based on several hundred locations in the Pacific predict that there may be 1.7 trillion tons of nodules there which, in addition to the manganese, contain 16.4 billion tons of nickel, 8.8 billion tons of copper, and 9.8 billion tons of cobalt.

The manganese nodules are authigenic, that is, they have grown in the place in which they occur. Therefore, they occur in areas where deposition of sediments is very slow, for once the nodule is buried, growth ceases. The abyssal regions of the oceans best fit this description. It is probable that nodules enlarge by the precipitation of elements, probably in colloidal form, from the surrounding solution. These colloidal particles have an electric charge which removes elements such as cobalt, nickel, zinc, copper, and others from solution. This acts as an enrichment process, since these elements are thereby accumulated in quantities far greater than their concentrations in seawater. Manganese freed by chemical weathering on the land and carried to the seas by rivers is probably a very important source of this element. Submarine volcanic activity has also been suggested as a source of manganese, and this source may be as important as its land-derived source. Decomposition of igneous outcrops on the ocean floors probably also contributes to the manganese supply. Manganese is among the 12 most abundant elements in the crust.

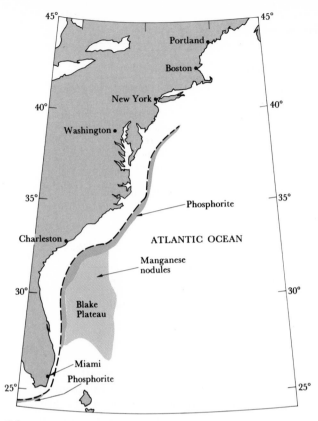

6-3

Most favorable area of potential reserves of phosphorite and manganese nodules along the Atlantic coast of the United States. The dashed line represents the edge of the continental shelf, which is at a depth of about 80 meters in the south and 140 meters in the north. The potential phosphorite deposits form a narrow band along much of the coast. The manganese nodules occur primarily on the surface of the Blake Plateau. (Adapted from K. O. Emery, Some Potential Mineral Resources of the Atlantic Continental Margin, *U.S. Geological Survey Professional Paper 525-C, 1965, Fig. 1.)*

Observational data as well as radiometric techniques have given us a general rate for the accumulation of nodules on the sea floor. In shallow water near a continental land mass, the rate is often faster than it is in the abyssal zone, but the quality of nodules is also much poorer. Accretion at about 1 mm per million years is the accepted rate for nodules on the ocean floor. Some estimates of accretion as rapid as 1 mm per thousand years have led to speculation that the growth of manganese nodules and the accompanying enrichment of other metallic elements occurs at a rate equal to that of our usage. Such a rate would represent a constantly replenished source; however, recent data do not support this.

Table 6-5 Tabulation of underground offshore mining operations

Location	Minerals	Maximum distance from shore
Finland	Magnetite Quartz Banded iron ore	1000 meters
Nova Scotia	Coal	5 miles
Taiwan	Coal	9000 feet
Turkey	Coal	1000 meters
Canada (Ontario)	Rock salt	2500 feet
Japan	Coal	
United States (Louisiana)	Sulphur	5 miles (drill holes)
United Kingdom (Cornwall)	Tin	1 mile
United Kingdom	Coal	5 miles
England	Potash	5 miles
Chile	Coal	4 miles
Alaska	Barite	1 mile

(From *Undersea Technology Handbook/Directory 1973*, Chapter 3, "Mining and Mineral Recovery," p. A/28 by Michael J. Cruickshank, National Oceanic and Atmospheric Administration, by permission of copyright owners, Compass Publications, Inc., Arlington, Va.)

Metalliferous muds In the mid-1960s it was discovered that hot, highly saline (240 parts per thousand) brines were causing the precipitation of metalliferous muds abnormally rich in certain metallic elements in three large basins in the Red Sea (Fig. 6-2). Some element concentrations in these brines were from 1,000 to 50,000 times greater than they were in average seawater. The water in the largest of the three basins, the Atlantis II Deep, was nearly ten times more saline than normal seawater and contained iron, manganese, zinc, lead, copper, silver, gold, and other metallic elements.

The muds below these brine pools have been cored to depths of 10 meters, and it has been found that the high enrichment of metallic elements persists even to this depth. Some seismic observations indicate that these metalliferous muds may extend as deep as 100 meters. These observations have led to estimates that muds in the upper 10 meters alone of the Atlantis II Deep, which has an area of about 60 square kilometers, have a potential value of 2.3 billion dollars just in copper, zinc, lead, silver, and gold. Although the conditions that culminate in this type of deposit are very complex and hence very rare, it has been suggested that other areas (the Gulf of Aden, the Gulf of Aqaba, the Gulf of California, and the East Pacific Rise) are potentially similar.

Underground offshore mining   A number of mining operations (Table 6-5) actually recovering mineral wealth from mines under the sea already exist. However, access to the mine is usually on shore and consequently, the mine does not extend a great distance from shore. This kind of operation is limited and offers little hope for easing the mineral problem, except in limited areas. On the other hand, we are presently incapable of dealing with the problems inherent in initiating and working a shaft-type mining operation on the sea floor itself.

SUMMARY

We hope that our consideration of the problem of the on-going depletion of mineral reserves has helped you recognize both the problem and some of the pitfalls and erroneous concepts on which some current discussions of mineral resources are based. Our mineral resources are finite, and the earth is not creating new resources to replace those being used. In considering mineral resources from the land, we have attempted to make you aware of the average concentration required for profitable exploitation of deposits. We also pointed out the geographic distribution of known reserves to show the inequitable nature of deposit location. Contrary to some beliefs, the sea will not provide for all our needs; however, new resources will be obtained from the sea to some extent, depending on new discoveries and the development of new technology. We have put all of these factors into a framework of current usage patterns and population growth so that we could then consider some alternatives open to us.

REFERENCES

"Another lost frontier?" *Forbes,* 15 August 1972: 29–31.

Bateman, A. M. *Economic Mineral Deposits,* 2d ed., New York: Wiley, 1965.

Burk, C. A. "Global tectonics and world resources," *The American Association of Petroleum Geologists Bulletin* 56, 2 (1972): 196–202.

Cloud, P. E., Jr. "Realities of mineral distribution," *Texas Quarterly* 11 (1968): 103–126.

————. "Mineral resources from the sea," in *Resources and Man,* National Academy of Sciences, National Research Council, San Francisco: Freeman, 1969.

Cruickshank, M. J. "Mining and mineral recovery," in *Undersea Technology Handbook/Directory 1973,* 3, Arlington, Va.: Compass, 1973.

Emery, K. O. *Some Potential Mineral Resources of the Atlantic Continental Margin,* U. S. Geol. Survey Professional Paper 525-C, 1965.

Fisher, J. L. and N. Potter. *World Prospects for Natural Resources,* Baltimore: Johns Hopkins Press, 1970.

Goldberg, E. D. "The oceans as a chemical system," in *The Sea,* ed. M. N. Hill, New York: Wiley, 1963, v. 2.

Gross, M. G. *Oceanography,* 2d ed., Columbus, Ohio: Merrill, 1971.

Gullion, E. A., ed. *Uses of the Seas,* Englewood Cliffs, N. J.: Prentice-Hall, 1968.

Hibbard, W. R., Jr. "Mineral resources: challenge or threat?" *Science* 160, 3824 (1968): 143–159.

Landsberg, H. H. *Natural Resources for U. S. Growth,* Baltimore: Johns Hopkins Press, 1967.

Lasky, S. G. "Mineral resources appraisal by the U. S. Geological Survey," *Colorado School of Mines Quarterly* 45, 1A (1950): 1–27.

———— . "How tonnage-grade relations help predict ore reserves," *Engineering Mining Journal* 151, 4(1950): 81–85.

Lovering, T. S. "Non-fuel mineral resources in the next century," *Texas Quarterly,* 11 (1968): 127–147.

———— . "Mineral resources from the land," in *Resources and Man, op. cit.*

McKelvey, V. E. and R. R. H. Wang. *Preliminary Maps—World Subsea Mineral Resources,* U.S. Geological Survey, Miscellaneous Geological Investigations Map I–632, 1970.

McKelvey, V. E. "Mineral resource estimates and public policy," *American Scientist* 60 (1972): 32–40.

Manheim, F. T., R. M. Pratt, and P. F. McFarlin. "Geochemistry of manganese and phosphate deposits on the Blake Plateau," Program for the meeting of the Geological Society of America, New Orleans, 1967, p. 139.

Meadows, D. H., D. L. Meadows, J. Randers, and W. W. Behrens III. *The Limits to Growth,* New York, Universe Books, 1972.

Mero, J. L. *The Mineral Resources of the Sea,* New York: Elsevier, 1965.

Park, C. F., Jr. *Affluence in Jeopardy—Minerals and the Political Economy,* San Francisco: Freeman, 1968.

Petersen, N. S., F. V. Carrillo, and R. C. Vars. *Materials Substitution Study—General Methodology and Review of U. S. Zinc Die-casting Markets,* United States Bureau of Mines Information Circular 8505, 1971.

Population Reference Bureau. "Population and resources: the coming collision," *Population Bulletin* **XXVI**, 2 (1970): 1–36.

Riley, C. M. *Our Mineral Resources,* New York: Wiley, 1956.

United Nations, Department of Economic and Social Affairs. *Mineral Resources of the Sea,* 1970.

United States Bureau of Mines. "Metals, Minerals and Fuels," *Minerals Yearbook* 2969, I–II, 1969.

———— . *Mineral Facts and Problems*, Bulletin 650, 1970.

"Vacuuming the Atlantic floor," *Science News* 98 (1970): 134–135.

Voskuil, W. H. *A Geography of Minerals*, Dubuque, Iowa: Wm. C. Brown, 1969.

Weyl, P. K. *Oceanography*, New York: Wiley, 1970.

In the nineteenth century man used crude oil primarily for making kerosene, which replaced whale oil used for lighting. (Photograph by George Sheng)

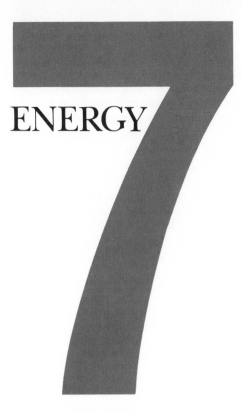

ENERGY 7 THE FOSSIL FUELS

INTRODUCTION

The occurrence of the Industrial Revolution in England, western Europe, and America during the period of about 1750 to about 1850 marked the beginning of man's demand for large quantities of energy. Energy, the capacity to do work, comes from a number of sources, which are broadly summarized in Table 7-1.

In the first century B. C., the horizontal waterwheel was man's primary means for generating power, aside from his own muscles and those of his animals, but the wheel generated only about 0.3 kilowatt of power, or the energy necessary to light three common 100-watt light bulbs. Later, the waterwheel was set vertically, and the power output was increased to about 2.0 kilowatts. Windmills were introduced as a source of power, and they generated as much as 12.0 kilowatts! Continued improvements in waterwheel design raised its generating power to about 56.0 kilowatts, which was its capacity just prior to the beginning of the Industrial Revolution. Until a little more than 200 years ago, the power derived from water, wind, and muscle was sufficient to meet man's energy needs. However, the introduction of power-driven machinery into the manufacturing process in England started man on his quest for ever-increasing amounts of energy which could be used to run his machines and do his labor. Since the start of the Industrial Revolution in England around 1750, the power output of basic energy-conversion devices has increased about ten thousand times, or four orders of magnitude. During the early and middle 1700s the steam engine, windmill, and water wheel could generate an average of 50 to 100 kilowatts of power, whereas modern steam and gas turbines generate an average of 500,000 to 1,000, 000 kilowatts of power.

Table 7–1 Energy and energy sources

Type of energy	Sources
Solar	The sun
Chemical	Petroleum, coal, and wood
Motion	Running water, including waterfalls, tidal basins, and man-made dams; wind
Geothermal	Geysers and hot springs; hot water at great depth
Nuclear	Uranium, thorium, and hydrogen; others

Wood was the fuel used to initiate industrialization in the United States, as it did in the other industrial centers of the world. As shown in Fig. 7–1, wood provided more than 90 percent of the energy used in the United States in 1850. However, rapid depletion of forests, combined with the need for a better energy source, thrust the use of coal as an energy source into the forefront, and by 1900 more than 70 percent of the energy consumed in the United States was provided by coal. The continuing need for energy led to the use of petroleum, and by 1970 petroleum and natural gas were supplying more than 75 percent of the energy in the United States. It seems apparent that our demand for ever-increasing amounts of energy will bring about still another change in the primary source of energy: that is, a change from fossil fuels to nuclear power.

In this chapter we will cover the fossil fuels, from which over 95 percent of our energy was derived in 1970. Chapter 8 will deal with other possible energy sources, e.g., nuclear, geothermal, and tidal power. We

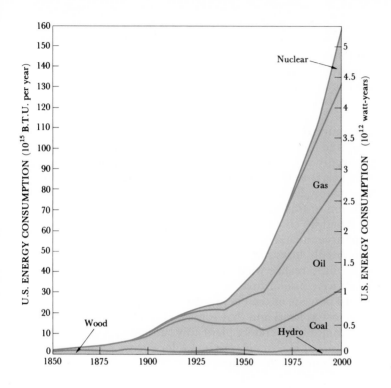

7-1
Energy consumption in the United States from 1850 and projected to the year 2000. Note that our energy consumption has multiplied some 30 times between 1850 and 1970. (Reprinted by permission from Chauncey Starr, "Energy and Power," Scientific American *224, 3, September 1971, p. 39. Copyright © 1971 by Scientific American, Inc. All rights reserved.)*

will also deal with some possible energy sources which have less obvious geologic relationships, such as solar energy and energy from solid organic waste. The conclusion of our discussion of energy will include some projections of future trends in energy usage for the United States and the world, as well as a tabulation of reserves of known energy sources.

The Fossil Fuels

The fossil fuels, which include petroleum, natural gas, and coal, are so named because they are thought to be the result of the distillation of plant and animal remains within the crust of the earth over a long period of geologic time. The chemical energy stored in fossil fuels originates in the process of photosynthesis, which in turn is a function of the energy radiating from the sun in the form of light. However, when we look at the energies involved in each individual stage of the process which takes solar energy, locks it up in fossil plants and animals, and releases it when petroleum or coal is burned, we see that using fossil fuels is a very inefficient way of providing energy.

Of the total amount of energy in the form of sunlight falling on an area of the earth covered by plants, about 2 percent is locked up as potential energy by photosynthesis. Perhaps only half of this, or 1 percent of the original solar energy, is available to animals who eat these plants. Since animals lose approximately 90 percent of their energy intake as heat or waste, at the conclusion of this entire process only 0.1 percent of the original energy is left. When we consider that coal is made from the remains of plants and that petroleum and gas are made from animal remains in a process which undoubtedly loses some of the potential energy locked up in these forms, we begin to realize what inefficient energy-storing devices fossil fuels are. When we consider further that the best fossil fuel plants convert only about 40 percent of the energy stored in these fuels into electricity (60 percent is lost as heat) and that some of this electrical energy is lost in transmission from the generating plant to your home, one is staggered by the implications of how much original solar energy was expended to light the 75-watt bulb in a reading lamp.

PETROLEUM

One of the world's most important sources of energy is petroleum. The use of petroleum derivatives such as gasoline, fuel oil, and natural gas, in addition to such nonenergy-producing products as asphalt and a host of petrochemicals, has increased dramatically in the past century. The petroleum industry is one of the biggest industries in the world; indeed, a great number of industries are associated in some way with the production, transportation, marketing, and use of petroleum and petroleum products. The manufactured substances in which petroleum is used are highly varied and include over 2000 different products. No other material has caused as intensive an examination of rocks, their makeup, their ages and fossils, structural attitudes, and types of layering as has petroleum. Because of both its extensive use in modern society and the enormity of the problem of finding and producing increasingly greater quantities of it, petroleum is one of the most important substances in the environment and in the science of geology.

7-2

The Signal Hill oil field, Los Angeles County, California, in 1922. (Photograph by W. S. W. Keen, U. S. Geological Survey)

Composition

Petroleum includes natural gas, crude oil, and a number of waxy or asphaltic solids and semisolids. The liquid portion of petroleum ranges from nearly colorless, light-weight oils to black, heavy oils rich in asphaltic materials. The various substances in petroleum are mostly (up to 98 percent) composed of hydrocarbons, with smaller amounts of other organic compounds, and traces of other, especially metallic, elements. The other organic compounds contain hydrogen, carbon, nitrogen, oxygen, and sulfur. Trace elements include about two dozen metals, the most important of which are nickel and vanadium.

Uses of Petroleum

When the first oil well was drilled by Edwin L. Drake near Titusville, Pennsylvania in 1859, petroleum did not have much practical use. The little petroleum that had been used until then was used mostly for heating and lighting and was obtained from naturally occurring seepages. After it was learned that man could obtain crude oil (that is, liquid petroleum in its natural state) from the ground through wells, oil was sought primarily for making kerosene to replace whale oil previously used in lamps. Wax and grease were among the minor products. After the turn of the century, when the invention of the internal combustion engine necessitated the production of gasoline to fuel the automobile, this became the most important substance obtained from crude oil.

Gasoline continues to be by far the most important petroleum product quantitatively, requiring about 45 percent of the crude oil. Another 45 percent of the crude oil and all the natural gas are also used for fuel. Natural gas and the lighter, liquid fractions of the paraffin group — so-called LPG — are used mainly for heating. Those fractions heavier than gasoline include kerosene, gas oil (fuel oil), and heavy residual oils. Among its other uses, kerosene is mixed with gasoline for use as jet-plane fuel. Gas oil is used for diesel fuel, home furnaces, industry, and for the heavy oils used in ships, steam generation, and heavy industry. The remaining 10 percent of natural petroleum is used for greases, lubricating oils, waxes, asphalt and tar, petrochemicals, and many other products. The petrochemical industry, especially, has become increasingly important in recent years and produces a very large number of commercial products.

The Occurrence of Petroleum

Petroleum is found in rocks of most geological ages and under a variety of conditions. A critical examination of the occurrence of petroleum reveals many limitations, however, and these limitations make possible an orderly exploration for it.

Petroleum occurs throughout the world, and its commercial production exists on every continent except Antarctica. However, in examining the geography of petroleum production and reserves, one discovers that petroleum is not at all evenly distributed. There are two reasons for this, the first of which is geological. As we shall see later, petroleum is found predominantly in certain kinds of rocks, under certain geological conditions, mostly in rocks of certain ages. The second reason is perhaps more apparent than real. Primarily for industrial and economic reasons, some areas of the world have been explored and developed more intensively than have others. This, however, does not necessarily mean that there is more petroleum in these areas. The large petroleum pools in Libya and the north slope of Alaska were undiscovered only a few years ago, mainly due to the difficulty of searching for oil in those places. Undiscovered oil fields elsewhere in the world may somewhat alter the petroleum distribution picture, but the overall distribution is reasonably well known.

Generally, the majority of the petroleum deposits of the world occur in two large belts (Fig. 7-3). One of these belts is the long Cordilleran region which extends from Alaska to Canada, through the western and Gulf Coast areas of the United States, to Venezuela, and through other South American fields to Argentina. The other belt is the Tethyan region, which runs along an east-west line from the Mediterranean, through the Middle East, to Indonesia. These belts rank as two of the greatest *geosynclinal*

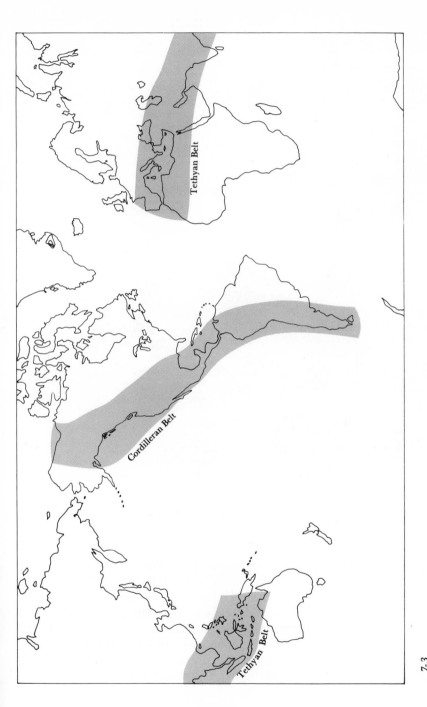

7-3
The major world petroleum belts.

7-4
Large drilling rig and platform in the Grand Isle area off the coast of Louisiana. (Photograph by E. F. Patterson, U. S. Geological Survey)

belts in the world. A *geosyncline* is a large trough, perhaps many hundreds or even thousands of miles long, that borders an uplifted area. Some geosynclines consist of an outer belt in which layers of sediments and associated volcanic deposits, ranging in thickness up to 50,000 feet, accumulate and an inner belt which has thinner deposits of sedimentary, but no volcanic, rocks. Island chains, similar to Japan and other island chains in the Western Pacific today, with volcanoes were present in the outer belts when the geosynclines were active. Still thinner deposits of sediments, called shelf deposits, which are only a few hundred or thousand feet thick, lie on the continent and border the inner geosynclinal deposits. Most oil deposits are found in shelf and inner geosynclinal areas. Of course, there are other areas in which there are significant petroleum deposits, but the majority of the occurrences are in these two large geosynclinal belts.

The major petroleum-producing countries in the western hemisphere are Venezuela, the United States, Canada, Mexico, Argentina, Columbia, and Trinidad. In the Middle East, nearly all countries produce some oil; the major producers are Kuwait, Saudi Arabia, Iran, and Iraq. Libya and Algeria are the major producers in Africa. Various countries in Europe produce some oil, but the USSR ranks higher than any other single country in the eastern hemisphere. Indonesia produces the most oil in the Pacific area. China's production of petroleum is not known, but is thought to be increasing continually. Petroleum reserves, at least proved reserves, are concentrated overwhelmingly in the Middle East. The Middle East is a huge oil province; since it probably will ultimately prove to be the largest of all, the political importance of the Middle East is obvious. Petroleum occurrences on the continental shelves are not as well known as are those of other locations, but exploration of the shelves is increasing. Undoubtedly, a great deal of petroleum will be discovered there (Fig. 7-5).

Although petroleum is found in rocks of all ages (Fig. 7-6), it is rare in both Precambrian and Recent rocks. Rocks of Tertiary age have yielded more oil than have those of other periods. Perhaps as much petroleum comes from Tertiary rocks as from rocks of all other ages combined. There may be a number of reasons for this. First, there are great masses of Tertiary rocks, especially of marine Tertiary rocks, in the world. Second, these rocks mostly lie near the surface, being covered, if at all, only by thin layers of Quaternary rocks, and therefore are readily accessible. Third, erosion, metamorphism, and other processes which could destroy the petroleum have not affected Tertiary rocks to the extent that they have the older rocks.

Rocks of Cretaceous age also contain large amounts of petroleum, furnishing nearly as much as the other periods combined (except, obviously, for the Tertiary). All the other geological systems do produce some petroleum, although the Cambrian and Triassic have a limited production. Thus, it is tempting to look for oil only in rocks of certain ages; but no rocks should be entirely omitted, although there is no economic justification for exploring Pleistocene sediments or Precambrian rocks.

Although there is some petroleum in igneous and metamorphic rocks, petroleum occurs almost entirely in sedimentary rocks. The rare occurrences of petroleum in igneous or metamorphic rocks are usually located near sedimentary rocks from which the petroleum could have migrated.

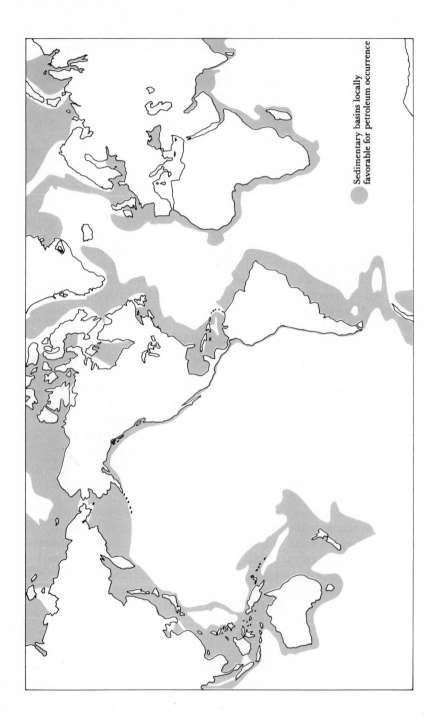

7-5 *Potential petroleum resources of the sea. (Source: U. S. Geological Survey Investigations Map I-632, 1970)*

Quaternary	
Tertiary	
Cretaceous	
Jurassic	
Triassic	
Permian	
Pennsylvanian	
Mississippian	
Devonian	
Silurian	
Ordovician	
Cambrian	
Precambrian	(Density of shading indicates abundance of petroleum)

7-6
Age distribution of petroleum.

No petroleum is found in the large Precambrian shield areas, which are comprised of crystalline rock.

Not all sedimentary rocks are equally productive of petroleum, since only certain types have the necessary porosity and permeability. The coarse-grained fragmental rocks, especially sandstone, meet this requirement, but the fine-grained rocks such as shale generally do not. Chemically precipitated sedimentary rocks productive of oil are almost always carbonate rocks such as limestone or dolomite. In short, rock type is related to petroleum occurrence, as is, apparently, the environmental type of the rock. Petroleum is vastly more abundant in rocks of marine origin than in rocks of continental, especially terrestrial, origin. Seemingly, the conditions under which marine rocks originate are more conducive to either, or both, the formation or preservation of petroleum.

Petroleum is found in *basins* formed as a result of the downwarping of the earth's crust and are usually located in shelf or geosynclinal rather than shield areas. As these basins sink, sediments from adjacent high areas are deposited in them, the thickest deposits normally being nearest the center of the basin. The basins are often structurally deformed later, forming traps in the reservoir rocks to catch and hold the oil which has been

7-7

An oil seep from sedimentary rock. The man is pouring oil he has collected from the seep. (Photograph by E. C. LaRue, U. S. Geological Survey)

generated in the sediments. A basin is a structural-sedimentary unit, complex in its interrelated rock types, fluid flow, geometry, and geologic history.

Origin of Petroleum

The origin of petroleum has been studied and debated, but is still mostly unresolved. Furthermore, it is possible that its origin may never be determined. Although the question may be important, it is largely academic, and failure to resolve it has scarcely hindered the discovery of the world's oil. Two schools of thought regarding the origin of petroleum exist. The great majority of scientists believes that petroleum forms from organic sources, while a few feel that the biological, chemical, and geological factors indicate an inorganic origin.

Sediment and soils that are forming today usually contain, and are often rich in, hydrocarbon compounds. These hydrocarbons are somewhat similar to petroleum hydrocarbons. Petroleum is intimately associated with rocks which formed in areas abundant in plant and animal life and in ages in which this life is known to have been plentiful. Virtually no petroleum occurs in Precambrian rocks, perhaps because life was not abundant at that time. Just how abundant life must actually be, other conditions being equal, is entirely speculative. Petroleum may form by either direct conversion of hydrocarbon compounds that organisms have incorporated into living tissue or conversion of organic matter after the organism has died. There is such a vast amount of biotic substance available that all the world's petroleum could form in either way.

As yet, there is no consensus on the exact kind of rock in which petroleum forms. Coarse-grained, well-sorted sediments, such as certain sands, and most carbonate sediments, including reef and some other limestones, are porous and well oxygenated, and their permeability allows the flow of water to wash away the organic material. This flushing and, especially, oxidation, would remove or destroy organic matter before it could be converted to petroleum. On the other hand, shale and calcareous mud (marl) are fine-grained and compactible. They are initially highly porous, so that they are likely to contain water and organic matter. Because their particles are fine, however, permeability is low, and the environment within the pore spaces becomes sufficiently isolated from the oxygen-bearing and freely moving overlying waters for stagnation to occur. Such conditions are thought to be enhanced if there is a rapid accumulation of sediment, which would quickly bury the organic material and thereby prevent its destruction by oxygen and by oxygen-breathing organisms, including bottom dwellers. The exclusion of oxygen creates a chemically reducing environment within the pore spaces of the sediment, which enables the organic matter, safe from aerobic destruction, to begin its conversion to petroleum (if it has not already done so).

The time it takes for petroleum to form is not known either. We do know of accumulations in some rocks less than three million years old. Perhaps a million years or even much less is the maximum time required for the formation of petroleum.

Assuming that petroleum does form in fine-grained sediments, it must then move to porous, more permeable, coarser-grained sediments, where it becomes concentrated into a commercially productive pool, that is, a place from which we can pump it. Probably, a *primary migration* occurs, in which the oil flows from the source beds to the reservoir beds (Fig. 7–8). As the sediments of the source beds accumulate, they begin to compact due to the increase in weight. As this takes place, fluids, including the petroleum, are squeezed upward or laterally out of the source beds. If there are no readily accessible reservoir rocks present, the fluids will return to the overlying waters, and the petroleum portion will be lost through oxidation or dispersal. If reservoir rocks do exist nearby, the fluids may move into them. The reservoir rock is composed of less compactible material and remains more porous and permeable than do adjacent compacting muds and clays. Fluids move down the pressure gradient toward the more open framework of the coarser-grained sediments.

It is probably not until the petroleum-bearing fluids reach the reservoir rock that the separation of the petroleum from water occurs. The petroleum apparently tends to move into the larger openings and also tends

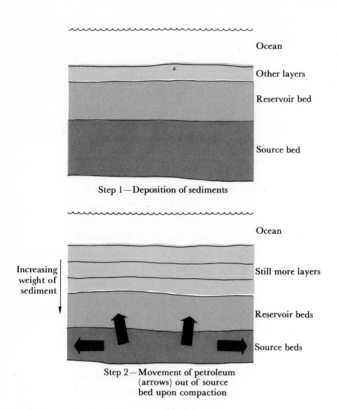

7-8
Primary migration of petroleum from source bed to reservoir bed.

to rise in the rocks, whereas the water stays in the finer openings and, because of its greater density, lies below the petroleum. Movement of the petroleum within the reservoir rock also is a result of pressure differentials. This movement might occur any time after the primary migration and includes movement of the petroleum either into or out of the *traps* which prevent the oil or gas from escaping.

Petroleum Traps

After petroleum forms, it permeates the rocks until it either escapes to the external environment, where it is destroyed or is trapped by some impermeable rock which prevents further migration. After it accumulates,

7-9

A small syncline exposed along banks of canal near Hancock, Maryland. (Photograph by C. D. Wolcott, U. S. Geological Survey 1897)

trapped in its reservoir rock, it may remain there until the trap is destroyed or until it is recovered by man.

There are many kinds of traps which provide barriers to continued migration. The particular conditions that cause traps to form may be quite complex, but for simplicity's sake we will follow a classification by A. I. Levorson, who classified traps as structural, stratigraphic, or fluid.

Structural traps include *folds, faults,* and *unconformities.* As petroleum moves up through permeable rocks, it moves out of downfolds (synclines) and into upfolds (anticlines). If an anticline lies underneath impermeable rock, the petroleum will accumulate in the top of the anticline (Fig. 7-11). The lightest part, natural gas, may segregate and rise above the heavier part, crude oil, to form a gas cap. Explorationists were quickly able to understand the anticlinal accumulation of oil; and for many years field geologists devoted their efforts to looking for dipping surficial rocks that indicated the presence of structural "highs." Some anticlines, however, are present wholly in the subsurface and are less readily discovered. Subsurface mapping, based on information from previously drilled wells and various geophysical methods, may lead to their discovery.

7-10
A small anticline exposed along the Potomac River in Maryland. (Photograph by I. C. Russell, U. S. Geological Survey)

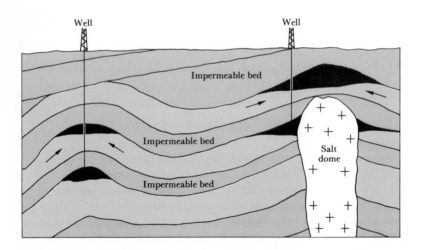

7-11
Anticlinal traps. Fold on right is caused by upward squeezing of low-density salt from salt bed at depth. Arrows show direction of movement of petroleum.

Faults, or ruptures along which rocks have been displaced, may truncate reservoir rocks and bring impermeable rocks into juxtaposition with them, forming a trap (Fig. 7–12). Some faults have open fissures that allow the petroleum to escape, but many of them are plugged with finely ground rock or precipitated mineral matter and thus form an impermeable barrier. Joints in rocks may be open and may contain petroleum. If fractured rocks are overlain or bounded laterally by impermeable rocks, they can become reservoirs.

Stratigraphic traps are caused by some lateral or vertical change other than the structural attitude of rock formations which tends to diminish the permeability. These traps could be the result of a decrease in pore space due to additional cementation, a "silting up" of a "clean" sandstone by an increase in the amount of clay or other fine particles among the sand grains, the disappearance of the reservoir bed by wedging out between other layers, or a change in which the lithology of the reservoir bed becomes different and less permeable (Fig. 7–13).

Sandstone lenses or other bodies that are surrounded by shale and might originate as river channels or off-shore bar sands can be considered stratigraphic traps. Reefs, composed of masses of limestone deposited by organisms such as coral and algae may be similarly surrounded by impermeable rocks and comprise stratigraphic traps (Fig. 7–14).

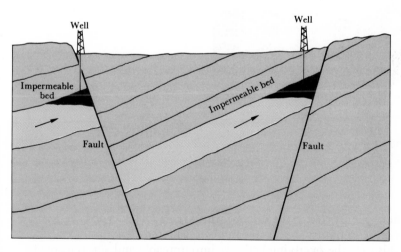

7-12
Fault traps. Petroleum is trapped between impermeable bed and fault.

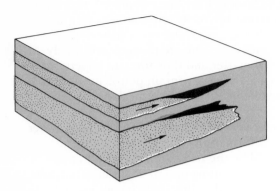

7-13
Petroleum traps caused by wedge-out of reservoir bed (top) and lateral change to fine-grained rock (bottom).

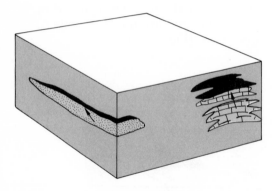

7-14
Petroleum traps in rock bodies surrounded by impermeable rocks. Left: elongate sandstone body such as bar or channel sand; right: lime-stone reef.

Another type of trap is the unconformity trap. This type is essentially stratigraphic in nature, but also can be considered structural (Fig. 7–15). In this case the overlying beds are impermeable and act as a covering that prevents migration farther up the dipping beds. Prehistoric soil layers may be present along an unconformity, and because they consist of compacted clays and other fine-grained materials may act as a barrier to migration.

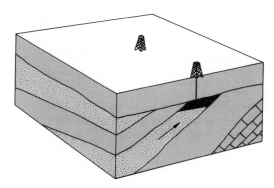

7-15
Petroleum trapped by angular unconformity.

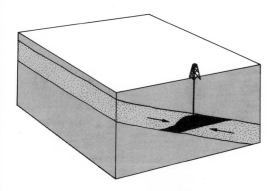

7-16
*Hydrodynamic trap. High pressure from strong
flow of water moving from left (arrow) down
pressure gradient prevents oil from rising further in reservoir bed.*

    The movement of fluids within reservoirs is a complex subject of
which our understanding has increased in recent years: we now recognize
that hydrodynamic forces, i.e., the force of moving water, may cause petroleum to accumulate in what may be called fluid traps. Since petroleum
is lighter than water, it moves upward through a reservoir bed. In a tilted
bed, water flows downward under the influence of the hydrodynamic gradient. When the two fluids meet, the force of the downward-moving water

may in some instances be sufficient to prevent further upward movement of the buoyant petroleum. In such cases the petroleum, unable to rise beyond a certain point, will accumulate in a pool (Fig. 7–16).

The Economics of Petroleum

Drilling for petroleum is both costly and risky. In the decade from 1959 to 1969, the average drilling cost per well in the United States rose from $53,500 to $88,554 (Fig. 7–17). Part of this cost is represented by the

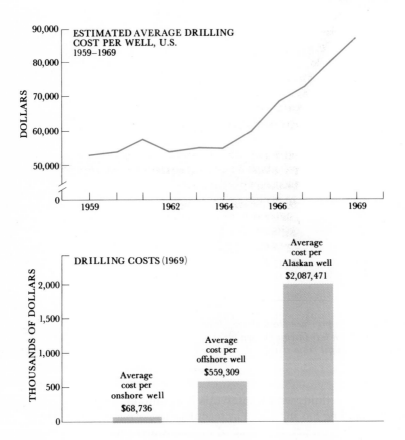

7-17
The steep rise in general drilling costs from 1950 to 1969 is shown graphically in the top figure; the marked difference in cost due to well location is shown in the bottom chart. (Reprinted by permission from Petroleum and Energy, *Sheet 7, Washington, D.C.: American Petroleum Institute, 1971.)*

Table 7-2 Materials necessary for the drilling
of a typical 10,000-foot well

14,000	feet of steel pipe
11,500	feet of steel casing
20	drill bits
5	reamers
1,050	tons of drilling muds and additives
4,850	sacks of cement
48,000	barrels of water
3,000	barrels of diesel fuel

(Reprinted by permission from *Facts About
Oil—A Handbook for Teachers*, Washington,
D.C., American Petroleum Institute, 1964,
p. 6.)

materials, shown in Table 7-2, while the remainder represents the cost of
labor of the teams of people such as the exploration crew, the geologists,
and the drilling crew, plus the office personnel necessary to support them
in the field.

The cost of drilling does not include the cost of the drilling rig, which
may have cost more than $500,000. In the period from 1960 to 1970, the
price of steel casing rose more than 17 percent, the cost of oil field machinery rose 26 percent, and hourly wages of production workers rose
more than 50 percent. This increase in the cost of drilling exploratory
wells is reflected in the almost yearly decline in the total number of wells
drilled in the United States since 1956, as shown in Fig. 7-18. The neces-

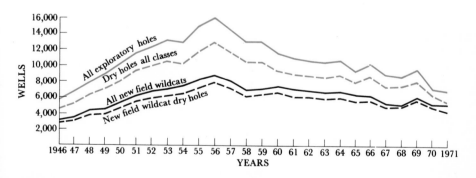

7-18
*Graph of exploratory wells drilled in the United States by year from 1946 to 1971. (Re-
printed by permission from Charles F. Iglehart, "North American Drilling Activity in
1971,"* American Association of Petroleum Geologists Bulletin **56**, 7, *July 1972, p. 1171.)*

sity of having to drill deeper for oil has added even more cost to the already inflationary cost of labor and material. This is graphically illustrated in Fig. 7–19. It appears that most of the near-surface oil has been discovered, and only the deeper reservoirs are now available to us. The deepest well to be drilled by early 1972 extends to a depth of 30,050 feet and is located in Beckham County, Oklahoma. Another well being drilled near Pampa, Texas, was expected to reach a depth of 31,000 feet. Contrast these depths with that of the first commercial oil well, drilled by Drake at Titusville, Pennsylvania, which struck oil at 69.5 feet.

The financial risk involved in drilling for oil is shown graphically in Fig. 7–20. It shows that of 100 new-field wildcats drilled in 1970, only nine *found* oil or natural gas, and *only two* had the one-million-barrel reserve of oil or its gas equivalent required to make it commercially profitable. Since an exploratory well may cost from $100,000 to $2,000,000, this search for oil represents a considerable financial risk. For example, the Chase Manhattan Bank has predicted that capital investments in the petroleum industry will have to rise from the $60 billion of the 1960–1969 period to a tremendous $150 billion during the 1970–1979 period (Fig. 7–21).

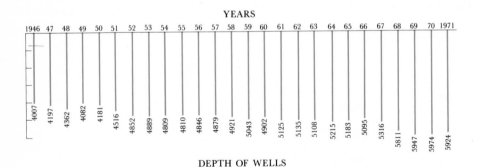

7-19
Average depth of new-field wildcat wells drilled in the United States by year from 1946 to 1971. (Reprinted by permission from Charles F. Iglehart, "North American Drilling Activity in 1971," American Association of Petroleum Geologists Bulletin 56, 7, July 1972, p. 1171.)

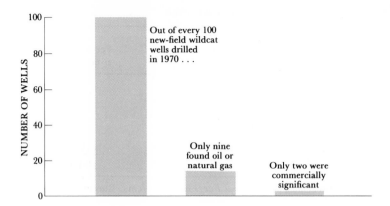

7-20
The high financial risks incurred in drilling for oil. (Reprinted by permission from Petroleum and Energy, *Sheet 8, Washington, D.C.: American Petroleum Institute, 1971.)*

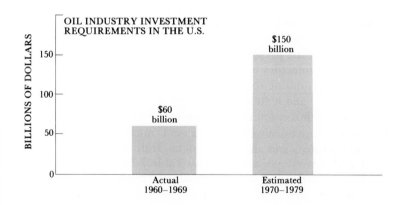

7-21
The massive capital requirements of the petroleum industry. (Reprinted by permission from Petroleum and Energy, *Sheet 9, Washington, D.C.: American Petroleum Institute, 1971.)*

Table 7-3 Use of natural gas by category of consumer, 1968

Natural gas consumers	Natural gas demand 1968 (billion cubic feet)	Percent of total demand
Industrial	8,530	45.0
Household-commercial	6,215	33.0
Electric generation	3,144	16.6
Transportation	591	3.1
Petrochemical and other nonfuel uses	441	2.3
	18,957	100.0

(Source: A. J. Warner, "Natural Gas," in *Mineral Facts and Problems*, Bureau of Mines Bulletin 650, 1970, p. 122.)

NATURAL GAS

Much of the discussion in the preceding section on petroleum is equally relevant to natural gas. The origin of both occurs by the same process; indeed, they are probably both generated simultaneously; therefore, their occurrence would be very similar, but not necessarily identical. Since migration of the gaseous portion to another area at a later time may tend to separate the two, it is possible to find one without the other. The structures by which gas is trapped and held are the same as those shown in Figs. 7–12 through 7–17.

Natural gas is made up of several light-weight hydrocarbons, those chemical compounds of hydrogen and carbon, the most commonly found of which are methane (the chief constituent), propane, ethane, and butane. Natural gas burns readily in the presence of air. In the process of burning, the carbon and hydrogen molecules break up into individual atoms of carbon and hydrogen, which combine readily with the oxygen in the air to form carbon dioxide (CO_2) and water (H_2O). This entire process of breaking up and recombining with oxygen is accompanied by the release of heat.

The heat released during burning is used to neat our homes, cook our food, and run our refrigerators and air conditioners. Industry, however, is the biggest consumer of natural gas, as shown by Table 7-3. In addition to its use as plant fuel by industry, natural gas is also used in the processing of metals, crude oil, and other raw materials; in the processing of food products; and for many other purposes. Relative to the consumption of the other fossil fuels for use in the generation of electricity, only a small amount of natural gas is used. Its total nonfuel uses, such as those in the petrochemical industry, account for little more than 5 percent of its consumption.

In the 25-year period between 1945 and 1970, natural gas usage grew at an average annual rate of 6.5 percent. During this period a high-pressure pipeline network, valued at over 17 billion dollars, was built to serve all of the continental United States. Vermont, the last state to be included in the network, received service in 1966, just at the beginning of a natural gas shortage, or as a report from the Federal Power Commission expressed it, "the end of natural gas industry growth uninhibited by supply considerations."

Although the economics of drilling for natural gas are similar to those of petroleum, there appears to be an additional factor which may drive the price of natural gas up: that is the location of potential future reserves. Apparently, perhaps two-thirds of our potential supply lies either at great depths (15,000 feet or more) or under the oceans in the continental shelf bordering the continent. Under these conditions, greater expense is incurred in both the exploration for, and the subsequent production of, gas.

OIL SHALE AND TAR SAND

Shale oil is petroleum derived from the destructive distillation of bituminous shales, generally called *oil shales.* Petroleum can also be recovered from certain bituminous sands called *tar sands.* The difference between oil shale and tar sand is that the bituminous matter in oil shale is a solid, whereas the bituminous matter in tar sand is a highly viscous liquid. Oil shale is a fairly resistant rock. Its organic matter is insoluble in organic solvents, such as carbon disulfide, and does not flow in response to heating until it is distilled in special retorts at temperatures of 800° to 900° F. As the viscous-liquid nature of the heavily asphaltic petroleum in tar sands permits the material to flow when heated with steam, the recovery of this petroleum differs from the recovery of the organic material in oil shale.

The composition of organic matter in oil shale is different from that of crude petroleum. Both their compositions are variable, but oil shale contains a small percentage of nonhydrocarbon substances such as oxygen and nitrogen. These substances are regarded as impurities and must be removed in the refining process. The organic matter of oil shale, often called *kerogen,* has been found to incorporate algal material, fungal and bacterial remains, transported wood fragments, spores, pollens, structureless globules or films of resin, other plant debris, and lesser amounts of recognizable animal remains. The substance has a dull to waxy luster, often has a curving fracture, and is black, brown, yellowish brown, or reddish brown. The inorganic portion of oil shale includes layers of clay and fine quartz in the form of silt or sand that has been washed in and,

7-22
Specimen of oil shale from Garfield County, Colorado. The specimen is moderately rich and is typically thinly laminated. (Photograph by W. H. Bradley, U. S. Geological Survey)

commonly, calcite crystals which probably grow in place. The material consists of thin laminations, the interlayering of "kerogen" and transported fragments perhaps being indicative of seasonal variation in deposition. Although similar to boghead and cannel coals, which contain plant spores and algal matter, oil shale has a greater inorganic content. Oil shale, cannel coal, and boghead coal form a related sequence of fossil fuels distinct from either petroleum or the peat-lignite-bituminous coal sequence.

Tar sand is a sandstone that contains heavy petroleum in its pore spaces. As this is too viscous to flow, it cannot be recovered in the manner in which oil is usually obtained through wells. The bituminous matter is, nevertheless, petroleum and is therefore different from the "kerogen" of oil shales. The origin of these deposits is not known.

Oil shale is found in a number of areas in the world, including New South Wales (Australia), southeastern Brazil, the Midland Valley of Scotland, Estonia, and various places in Russia, including the Domanik deposits which are widespread from the Caspian to the Arctic. In the United States, the best known and most extensive deposits are in the Green River Formation, which is a sequence of lime muds and other rocks that were deposited in a large lake in the Eocene Epoch, 40 to 60 million years ago. This lake occupied areas in southwest Wyoming (Green River Basin), (Fig. 7–24), although the richest deposits of oil shale are in the Colorado region. The oil shale has been produced away from the shoreline of a quiet-water lake environment. It is believed that oil shales such as those of the Midland Valley were formed as a result of the accumulation of organically rich ooze on the bottom of shallow estuaries and lagoons which were for the most part barred from the sea, whereas the Domanik deposits of Russia are wholly marine. There seems to be a rather wide range of depositional environments in which oil shale may have formed.

Probably the world's largest and best-known tar sand deposits are those of northern Alberta. These deposits occur in sandstones of Early

7-23
Aerial view of Roan Cliffs along Parachute Creek, Colorado, oil shale area. (Photograph by J. A. Donnell, U. S. Geological Survey)

Cretaceous age in a belt extending approximately 600 miles from Peace River, Alberta, to Lloydminster, Saskatchewan, the larges deposit being the Athabasca deposit centered at Fort McMurray.

The amount of energy that can be obtained from oil shale and tar sands may compare favorably with that from conventional oil and gas pools. Brian Skinner lists potential world resources of oil and gas as 2500 billion barrels, W.P. Ryman suggests 2090 billion barrels, and Moody estimates 1800 billion barrels. According to Skinner, there may be an additional 1000 billion barrels of shale oil and perhaps more than 600 billion barrels of oil from tar sands. M. K. Hubbert indicates that there may be 2000 billion barrels of shale oil, but that probably only 190 billion barrels can be recovered under present conditions. He also says that there are about 300 billion barrels of the Canadian tar sand oil reserves; this figure should be compared with 750 billion barrels, the amount actually in place in Canada, according to Vigrass. In the Uinta Basin there may be about 120 billion barrels of oil in shale which yield 25 gallons or more per ton, and 290 billion barrels of oil in shale yielding 15 gallons or more per ton. Other estimates suggest there may be about 600 billion barrels of recoverable 25-gallon per ton oil in the Green River shale, most of it located in the Piceance Basin.

Recovery of petroleum from oil shale involves a number of environmental and economic factors not encountered in ordinary recovery of oil and gas through wells. The solid or viscous-liquid nature of the "kerogen" and asphaltic petroleum has already been noted. Since the hydrocarbons cannot be pumped out of the rock, the entire rock must be dug up and then treated; this involves the removing and processing of enormous quantities of rock. Some of the rock is at the surface, and some lies at depths of

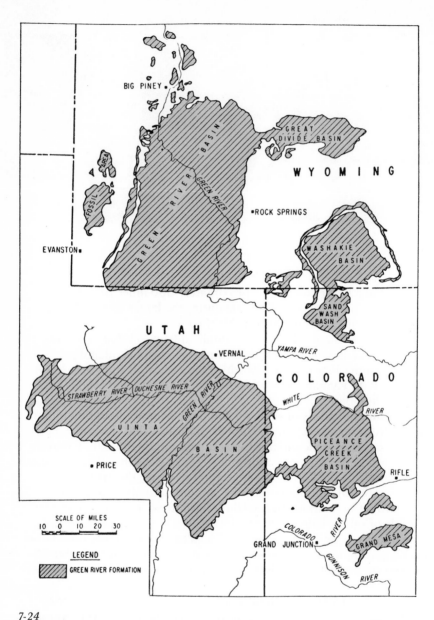

7-24
Location of basins containing oil shale deposits in the western United States. (From Glenn L. Cook, "Oil Shale – An Impending Energy Source," Journal of Petroleum Technology **24**, *November 1972, pp. 1325–1330. Reprinted by permission of the Society of Petroleum Engineers of the American Institute of Mining, Metallurgical, and Petroleum Engineers.)*

up to a couple of thousand feet. Out of economic necessity, strip mining will have to be the method used to retrieve these huge amounts of rock.

Inherent in the recovery of petroleum from shale are the problems of depth, amount of overburden (overlying waste rock), disposal of overburden, and disposal of the spent material after the shale has been treated. The disturbance of the landscape and the dumping of waste pose great ecological problems. The actual price of producing oil from oil shale does not differ greatly from the price of producing oil in the usual way; in fact, it may soon be competitive. At present, however, it appears that the greatest bulk of our crude oil will continue to be produced from conventional sources. When these supplies dwindle, we will then, out of necessity, have to turn to oil shale. Canadian tar sands are already being commercially worked. Deposits in the Fort McMurray area were yielding a few tens of thousands of barrels per day by 1970. Continued expansion of this production is expected.

COAL

Coal is one of the most important energy sources in the world. It is formed through chemical and physical action on the organic substances of once living organisms. Early man used sunlight as his prime source of energy and only later changed to wood. Coal was used next, and it was certainly the first of the fossil fuels to be used as a major source of energy. Oil and natural gas have supplanted it for some purposes, but more coal is used in the world today than ever before, and still furnishes about one-quarter of the industrial energy of the United States. The production of coal continues to increase nearly everywhere, in spite of the increasing usage of other energy sources. Since it is likely that the production of gasoline and natural gas, both of which can be made from coal, will increase, it seems obvious that coal production will continue to increase.

Coal has long been known as a source of energy. The record of its first use is lost in antiquity, but undoubtedly early man used it on occasion. People in the northeast of England used coal as early as the ninth century. They called the material *sea coals* because they found the lumps of coal along the coast. The word "coal" comes from the Anglo Saxon *col,* the approximate meaning of which is a glowing or burning rock. For a few hundred years then, coal was used in much of Europe for domestic heating and cooking. The Industrial Revolution began, and in England the geology of coal occurrence received attention at least as early as 1719, when John Strachey reported on coal strata to the Royal Society. In the United States coal was first mined in Virginia as early as the 1740s. William Smith, known as the father of stratigraphy and publisher of the first geologic map, was a coal-mine surveyor in England whose work in the con-

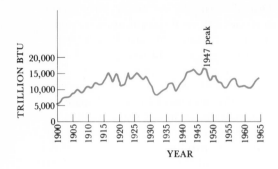

7-25
Production of bituminous coal and lignite in the United States. (Source: U. S. Bureau of Mines Yearbook, 1965)

struction of the Somerset Coal Canal led him to his contributions to geology. This work was done at the beginning of the nineteenth century, when England was busily industrializing; from the middle of the nineteenth century on, the use of coal increased markedly. Today, the use of coal is still widespread.

Uses and Production of Coal

Coal is burned to release heat, and one of its main uses is to provide heat for driving steam turbines which generate electric power. About half of the coal used in the United States is used for this purpose, and much of the rest is used in making steel. The production of coal, as we have already noted, has increased greatly since industrialization began. In 1870 the world production of coal was a bit under one-quarter billion metric tons; it is estimated that in 1970 its production was about three billion tons. Thus, in only 100 years, there has been a twelvefold increase in the world's coal usage. In the United States coal became a prime energy source in 1860, reaching a peak of 630 million tons in the year 1947 (Fig. 7–25). Presently, the United States is continually increasing its production of coal and is approaching a new peak. Production in 1970 was approximately 590 million tons. M. K. Hubbert has pointed out that half of the coal produced in the last 800 years has been mined in the last 31 years.

A number of attempts have been made to estimate how much coal is left in the world, how much of it is recoverable, and how long it will last. In 1969 Paul Averitt of the United States Geological Survey estimated that there existed in the world about 16.8 trillion short tons of coal, including coal which had already been produced. Of this, perhaps half will be recoverable. According to Hubbert's figures, if the production doubles only three more times, the peak rate of production will occur in about 170 to 200 years, after which a decline must occur. Nearly all the coal will be mined within 200 years after that. The supplies of coal in the world, then, are obviously finite and will not last more than a fraction of the time of human history.

Occurrence of Coal

Coal is a familiar, widespread, but not overly common rock. However, like many other resources, coal is not distributed uniformly throughout the world. There are extensive coal deposits in the U.S.S.R., the United States, and China. Canada, India, Japan, and various European countries, particularly Great Britain and Germany, also have fairly large amounts of coal. Compared to the other continents, South America, Africa, and Australia are relatively poor in coal. According to Averitt, Asia has more coal reserves than do all other continents combined, while North America has well over half of the remainder.

In the United States coal occurs in three regions (Fig. 7–26): the Appalachian Mountains, the central region, and the Rocky Mountains, the latter extending northwest into Canada. New England, the Atlantic Coast, the Canadian Shield, the Great Basin, the Columbia Plateau, and the High Plains from South Dakota through west Texas are mostly lacking in coal. A summary of the United States' coal production for 1967 (Table 7–4) gives some idea of the distribution of bituminous coal in this country and

Table 7–4 Production of bituminous coal in the United States, 1967 (in thousands of tons)

West Virginia	152,500
Kentucky	99,500
Pennsylvania	79,000
Illinois	65,000
Ohio	46,000
Virginia	38,000
Indiana	19,000
Alabama	15,000
Tennessee	7,000
Colorado	5,000
Utah	4,000
North Dakota	4,000
Wyoming	4,000
New Mexico	3,500
Missouri	3,000
Others	6,000
	550,500

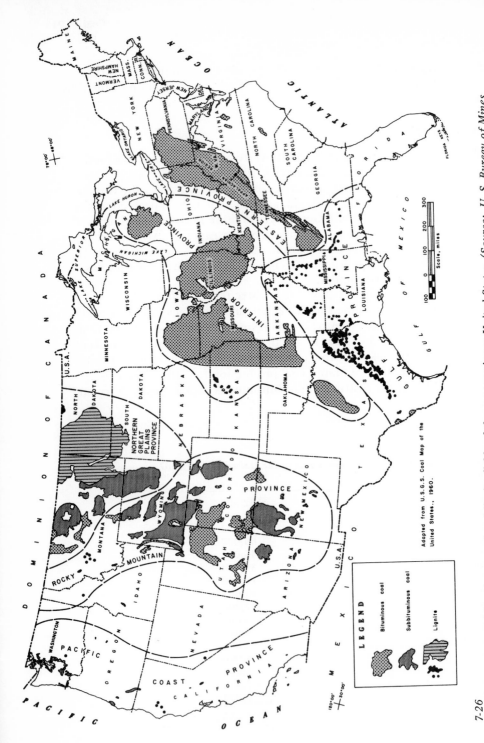

7-26
Bituminous and subbituminous coal and lignite fields of the conterminous United States. (Source: U. S. Bureau of Mines Information Circular 8535, U. S. Department of the Interior, 1972)

LEGEND

Bituminous coal

Subbituminous coal

Lignite

Adopted from U.S.G.S. Coal Map of the
United States, 1960.

Scale, miles
100   0   100   200   300

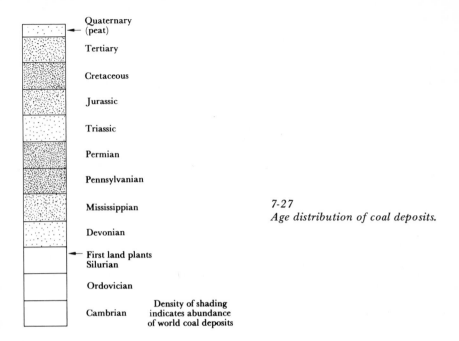

7-27
Age distribution of coal deposits.

indicates that the Appalachian area is by far the most important of the coal-producing areas. Virtually all the anthracite is produced in Pennsylvania.

Coal occurs in every geological system since the Silurian (Fig. 7-27). The distribution of coal in time is not at all uniform: some systems have an abundance of coal, while others have little. According to N. Strakhov, Devonian coal is known to exist because it was found in the Soviet Union; yet even so, there is very little of it. Coal seems to have originated from plants in swamp environments, and the earliest known nonmarine plants are found in Upper Silurian rocks. The development of land plants proceeded rapidly, but it was not until the Carboniferous that the combination of abundant plant life and proper environment led to the formation of coal in great quantities. Most of the Carboniferous coal is in the Pennsylvanian (Upper Carboniferous) of the United States, western Europe, and the U.S.S.R. Strakhov attributes even more coal to the Permian, most of which is found in the U.S.S.R., China, South Africa, India, and Australia. Cretaceous coal is quite abundant, but Tertiary coal is scarcer and is mostly of low rank.

Coal occurs in stratified deposits in sedimentary basins. The individual coal beds, usually referred to as *seams,* are usually thin, their thickness rarely exceeding 10 feet, and are laterally limited to a few miles. Some of the economically important beds of coal are only a few tens of feet thick but underlie hundreds of square miles. The notable Pittsburgh coal seam underlies an area of 22,000 square miles. Seams characteristically interfinger with other sedimentary strata, especially with thin beds of shale or claystone.

Petrology and Composition

Coal is composed primarily of carbon, with smaller amounts of hydrogen and oxygen. Some sulfur and nitrogen are commonly present, though sulfur, in particular, is an undesirable impurity. In addition to the organic materials in coal, there are traces of inorganic mineral matter. Many sedimentary minerals, including quartz and clay, have been found in coal. Chemically precipitated materials include pyrite, marcasite (both of which are iron sulfides), and salts such as calcite and gypsum. A great number of metallic trace elements have been reported in coal, including germanium, gallium, vanadium, and nickel.

Classification

Coal can be classified in terms of two different categories. First, coal is classified, as are other rocks, according to its petrological characteristics. These are mostly physical characteristics and lead to the recognition of *lithotypes* such as vitrain and fusain. The classifications of these lithotypes is amplified by microscopic descriptions of the various materials which comprise the lithotypes. Alternatively, coal can be classified by *rank*. This classification is based on the degree of coalification and ranges from lignite, the lowest rank coal, to anthracite, the highest. Rank is the degree of carbonification of a given coal. As the progressive alteration from lignite to anthracite occurs, the percentage of *fixed carbon* (carbon not chemically combined) and heat value increases. Table 7–5 shows the ranks of coal. The American Society for Testing Materials has arranged ranks of coal on the basis of fixed carbon (dry) for high ranks and heat value (moist) for low ranks. Bituminous coal, which has a higher heat value than do the ranks below it, even has a slightly higher heat value than does anthracite.

Origin

Coal forms from the progressive alteration of organic material which comes from plants which grew in swampy areas possibly similar to the Dismal

Table 7–5 Classification of coal by rank

Approximate percentage of fixed carbon		Rank		Heat value in British Thermal Units (BTU) per pound
As high as 96		Anthracite		15,000
92			Semi-anthracite	15,000
86	Bituminous coal		Low-volatile bituminous coal	15,000
78			Medium-volatile bituminous coal	15,000
69			High-volatile bituminous coal	
				13,000
60		Subbituminous coal		8,300
As low as 38		Lignite		
		Brown coal		8,300

Rank names are based on ASTM Classification of Coal (commercial)

Swamp of Virginia and North Carolina. A rather delicate balance of physical and chemical conditions must exist for the organic matter to survive through the coalification process to the final stage of coal formation. Drainage must be retarded so that the plants which die will be buried under water and debris of other plants. Such burial prevents the destructive oxidation of the organic matter and allows toxic products to accumulate. Through-flowing streams might erode the plant debris and would possibly provide enough oxygen to eliminate necessary reducing conditions; therefore, the swamps must be mostly stagnant. Sufficient precipitation has to occur, but it must be distributed uniformly throughout the year so that dry or excessively wet spells cannot occur. If the water table, or level of the water, in the swamp is lowered, the reducing conditions will be lost as the organic matter is exposed to the air.

Temperatures in this swamp environment apparently can be cool but not frigid. The great accumulations of modern peat bogs are found in cool, moist, temperate areas of the world, such as Scotland, Newfoundland, and Tierra del Fuego. The plant life of the coal swamps must have been extremely varied; about 3000 species have been reported from the Carboniferous alone. Many of the large plants were giant club mosses that grew as high as 100 feet and were as much as 3 feet in diameter. In addition to the giant club mosses, there were also horsetails, rushes, sedges, ferns, seed ferns, mosses, and primitive coniferlike Cordaitales, some of which were also large. Certain rushes reached 30 feet in height; some giant horsetails and Cordaitales reached nearly 100 feet. The seed ferns included treelike forms. The ferns and seed ferns were typical of the Paleozoic, as were giant horsetails, the Cordaitales, and giant club mosses. More modern forms, including the conifers and angiosperms, were typical of the Meso-zoic and Tertiary.

A great number of fungal and algal remains have also been reported in coal. Spores of all sorts, including those from plants living in the swamps and those from nonswamp dwellers, are present in coals and are quite important, both quantitatively and for purposes of correlating coal beds. Cannel and boghead coal consist mostly of these materials.

Although there is a relationship between coal swamps and sedimen-tary basins, the swamps apparently were not themselves freely connected with marine waters during the critical times of organic accumulation. Some may have been quite distant from the sea; others were undoubtedly separated from salt water by only a baymouth bar, for example. The plants found in coal are fresh- or brackish- water inhabitants. The sea, however, invaded many areas, submerging the peat bogs and spreading layers of sediment over them. In some places, the sea's tidal movements led to the formation of peculiarly cyclic sequences of coal and associated nonmarine and marine beds called *cyclothems*. The coal itself remained nonmarine in origin.

Masses of tree trunks and limbs, leaves, pollens, spores, and other plant material accumulated in the quiet water of the coal swamps. As the mass of material piled up on the bottom of the swamp, decay began, but soon ceased as oxygen supplies were stifled by burial under water and un-der still more debris. Moldering, then putrefaction took place, and the wet mass of debris became peat. Peat is a yellowish brown to black, usu-ally fibrous, material in which fragments of plants are still recognizable. If the floor of the swamp continued to sink and debris continued to pile on top of the peat, the gradual transformation of the peat continued. The peat changed into brown coal, and the brown coal into lignite, continuing to some terminal point in the stages of coalification. The ultimate stage is usually a high-rank bituminous coal, or if diastrophic forces squeeze the beds, anthracite. As the gradual changes occur, cellulose, which is the dominant material of plants, is converted chemically to carbon dioxide, water, methane, and carbon. The first three of these products are volatile and are expelled, whereas the carbon remains. Thus, relative to the other substances, the carbon content increases, and the higher the fixed carbon, the higher the rank of coal. Under the most extreme conditions of meta-morphism, graphite (pure carbon) may form. Graphite is not coal, but is a metamorphic mineral that forms under conditions beyond the stages of coalification.

REFERENCES

American Petroleum Institute. *Facts About Oil — A Handbook for Teachers,* Washington, D. C.: API, 1964.

———— *Petroleum and Energy,* Washington, D. C.: API, 1971.

Averitt, P. *Coal Resources of the United States - January 1, 1967,* U. S. Geological Survey Bulletin 1275, 1969.

Cook, G. L. *Oil Shale - An Impending Energy Source,* paper prepared for the Northern Plains Section Regional Meeting of the Society of Petroleum Engineers of AIME, Omaha, Nebraska, May 18-19, 1972.

Ellison, S. P., Jr. "Toward a national policy on energy resources," *American Association of Petroleum Geologists Bulletin* 56, 9 (1972): 1597–1598.

Federal Power Commission, Bureau of Natural Gas. *National Gas Supply and Demand: 1971-1990,* Staff Report No. 2, 1972.

Hubbert, M. K. "Energy resources," in *Resources and Man,* National Academy of Sciences, National Research Council, San Francisco: Freeman, 1969.

Iglehart, C. F. "North American drilling activity in 1971," *American Association of Petroleum Geologists Bulletin* 56, 7 (1972): 1145-1174.

McKelvey, V. E. and F. F. H. Wang. *Preliminary Maps - World Subsea Mineral Resources,* U. S. Geological Survey Miscellaneous Investigations Map I-632, 1970.

Pearse, C. R. "Athabasca tar sands," *Canadian Geographical Journal* LXXVI, 1 (1968): 2-9.

"Plenty of gas, but costly," *Chemical and Engineering News* 49 (June 7, 1971): 11-12.

Sparling, R. C., N. J. Anderson, and J. G. Winger. *Capital Investments of the World Petroleum Industry — 1970,* Report of the Energy Economics Division, Energy Group, Chase Manhattan Bank, 1971.

Starr, C. "Energy and power," *Scientific American* 225, 3 (Sept. 1971): 37-49.

United States Department of the Interior, Bureau of Mines, Oil Shale Program. *Some Results from the Operation of a 150-ton Oil Shale Retort,* by A. E. Harak, L. Dockter, and H. C. Carpenter, Technical Progress Report 30, 1971.

Vigrass, L. W. "Geology of Canadian heavy oil sands," *American Association of Petroleum Geologists Bulletin* 52, 10 (1968): 1984–1999.

Warner, A. J. "Natural gas," in *Mineral Facts and Problems,* Bureau of Mines Bulletin 650, 1970, pp. 111-136.

The sun releases enormous amounts of energy, which man is trying to harness to meet his increasing need for new energy sources. (Photograph by George Sheng)

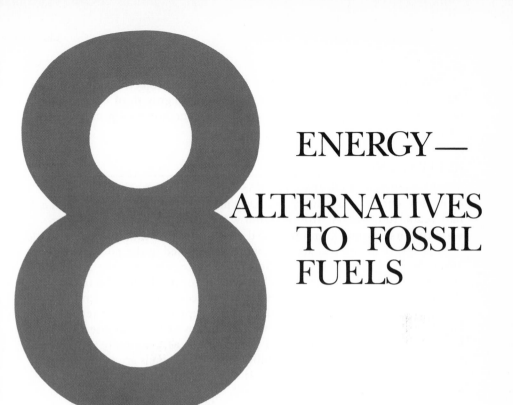

ENERGY—
ALTERNATIVES TO FOSSIL FUELS

INTRODUCTION

Having dealt with the fossil fuels as sources of energy to supply the world's increasing demands for energy, we shall now turn to alternative sources of energy which may help us meet the enormous demands for energy which are being projected for the future. Since these projections indicate that the fossil fuels will not be adequate to fill the predicted demand, we may have to turn to some of the alternative sources discussed here. We must either turn to alternative sources of energy or put a ceiling on the amount of energy we can use. In this chapter we will explore these alternatives, and rank them according to their potential ability to provide significant amounts of energy in the next two to three decades. Undoubtedly, we will find some disagreement about the relative rank assigned to the various potential sources, but this is our assessment based on present technology.

The last section of this chapter is devoted to projections of future energy demands, known fuel reserves, and the rapid growth of energy usage rates. We will attempt to summarize this vast and complex picture of energy and its usage today.

NUCLEAR ENERGY

The first self-sustaining chain reaction and consequent controlled release of nuclear energy was achieved on December 2, 1942. About two and a half years later the first atomic bomb was detonated, and shortly afterward the first atomic bomb was used in war. At that time the entire world became aware of the vast store of energy contained in the nucleus of the atom. In 1954 the Congress of the United States amended the Atomic Energy Act to permit the use of nuclear energy for producing electricity. Now that we understand that nonrenewable resources of petroleum and coal are being rapidly used and can even discern the depletion of these resources based on current projections, we are turning toward nuclear energy at least as a supplementary source of power, if not as a complete replacement for fossil fuels. The energy possibilities in nuclear fuels are tremendous: one cubic foot of uranium contains the energy equivalent of 1.7 million tons of coal or 7.2 million barrels of oil.

In 1970 only 1 percent of this nation's electrical power was obtained from nuclear energy; perhaps 25 percent will be obtained from this source by 1980. Since the demand for electricity doubles about every 10 years, this will represent a 50-fold increase. After 1980 the percentage may rise until over half of our energy needs are satisfied by nuclear energy. Just what the ultimate percentage will be is a matter of speculation and depends on a number of factors such as pollution, development and use of other forms of energy, safety, and the health hazards of radiation.

Producing Nuclear Energy

Nuclear energy can be generated in two fundamentally different ways: fusion or fission. In fusion energy is released in the process of fusing the nuclei of two relatively lightweight atoms to form a heavier atom. In fission, energy is given off when heavier atoms are split into lighter atoms.

Fission   One way that energy is released from the atom is through a phenomenon called "radioactivity," whereby various particles in the nucleus are shed or expelled due to the spontaneous decay of certain elements; energy is liberated simultaneously. In nature this phenomenon occurs with the decay of elements such as uranium and thorium. It was discovered that man himself could "smash" the atom by bombarding the nucleus with highly energized particles. This procedure always involves a slight loss of

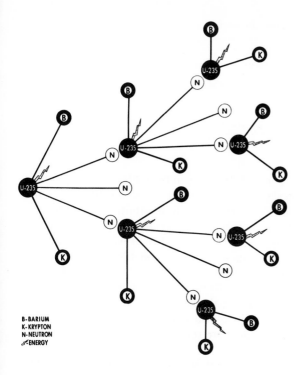

B-BARIUM
K-KRYPTON
N-NEUTRON
⚡-ENERGY

CHAIN REACTION

8-1
Chain reaction in uranium fission. The splitting of uranium produces
energy, neutrons, and isotopes of barium and krypton. Neutrons
then bombard other uranium atoms, producing similar results.
(Source: U. S. Atomic Energy Commission)

mass from the mass expected from the sum of the mass of the nucleus plus the mass of the particle shot into it. Albert Einstein showed theoretically that mass could be converted to energy; therefore, the energy released when an atom is "smashed" must be produced from the lost mass.

The products of nuclear splitting are energy and other elements that have lighter atomic weight. One common example of this is the splitting, or fission, of one isotope of uranium, U-235. If the nucleus is bombarded with a neutron, the atom may split into barium-141, Krypton-92, neutrons, and energy. Because the released neutrons are very fast-moving particles, they penetrate the nuclei of adjacent uranium atoms and cause the reaction to continue in a similar manner, a process known as a *chain reaction* (Fig. 8-1). Chain reactions will occur only if sufficient uranium is present; hence, we refer to this amount as the *critical mass*.

Table 8–1 Isotopic composition of naturally occurring uranium

Isotope	Percent of naturally occurring uranium
U-238	99.238%
U-235	0.711%
U-234	0.006%
	100.000%

The production of energy through fission can be achieved in any one of three basic atomic reactor designs—burners, converters, and breeders. In a burner reactor all the fissionable material, usually uranium-235 and plutonium-239, is consumed by the end of one fuel-consumption cycle. In a converter reactor uranium-238 or thorium-232, neither of which is readily fissionable, is fed into the reactor, along with the fissionable material contained inside this quantity of the uranium-238 and thorium-232. As the readily fissionable uranium-235 and plutonium-239 decay, they emit high-speed neutrons. These neutrons bombard the uranium-238 and the thorium-232, which are then transmuted into plutonium-239 and uranium-233, respectively. Both of these end products are readily fissionable and usable as fuel.

If the fissionable isotopes are produced in amounts less than the original quantity of U-235 fuel, the reactor is called a converter, but if the amount is greater, it is called a breeder reactor. By fueling one or two breeder reactors with natural U-235, sufficient fissionable isotopes can be produced to fuel other breeder reactors. However, no commercial breeder reactors exist at this time. Federal funds have been committed to develop such a reactor by 1980, and it seems imperative that we do so.

Uranium occurs in nature as three principal isotopes, as shown in Table 8–1. It is obvious from this table that burner reactors, capable of using only U-235, would soon deplete our supply of this isotope unless it was available in vast quantities. In addition, we can increase our fuel supply nearly a hundredfold by using breeding reactors.

Although there are a number of naturally occurring radioactive elements, uranium is the most common and the one most frequently used as a source of atomic energy. With a crustal abundance of only 1.8 parts per million, uranium is not an abundant element. It is most common in granite, where it may have about twice the crustal abundance and sometimes occurs in quantities of almost four grams per ton. It is progressively less common in less siliceous igneous rocks, ranging to an amount of less than one gram per ton in basalt. It is also common in marine carbonaceous shales.

Chemically, uranium, as well as thorium, has a pronounced affinity for oxygen and tends to concentrate in the upper part of the surface crust.

Uranium and thorium are the chief contributors to the radioactivity which produces heat in the upper mantle and crust, particularly the continental crust.

Uranium ore minerals are principally uraninite (pitchblende), UO_2, and carnotite, $K_2(UO_2)_2(VO_4)_2 \cdot n\,H_2O$. Uraninite occurs in granite and related rocks, particularly pegmatite. It is virtually never pure UO_2, but is usually partially oxidized, with UO_3 being mixed with UO_2. The mineral usually contains other elements, too, including thorium and lesser quantities of lead, radium, and various gases and rarer metals. Carnotite, along with perhaps nearly 100 less common uranium minerals, forms as a secondary mineral from the alteration or decomposition of uraninite. Many of these minerals are common in sedimentary rocks, where they are precipitated under conditions of low temperature and pressure.

Uranium ores occur in many countries; important producers include the United States, Canada, Zaire (Congo), and Czechoslovakia. Ranked according to abundance, the measured reserves of uranium in the free world are located in Canada, the United States, the Republic of South Africa, and France. The deposits near Great Bear Lake and in other areas of Canada, the province of Katanga in Zaire (Congo), and Joachimstal, Czechoslovakia are in *hydrothermal veins,* deposited there by hot solutions. The deposits of the Colorado Plateau area of the United States, by contrast, are of secondary origin and consist mainly of carnotite disseminated in sandstone of the Chinle and Morrison Formations. The Colorado Plateau deposits have yielded by far the majority of uranium produced in this country. Although uranium is abundant in other areas, it is often too finely disseminated for these to be considered major sources (Fig. 8-2).

Uranium ore is processed so that the uranium becomes a concentrate of the oxide mixture U_3O_8, popularly called "yellow cake." The U_3O_8 is purchased mostly by the government but is occasionally also bought by private users. The price for the oxide concentrate in the United States has been about $8 per pound, and at this price there will be enough uranium to last only a few years, depending on the development of the nuclear power industry. At higher prices, lower-grade ores could be used.

Burner-type reactors are the only type commercially used today, and almost all of those in use or on order are light-water reactors (LWR) or those using ordinary H_2O in their internal system. Light-water reactors are of two types: boiling-water reactors (BWR) and pressurized-water reactors (PWR). Approximately 60 percent of the current nuclear power generation market uses pressurized-water reactors, while the remaining 40 percent uses boiling-water reactors. Figures 8-3 and 8-4 show schematics of the operation of the boiling-water reactor and the pressurized-water reactor in a power-generating station.

The high-temperature gas-cooled reactor (HTGR) fills the gap between the burner reactors and the fast breeder reactors. The HTGR is desirable

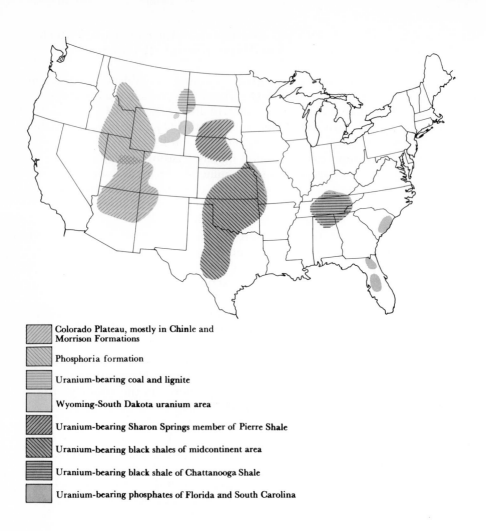

Colorado Plateau, mostly in Chinle and
Morrison Formations

Phosphoria formation

Uranium-bearing coal and lignite

Wyoming-South Dakota uranium area

Uranium-bearing Sharon Springs member of Pierre Shale

Uranium-bearing black shales of midcontinent area

Uranium-bearing black shale of Chattanooga Shale

Uranium-bearing phosphates of Florida and South Carolina

8-2
Uranium deposits of the United States.

because its initial fuel requirements are more than 25 percent below those
for light-water reactors and slightly less than 25 percent during the first
seven or eight years of operation. At the end of this period the U-233
(U-233 comes from thorium-232) recycle begins, and the net additional ore
required drops even lower, to about 63 percent of that required by light-
water reactors. However, there are disadvantages to these reactors: the

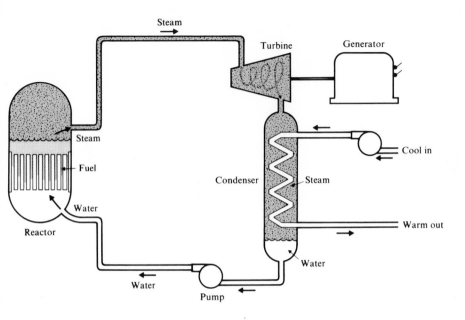

8-3
Schematic of a boiling-water reactor (BWR), showing that the water which forms steam to drive the turbine circulates in direct contact with the radioactive core. (David Ritten-house Inglis, Nuclear Energy — Its Physics and Its Social Challenge, *Reading, Mass.: Addison-Wesley, 1973, p. 97. Reprinted by permission.)*

production of a given rate of heat generation requires a larger high-temperature gas-cooled reactor than light-water reactor, and since gas does not remove heat very efficiently, the heat generation per unit volume of reactor must be fairly low.

Fusion Another process by which we obtain energy from the atom is fusion, or thermonuclear reaction, in which nuclei of light atoms are combined to form heavier atoms with a concomitant release of energy. A typical, and perhaps most common, example is the fusion of the various isotopes of hydrogen to form helium. Natural hydrogen consists of three isotopes: protium, having a single proton in the nucleus; deuterium (heavy hydrogen), having a proton and a neutron in the nucleus; and tritium, having a proton and two neutrons in the nucleus. Of the various methods that can be used to fuse isotopes of hydrogen to form helium, perhaps the most common is illustrated by the following reaction: $H^3 + H^2 \longrightarrow He^4 + n + E$, where

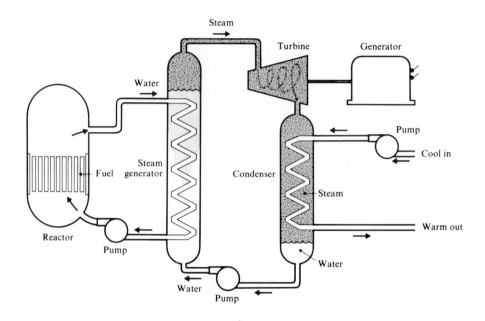

8-4
Schematic of a pressurized-water reactor (PWR), showing how the water which is heated by contact with the reactor core is confined by a pressurized system which is separate from the system containing the steam used to drive the turbine. (David Rittenhouse Inglis, Nuclear Energy — Its Physics and Its Social Challenge, *Reading, Mass.: Addison-Wesley, 1973, p. 96. Reprinted by Permission.)*

H^2 represents deuterium, H^3 an atom of tritium, He^4 a helium atom, "n" an escaping neutron, and "E" nearly 25 million electron volts of energy freed in the process.

The fusion of deuterium and tritium which produced the hydrogen bomb is accomplished by the same process as the sun's "burning." This, however, is an uncontrolled reaction, and while controlled fusion has not yet been achieved, research is under way. The fusion of hydrogen isotopes to form helium produces enormous amounts of energy; therefore, when this process can finally be controlled, the problem of an energy source may well be solved. Considering that approximately one out of every 5000 hydrogen atoms is deuterium, the ocean is obviously an almost inexhaustible fuel source. The amount of energy obtainable from one cubic yard of sea water is nearly equivalent to that obtainable from 1000 barrels of oil. The energy of the deuterium in a cubic mile of sea water is of the same order of magnitude as is the energy of the world's entire petroleum supply.

Much work is being done on achieving controlled nuclear fusion. In order to fuse deuterium nuclei (or deuterium and tritium nuclei), temperatures of millions of degrees, enormous magnetic field confinement, and the right quantities of nuclei are required. A group of hot, ionized nuclei—called a *plasma*—must be confined in a strong magnetic field for a few milliseconds. The Russians have achieved the feat of holding a plasma, under conditions almost long enough and hot enough, in a doughnut-shaped magnetic chamber they call a "Tokamak." American experiments have achieved some success using similar Tokamak devices. The plasma in these devices is of low density, as relatively few nuclei are present.

The design of any fusion reactor will present some difficult engineering problems. For greatest efficiency, the fusion process will have to take place within a very strong magnetic field, and superconducting magnets will have to be maintained at very low temperatures. Thus, it will be necessary to create a device which could have temperatures as high as 100,000,000° C at the center and almost as low as absolute zero only two meters away. This is only one of the numerous problems involved.

Another approach to achieving controlled thermonuclear fusion is to produce a high-density plasma, that is, a greater concentration of nuclei in a given space, by pulsing. This method involves the quick heating of a high-density plasma by such means as the shock and compression of a magnetic implosion (opposite of "explosion," which is an expansory phenomenon). Efforts toward harnessing fusion continue along both lines.

A more recent advance calls for directing a laser beam (or several beams at once) on a small pellet of deuterium and tritium to reach the 100 million degrees C necessary to ignite fusion. The laser must be capable of delivering its energy in one-billionth of a second or less, and recent advances in laser research suggest that this is feasible. So far, much of the work has been theoretical, using computer simulations of the very complex physical effects that take place when a fuel pellet and laser interact. Some scientists working in this area feel that the feasibility of laser fusion will be demonstrated by 1976.

In summary, nuclear energy can, and in the future undoubtedly will, furnish man with a very large share of his energy requirements. In general, there are three methods of producing nuclear energy: using conventional burner-type reactors, which includes virtually all those in present use; using the vastly more efficient breeder-type reactors; and using thermonuclear or fusion-type reactors, which have not yet been (and just possibly might never be) made to work. Each process is successively more difficult to develop and perfect. As of June 1971, there were 22 nuclear power plants

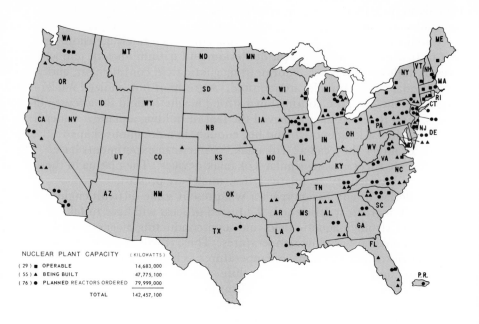

NUCLEAR PLANT CAPACITY   (KILOWATTS)
(29) ■ OPERABLE   14,683,000
(55) ▲ BEING BUILT   47,775,100
(76) ● PLANNED REACTORS ORDERED   79,999,000

                    TOTAL   142,457,100

8-5
Nuclear power plants in the United States at the end of December 1972. (Source: U. S. Atomic Energy Commission)

in operation, 55 under construction, and 44 on order (Fig. 8-5). Table 8-2, compiled two years earlier, gives details for most of these. Table 8-3 gives the world picture. A small number of breeder reactors, mostly experimental, were in existence at the beginning of the 1970s, and regular usage of commercial breeder reactors is expected to start during the 1980s.

What of the development of fusion energy? Eventually, it probably will be a reality. But it does not appear likely that we will use fusion as a major source of energy until the first years of the twenty-first century. There is certainly a sufficient supply of deuterium in the oceans; however, as long as we must use uranium to fuel fission-type reactors, we must continue to explore for, and develop adequate supplies of, uranium. This search for uranium is particularly urgent as long as we depend on burner reactors, which waste useful uranium. The use of fission reactors involves the additional problem of what to do with the radioactive waste materials, some of which present difficult disposal problems. We shall explore this problem elsewhere.

Table 8–2 Domestic civilian nuclear power plants operating, under construction, and planned in the United States

Reactor	Location	Type [1]	Electrical capacity, megawatts (Mwe)	Initial criticality
OPERABLE				
Shippingport Atomic Power Station	Shippingport, Pa	PWR	90	1957
Dresden Nuclear Power Station, Unit 1	Morris, Ill	BWR	200	1959
Yankee Nuclear Power Station	Rowe, Mass	PWR	175	1960
Big Rock Point Nuclear Plant	Big Rock Point, Mich	BWR	70	1962
Elk River Reactor	Elk River, Minn	BWR	22	1962
Indian Point Station, Unit 1	Indian Point, N.Y	PWR	265	1962
Enrico Fermi Atomic Power Plant	Lagoona Beach, Mich	FBR	61	1963
Humboldt Bay Power Plant, Unit 3	Eureka, Calif	BWR	69	1963
Peach Bottom Atomic Power Station, Unit 1	Peach Bottom, Pa	HTGR	40	1966
San Onofre Nuclear Generating Station	San Clemente, Calif	PWR	430	1967
LaCrosse Boiling Water Reactor	Genoa, Wis	BWR	50	1967
Connecticut Yankee Atomic Power Plant	Haddam Neck, Conn	PWR	575	1967
Oyster Creek Nuclear Power Plant, Unit 1	Toms River, N.J	BWR	515	1969
Nine Mile Point Nuclear Station	Scriba, N.Y	BWR	500	1969
Robert Emmett Ginna Nuclear Power Plant, Unit 1	Ontario, N.Y	PWR	420	1969
Total operable capacity			3,482	
UNDER CONSTRUCTION				
Dresden Nuclear Power Station, Units 2 and 3	Morris, Ill	BWR	1,618	1970
Millstone Nuclear Power Station, Unit 1	Waterford, Conn	BWR	652	1970
H. B. Robinson S.E. Plant, Unit 2	Hartsville, S.C	PWR	700	1970
Palisades Nuclear Power Station, Unit 1	South Haven, Mich	PWR	700	1970
Monticello Nuclear Generating Plant	Monticello, Minn	BWR	545	1970
Quad-Cities Station, Units 1 and 2	Cordova, Ill	BWR	1,618	1970–71
Point Beach Nuclear Plant, Units 1 and 2	Two Creeks, Wis	PWR	994	1970–71
Surry Power Station, Units 1 and 2	Gravel Neck, Va	PWR	1,560	1970–71
Oconee Nuclear Station, Units 1, 2, and 3	Seneca, S.C	PWR	2,613	1970–72
Vermont Yankee Generating Station	Vernon, Vt	BWR	514	1971
Fort Calhoun Station, Unit 1	Fort Calhoun, Nebr	PWR	457	1971
Pilgrim Station	Plymouth, Mass	BWR	625	1971
Fort St. Vrain Nuclear Generating Station	Platteville, Colo	GCR	330	1971
Cooper Nuclear Station	Brownville, Nebr	BWR	778	1971
Browns Ferry Nuclear Power Plant, Units 1, 2, and 3	Decatur, Ala	BWR	3,194	1971–72
Turkey Point Station, Units 3 and 4	Turkey Point, Fla	PWR	1,303	1971–72

See footnotes at end of table.

Table 8-2 (cont'd)

Reactor	Location	Type [1]	Electrical capacity, megawatts (Mwe)	Initial criticality
UNDER CONSTRUCTION—Continued				
Peach Bottom Atomic Power Station, Units 2 and 3	Peach Bottom, Pa	BWR	2,130	1971–72
Indian Point Station, Units 2 and 3	Indian Point, N.Y	PWR	1,838	1971–73
Zion Station, Units 1 and 2	Zion, Ill	PWR	2,100	1971–73
Diablo Canyon Nuclear Power Plant, Unit 1	Diablo Canyon, Calif	PWR	1,060	1972
Maine Yankee Atomic Power Plant	Wiscasset, Maine	PWR	790	1972
Kewaunee Nuclear Power Plant	Carlton, Wis	PWR	527	1972
Crystal River Plant, Unit 3	Red Level, Fla	PWR	858	1972
Rancho Seco Nuclear Generating Station, Unit 1	Clay Station, Calif	PWR	800	1972
Edwin I. Hatch Nuclear Plant, Unit 1	Baxley, Ga	BWR	786	1972
Arkansas Nuclear One	London, Ark	PWR	850	1972
Calvert Cliffs Nuclear Power Plant, Units 1 and 2	Lusby, Md	PWR	1,600	1972–73
Donald C. Cook Plant, Units 1 and 2	Bridgman, Mich	PWR	2,114	1972–73
Three Mile Island Nuclear Station, Units 1 and 2	Goldsboro, Pa	PWR	1,641	1972–73
Salem Nuclear Generating Station, Units 1 and 2	Salem, N.J	PWR	2,100	1972–73
Prairie Island Nuclear Generating Plant, Units 1 and 2	Red Wing, Minn	PWR	1,060	1972–74
Total capacity under construction			38,455	
PLANNED				
Beaver Valley Power Station, Unit 1	Shippingport, Pa	PWR	847	1972
Hutchinson Island, Unit 1	Fort Pierce, Fla	PWR	800	1973
Duane Arnold Energy Center, Unit 1	Palo, Iowa	BWR	545	1973
James A. FitzPatrick Nuclear Power Plant	Scriba, N.Y	BWR	821	1973
Millstone Nuclear Power Station, Unit 2	Waterford, Conn	PWR	828	1973
North Anna Power Station, Unit 1	Mineral, Va	PWR	845	1973
Diablo Canyon Nuclear Power Station, Unit 2	Diablo Canyon, Calif	PWR	1,060	1973
Enrico Fermi Atomic Power Plant, Unit 2	Lagoona Beach, Mich	BWR	1,126	1973
Sequoyah Nuclear Power Plant, Units 1 and 2	Daisy, Tenn	PWR	2,248	1973–74
Brunswick Steam Electric Plant, Units 1 and 2	Southport, N.C	BWR	1,642	1973–75
Malibu Nuclear Plant, Unit 1	Corral Canyon, Calif	PWR	462	1974
Trojan Nuclear Plant, Unit 1	Rainier, Oreg	PWR	1,106	1974
Davis-Besse Nuclear Power Station	Oak Harbor, Ohio	PWR	872	1974
Joseph M. Farley Nuclear Plant	Dothan, Ala	PWR	829	1974
Consolidated Edison Co	Verplanck, N.Y	BWR	1,115	1975
Shoreham Nuclear Power Station	Brookhaven, N.Y	BWR	819	1975
William H. Zimmer Nuclear Power Station, Units 1 and 2	Moscow, Ohio	BWR	1,680	1975–76
Philadelphia Electric Co., Units 1 and 2	Pottstown, Pa	BWR	2,130	1975–77
Public Service Electric & Gas Co., Units 1 and 2	Newbold Island, N.J	BWR	2,200	1975–77
Oyster Creek Nuclear Plant, Unit 2	Toms River, N.J	PWR	1,100	1976
Bailly Generating Station	Dunes Acres, Ind	BWR	515	1976
Carolina Power & Light Co	North Carolina	BWR	821	1976
Pennsylvania Power & Light Co., 2 units	Not determined	BWR	2,104	1976–78
Bell Station	Lansing, N.Y	BWR	838	Not given
Duke Power Co., 2 units	Not determined	PWR	2,200	1977–79
Seabrook Nuclear Station	Seabrook, N.H	PWR	860	Not given
Total planned capacity			30,413	
Grand total			72,350	

[1] BWR   Boiling light water cooled, light water moderated reactor.
FBR   Fast Breeder Reactor.
GCR   Gas cooled, graphite moderated reactor.
HTGR High temperature gas cooled, graphite moderated reactor.
PWR   Pressurized light water moderated and cooled reactor.

Source: Adapted from "Nuclear Reactors Built, Being Built, or Planned in the United States as of Dec. 21, 1969," AEC Division of Technical Information, TID-8200 (21st Rev.) pp. 7–9.

(Source: *Minerals Yearbook*, Vol. I-II, 1969, pp. 1113–1114)

Table 8-3 Nuclear power plants operating, under construction, or planned in the world, as of June 30, 1969

Country	Operating Mwe	Operating No.	Operating Type [2]	Under construction Mwe	Under construction No.	Under construction Type [2]	Planned Mwe	Planned No.	Planned Type [2]
Canada	230	2	BHWR PHWR	2,282	5	4 PHWR 1 HWLWR	3,350	5	4 PHWR [3] 1
France	1,622	8	6 GCR 1 HWGCR 1 PWR	1,311	3	2 GCR 1 FBR(250)	1,480	2	([3])
Germany, West	842	4	2 BWR 1 PGWR 1 PWR	2,570	6	2 BWR 2 PWR 1 HTGR 1 HWGCR	9,040	16	1 FBR(300) 1 HTGR [3] 14
India	380	2	BWR	600	3	PHWR	200	1	PHWR
Italy	597	3	BWR GCR PWR	40	1	HWLWR	1,300	2	([3])
Japan	158	1	GCR	2,402	5	3 BWR 2 PWR	9,856	16	6 BWR 1 FBR(300) 1 HWLWR 3 PWR [3] 5
Netherlands	52	1	BWR	400	1	PWR	------	--	----------
Spain	153	1	PWR	920	2	1 BWR 1 GCR	3,000	6	1 PWR [3] 5
Sweden	------	--	----------	1,295	3	2 BWR 1 BHWR	3,619	6	2 BWR 2 PWR [3] 2
Switzerland	350	1	PWR	656	2	1 BWR 1 PWR	1,550	3	([3])
U.S.S.R.	1,165	10	8 LWGR 1 BWR 1 PWR	6,775	11	5 PWR 4 LWGR 2 FBR(750)	760	2	PWR
United Kingdom	4,131	26	24 GCR 1 AGR 1 HWLWR	6,380	11	8 AGR 2 GCR 1 FBR(250)	5,000	8	AGR
Others	------	--	----------	1,354	[4] 5	2 PHWR 2 PWR 1 HWGCR	9,944	[5] 24	7 PWR 1 PHWR 1 HWGCR [3] 15
Total	9,680	59	32 GCR 8 LWGR 7 BWR 6 PWR 2 PHWR 1 AGR 1 BHWR 1 HWGCR 1 HWLWR	26,985	58	13 PWR 9 BWR 9 PHWR 8 AGR 5 GCR 4 FBR 4 LWGR 2 HWGCR 2 HWLWR 1 BHWR 1 HTGR	49,099	91	15 PWR 8 AGR 8 BWR 6 PHWR 2 FBR 1 HTGR 1 HWGCR 1 HWLWR [3] 49

[1] Excluding United States and experimental power reactors (output below 20 Mwe).
[2] AGR   Advanced gas cooled, graphite moderated reactor.
  BHWR Boiling heavy water cooled, heavy water moderated reactor.
  BWR   Boiling light water cooled, light water moderated reactor.
  FBR   Fast Breeder Reactor.
  GCR   Gas cooled, graphite moderated reactor.
  HTGR High temperature gas cooled, graphite moderated reactor.
  HWGCR Heavy water moderated, gas cooled reactor.
  HWLWR Heavy water moderated, boiling light water cooled reactor.
  LWGR Light water cooled, graphite moderated reactor.
  PHWR Pressurized heavy water moderated and cooled reactor.
  PWR   Pressurized light water moderated and cooled reactor.
[3] Not stated.
[4] Includes Argentina, Bulgaria (2), Czechoslovakia, and Pakistan.
[5] Includes Argentina (1), Australia (1), Austria (1), Belgium (3), Brazil (1), Czechoslovakia (1), Finland (2), Greece (1), Hungary (2), Israel (1), Korea, Rep. of (1), Mexico (1), N. Zealand (1), Norway (1), Pakistan (2), So. Africa, Rep. of (1), Taiwan (1), Thailand (1), and United Arab Republic (1).

Source: Power and Research Reactors in Member States. September 1969 Edition. International Atomic Energy Agency, Vienna, 1969, 82 pp.

(Source: *Minerals Yearbook*, Vol. I-II, 1969, p. 1122)

8-6
The Enrico Fermi nuclear power plant, near Detroit, Michigan, was the first experimental breeder plant. (Photograph courtesy Power Reactor Development Co.)

GASIFICATION

Synthetic "natural gas," composed largely of methane and having the same heating characteristics as natural gas, can be made from oil, naphtha, and coal. Two conditions have caused us to intensify our interest in making synthetic gas. One is the shortage of natural gas, and many people first became aware of this during the winter of 1972–1973, when many areas of the United States experienced a shortage of natural gas. Known reserves of natural gas in the United States are sufficient to satisfy demand for only about the next 13 years. The second factor that generates interest in this process is the probable high cost of finding and producing the estimated potential reserves of natural gas. The depth of the average reservoir of estimated reserves in Texas, for instance, has been calculated to be about 20,000 feet, or about double the average depth of reservoirs now being produced there. Estimates indicate that it will cost ten times more to recover these estimated reserves than it is presently costing to produce the known reserves. If these predictions hold true nationwide (and they

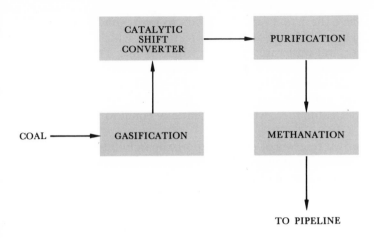

8-7
*Generalized flow chart for the gasification of coal, showing the
basic steps of the process. The process for naphtha is similar to
this.*

probably do), the average price of natural gas at the well-head in the
United States will go from the $0.17/1000 cubic feet in 1970 (in Texas)
to $1.70/1000 cubic feet; this increase may price natural gas out of the
market. Since synthetic gas in the form of methane can probably be
produced for about $1.00 to $1.25 per 1000 cubic feet, it appears to be a
viable alternative.

Gasification (Fig. 8–7) is really a simple process: carbon from oil,
naphtha, or coal is combined with water at a high temperature and forms
methane, which constitutes more than 90 percent of the natural gas serving
domestic and industrial needs. Gasification of naphtha is much simpler
than is the gasification of coal; however, because of the large reserves of
coal, import quotas set on naphtha, and increasing demand for naphtha as a
raw material in the chemical industry, the gasification of naphtha will not
provide as useful an alternative to natural gas as will the gasification of coal.

There are four basic steps in the coal-gasification process. Coal, after
being suitably prepared, is introduced into a gasifier. This unit, whose
purpose is to convert the solid coal into gases, operates at high temperatures
and pressures. These gases are fed into a second unit, the catalytic shift
converter, to be processed in order to produce the proper hydrogen-to-
carbon monoxide ratio necessary for the formation of methane at a later
stage. The ratio of carbon to hydrogen is about 1 to 0.8 in bituminous coal
and 1 to 0.7 in lignite. It is necessary to introduce hydrogen from some

other source to achieve the 1-to-4 ratio of carbon to hydrogen found in methane. The third step in the process is purification, in which excess CO_2, H_2S, organic sulfides, and water are extracted from the gas. The final step is the catalytic methanation process in which methane is formed. After the removal of some excess water formed during this process, the synthetic gas is 95 to 98 percent methane, having an energy content that averages around 1000 BTU per cubic foot, which is essentially equal to that of natural gas.

The present known reserves of coal are approximately 5 trillion tons and should last for several hundred years, even at current rates of increase in usage. The relative abundance of coal to feed this process, the critical need for natural gas, and the available technology combine to make this an attractive source for future energy. The gasification of coal is already being used commercially in Europe, and in the United States there are currently several coal gasification plants either under construction or in the planning stage.

GEOTHERMAL ENERGY

Geothermal energy, produced from the heat of the earth's interior, has recently been receiving renewed attention by those who are searching for new sources of energy. The United States Geological Survey estimates that as much as 6×10^{24} calories of energy, equal to the heat content of 900 trillion short tons of coal, are stored as heat in the crustal rocks, extending to a depth of ten kilometers, under the United States. This is an enormous amount of energy, equalling nearly 350,000 times the amount of energy from all sources used by the United States in 1970. Or, to put it another way, it is enough energy to supply the United States at the 1970 rate for roughly one-third of a million years. Why, then, do we not avail ourselves of this source of energy? Unfortunately, like other of our needed resources, it is sparsely distributed through most rocks, making most of the crustal material an extremely low-grade source of geothermal energy. There are, however, concentrations of heat in selected areas, and these are indeed promising power sources. The Congress of the United States adopted the Geothermal Steam Act in 1970, establishing as a national goal the development of such areas in the United States.

The heat within the interior of the earth is produced primarily by radioactive decay and friction. The temperature in the crust increases proportionately with depth until it begins to level off at a depth of about 50 miles. The mantle maintains a temperature of about 1000° C. These areas of more or less normal geothermal gradients may be responsible for a type of geothermal field which has only recently begun to receive attention. These are fields of low-temperature water, of 120° to 180° F, which are found in association with sedimentary deposits. Water of such a low temperature is not usable in the generation of power, but can be used very effectively, as it is in Iceland, as space heating in homes, industry, and greenhouses.

The geothermal energy usable as a source of power occurs in two ways: dry steam and wet steam. Dry-steam fields occur in geologic situations where the temperature is high and the water is not under much more than atmospheric pressure. Under these circumstances the water boils underground and generates steam. If this continues for some time, any natural porosity in the rock will become filled with steam, as will any fractures, and this steam can be tapped directly by wells drilled into the field. This is the situation at The Geysers, a field about 90 miles north of San Francisco, where the first geothermal power plant in the United States was commissioned in 1960. By the end of 1971, the generating capacity had risen above the original 12,500-kilowatt capacity.

There are at least five major dry-steam fields: The Geysers field in northern California, the Larderello field in northern Italy (which started producing in 1904), the Valle Caldera field in the Jemez Mountains near Los Alamos, New Mexico, and two fields in Japan. The extent of the Valle Caldera field is still to be assessed by a drilling program, but the structure of the area suggests a potential greater than 1000 megawatts. Neither the duration nor the extent of dry-steam fields is known, but the experience of the Italians suggests a minimum field-life expectancy of 30 to 60 years and a maximum of 1000 to 3000 years. On the basis of field discoveries to date, wet-steam fields may be as much as 20 times more abundant than dry-steam fields.

Wet-steam fields differ radically from dry-steam fields. Reservoirs in the wet-steam fields are under high pressure, and the temperature of the water may reach as high as 350° to 700° F without boiling. The temperature at which water boils increases with increased pressure, as is shown by Fig. 8-8. Curve "A" shows the theoretical increase in boiling point in relation to depth below the surface. Note that the increase in boiling point is not uniform, but increases 230° F in the first 1000 feet and only about 170° F in the next 4000 feet. Curve "B" shows the increase where water has circulated to hot rocks at great depth, as is the case with the hot springs area at Yellowstone National Park and Wairakei, New Zealand. Curve "C" shows a more favorable situation, such as the one found at Larderello, Italy, where impermeable rocks above the reservoir insulate it and allow deep waters to be even hotter than those under the conditions represented by curve "B." Figures 8-9 and 8-10 show the two types of wet steam fields: the field which has some upward leakage to form hot springs, and the field which is confined and has little or no leakage. In the latter type the superheated water "flashes" to steam as it comes to the

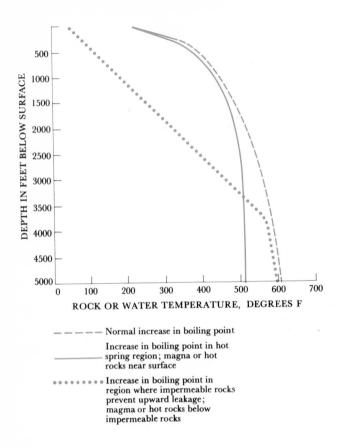

8-8
Water temperature variations with depth. (Source: "Natu-
ral Steam for Power," U. S. Geological Survey, 1968)

surface in the well. As the water nears the surface, its heat remains essentially the same, but the confining pressure is greatly reduced. At some
point the pressure falls below that necessary to keep the water liquid, and
steam is formed in a spontaneous reaction. In reality, only 10 to 20 percent of the discharge, by weight, is steam, while the remainder is very hot
water.

    In past operations the steam was used to generate power, and the hot
water and condensed steam were discharged as waste into some convenient
stream or lake. However, this wasteful practice is slowly being discontinued.
At some installations today the hot water is routed through a multiple-

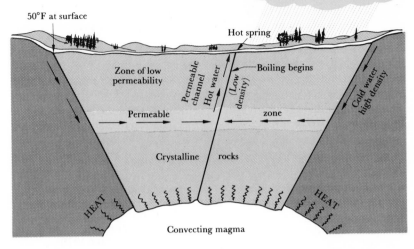

8-9
The hot-spring type of wet-steam field, where some of the hot water escapes to the surface. (Source: "Natural Steam for Power," U. S. Geological Survey, 1968)

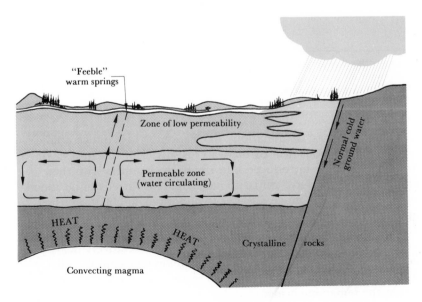

8–10
A dry-steam field, which is insulated from the surface by a layer of impermeable material, thus allowing little or no hot water to escape to the surface. (Source: "Natural Steam for Power," U. S. Geological Survey, 1968)

Table 8–4 Tabulation of selected geothermal fields being used in late 1972.
The expected development of electric-power capacity at those and other fields
to 1980 is given in the last column.

Nation	Field	Installed capacity late 1972, Mw	Expected development, Mw
El Salvador	Ahuachapán	—	30 by 1975; 60 by 1980
Iceland	Hengill (Hveragerdi)	—	Up to 32 by 1980
	Namafjall	3	None known
Italy	Larderello	365	15-percent increase possible
	Monte Amiata	25	
Japan	Hachimantai	—	Perhaps 10 by mid-1970s
	Matsukawa/Takinokami	20	Perhaps 60 by 1980
	Onikobe	—	Perhaps 10 by 1980
	Shikabe	—	7; salt-recovery works planned for 1970s
	Otake/Hachobaru	13	Perhaps 60 by 1980
Mexico	Cerro Prieto	75	150 by 1980
New Zealand	Kawerau	10	None planned
	Wairakei	160	None planned
U. S. S. R.	Pauzhetsk	6	Up to 25 by 1980
	Kunashir	—	Up to 13 by 1980
United States	The Geysers	302	110 per year through 1980, to 1180
	Imperial Valley	—	Demonstration, for desalination; power station by 1980

(James B. Koenig, "Worldwide Status of Geothermal Resources Development," in
Geothermal Energy: Resources, Production, Stimulation, ed. P. Kruger and C. Otte,
Stanford: Stanford University Press, 1973, p. 56. Reprinted by permission.)

flash distillation unit (see "Desalination" in Chapter 3) to produce usable
fresh water. The indigenous heat of the water is sufficient to cause steam
to form when the water is introduced into the flash chambers with their
less-than-atmospheric pressure. Since no additional heat needs to be added
to the water, this process is much cheaper than normal desalination using
the multiple-flash method. This process is being further refined by the
Chilean government at a recently discovered wet-steam field located at
El Tatio, Chile. The water in this field was found to be relatively heavily
mineralized, and therefore the government is developing a complex that
will generate electricity, produce desalinized water, and extract valuable
minerals from the remaining brine.

Most areas that have geothermal power (Table 8–4 and Fig. 8–11) are
situated in the well-known volcanic regions of the world. The geothermal

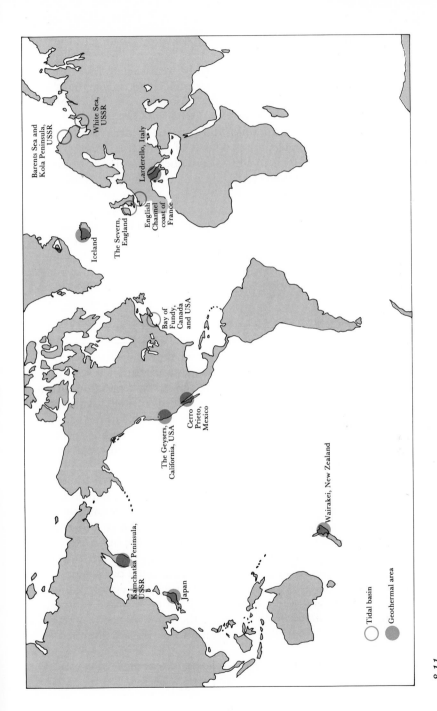

8-11
Tidal and geothermal power sites.

Barents Sea and
Kola Peninsula,
USSR

White Sea,
USSR

Larderello, Italy

Iceland

The Severn,
England

English
Channel
coast of
France

Bay of
Fundy,
Canada
and USA

The Geysers,
California, USA

Cerro
Prieto,
Mexico

Wairakei, New Zealand

Kamchatka Peninsula,
USSR

Japan

◯ Tidal basin

● Geothermal area

power installations of California, Mexico, New Zealand, Japan, and the Kamchatka Peninsula of the USSR are all on the "Ring of Fire" circum-Pacific volcanic belt (Fig. 10–10). Those of Italy are associated with the Mediterranean volcanic area, and the hot water of Iceland is over the Mid-Atlantic Ridge, also a belt of volcanic activity. Use of geothermal energy for producing electric power dates from 1904, when the first such power station went into operation at Larderello, Italy, about 50 miles southwest of Florence. Italy remains the leading producer of geothermal energy.

There seem to be many areas in the volcanic belts that look promising for the development of geothermal energy; recent work in Mexico and in California's Imperial Valley suggests that these areas may be suitable. More exploration will be done in the future as man attempts to develop every possible source of power. The utilization of geothermal energy can be increased somewhat, but it is not known how much at this time. Geothermal wells will provide excellent local sources of inexpensive, nearly pollution-free power. For this reason we should continue to pursue our geological exploration for favorable areas.

SOLAR ENERGY

The sun is continually converting hydrogen into helium, releasing huge amounts of energy in the process. Every second, approximately four million tons of mass are converted into energy by the sun and radiated out into space. The earth intercepts only one two-billionths of the total amount radiated. This amounts to about 261,448 trillion kilowatt-hours every day. This is roughly equivalent to the energy in some 500 trillion tons of coal. The total production of electricity for the world in 1969 was only about 4.4 trillion kilowatt-hours!

The supply of energy at the earth's surface is less than what the earth intercepts because of absorption and reflection by the atmosphere and because the amount of energy striking a unit area is decreased geometrically as that surface changes orientation from being perpendicular to the sun's rays to receiving the sun's rays at some other angle. Taking these factors into account, it has been calculated that the daily energy on a horizontal surface at low latitudes averages six to eight kilowatt hours per square meter. If we use the larger of the two figures, we find that energy equivalent to about 3.8 times that used by the United States, from all sources, in 1970 falls on the United States daily. Why, then, are we not utilizing this source of energy now? Is it likely to become a significant source of energy in the near future?

Solar energy *is* currently being used to some extent and for certain processes.

Solar evaporation   This is an important industrial process in the United States as well as in many of the developing countries. In areas where evaporation exceeds rainfall, shallow ponds filled with brine are allowed to evaporate dry. The crystallized salts are then harvested. The United States produces about one million tons of salt annually using this method.

Solar distillation   This is perhaps one of the oldest uses for which man has used solar energy. Salt water is put into a shallow receptacle having a large surface area relative to its volume. This receptacle is then covered by a transparent dome. The water evaporates, leaving behind the salt, then condenses on the dome and drips off the edges to be caught and used.

Water heating   This technique is used in Israel, Japan, Australia, and southern Florida. Water circulated through a solar collector is heated by the sun's energy. It is then held in an insulated tank until needed. This use applies mainly to individual household heaters, but there is also a small amount of commercial usage.

Solar furnace   In this application, solar energy is concentrated with the aid of a parabolic reflector to provide very high heat at a single point. According to Lucian, a not-too-reliable Greek writer of about A.D. 150, this principle was used by Archimedes during the seige of Syracuse by the Roman, Marcellus, in 212 B.C. Marcellus assailed Syracuse by land and by sea and was temporarily defeated when, among other defense measures, Archimedes set his fleet afire by using great concave mirrors to concentrate the sun's energy on his ships.

In 1774 the same principle was used in Paris by a French chemist named Lavoisier, who used a 52-inch lens to conduct chemical studies at high temperature. Currently, the largest solar furnace in existence is located in France at Mont-Louis in the Western Pyrenees. Sixty-three individual reflectors, capable of tracking the sun, reflect the sun's rays onto a parabolic reflector 140 feet high. The parabolic reflector, set in the side of a ten-story concrete office and laboratory building, collects the rays and focuses them on the oven, where temperatures up to 6300° F can be reached. The furnace is used to produce large, ultrapure crystals and extremely high-temperature refractory materials.

Presently, a number of new applications for solar energy are in various stages of experimentation; these include space heating and cooling, solar refrigeration, solar cooking, and the generation of electrical energy. It is the latter in which we are most interested.

There are currently two difficulties associated with power generation from solar cells: the cost of power generation by this method and the low efficiency of solar cells now in use. The conversion of solar radiation to electrical energy has had its chief function as a power source for spacecraft

8–12
Solar energy collectors (foreground) reflect energy onto a large parabolic mirror (background) which focuses energy into oven in white building. (Photograph courtesy French Embassy Press & Information Division)

in the Russian and American space programs. To date, more than 1000 spacecraft from these two programs have used solar cells as primary sources of electrical power; indeed, the space program has been nearly the sole user of solar cells. This relatively small market, with no forseeable increase from other demands, has kept the cost of solar cells high. The very stringent requirements imposed on the cells used in satellites and the relatively small demand for these cells have both worked to exclude the introduction of mass-production methods into their fabrication.

How do these solar cells work? It has long been known that certain materials, principally selenium, cadmium sulfide, and silicon, produce electrical energy when exposed to light. In 1954 the Bell Telephone Laboratories began making solar cells using silicon, and this is still the basic material of most modern solar cells. The cells are specially made so that the electrons in the molecular structure will move only in a certain way when the silicon absorbs light. Basically, the cell is composed of three parts: one layer of material which is deficient in electrons, one layer of material which has an excess of electrons, and a barrier layer between them. In the presence of light, photons (units of light) are absorbed in the layer facing

8-13
Solar energy collector at Mont-Louis in the French Pyrenees. (Photograph courtesy French Government Tourist Office)

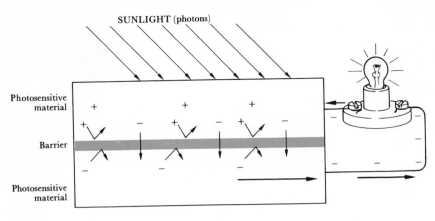

8-14
A typical solar cell consists of a "positive-rich" layer, a "negative-rich" layer, and a barrier layer. The flow of electrons is directional through the barrier layer.

the light, thereby upsetting the electron equilibrium of the material. Electrons are displaced by photons and are then free to move about. The barrier layer is so designed that electrons can pass readily through it, but in only one direction (Fig. 8-14), thus giving direction to the electron flow. The positive "holes" left by the exit of the electron cannot pass through the barrier; eventually, one side of the barrier becomes positively charged, and the other side becomes negatively charged. If an external path is established between the two sides, a flow of electricity occurs. For this flow to be useful, an electric motor, lightbulb, radio transmitter, or other electrical device, must be placed in this external circuit to intercept the flow of electricity.

It has been estimated that if the same devices presently being used in the Skylab power unit were to be installed and made operational in a commercial generating plant, the cost would be about \$2,000,000 per kilowatt, compared with \$463 for a coal-burning plant, \$534 for an oil- or gas-burning plant, and approximately \$750 for a nuclear plant. Realistically, this figure could be reduced to about \$15,000 per kilowatt simply because the manufacturing specifications for a surface generating station would be much less severe than those used in the Skylab program. Even so, the cost differential is excessive. The figure might be reduced to around \$2500 if mass-produced cells using cadmium sulfide as the photosensitive material are developed from recent techniques of applying this material as a very thin film coating.

Part of the reason for such high cost per kilowatt of generating power is the low efficiency of present solar cells. Solar cells using silicon as the

photosensitive material are the most widely used and most permanent.
They provide the power for the long-range satellites of both Russia and the
United States. However, the conversion efficiency of the silicon cell is only
about 10–12 percent. Therefore, other compounds, e.g., gallium arsenide,
indium phosphide, gallium phosphide, aluminum-gallium arsenide, cadmium
sulfide, and cadmium telluride have also been used. However, most of these
compounds have efficiencies in the 6–8 percent range. Improvements that
might double its efficiency to 20–22 percent have been suggested for the
silicon cell. Efficiencies higher than this are not currently being discussed.

There are two geologic aspects that relate to solar energy: discovery
and mining of the elements needed in the manufacture of solar cells, and
site analysis of the proposed locations for the solar receptor arrays. In order
to generate the 4.4 trillion kilowatt hours of electricity produced by the
world in 1969, it would take at least 1300 square miles of solar receptors,
if we assume a 10 percent overall efficiency for conversion of solar energy
to electricity. A single power station generating ten million kilowatts would
require a receptor area of some square 25 miles. Obviously, the geologist
should be consulted when choosing such extensive sites.

To circumvent this problem of having to use large parcels of land for
the generation of electricity by solar energy, it has been suggested that a
large array of receptors be placed in a permanent parking orbit 22,300 miles
above the earth to form a space power plant. The station would be a solid
bank of solar panels approximately 7.5 miles long by 3 miles wide. Because the station would be above the earth's atmosphere, it would receive
more radiation per unit area than if it were on the earth's surface. Therefore, a station of these dimensions should generate five billion watts of
electrical power. The energy collected would be converted to microwaves
and beamed to earth, where it would be converted to electrical energy. The
efficiency of converting beamed energy to electrical energy is approximately
80 percent, which is relatively high.

TIDAL ENERGY

Energy can be developed from the oscillatory flow of ocean tides. It is well
known that moving water possesses much potential energy, and for many
years man has used the energy of waterfalls to provide electrical energy.
He has even built dams in order to obtain energy from man-made waterfalls.
Since tides are moving water, obtaining energy from tidal flow is essentially
equivalent to obtaining it from waterfalls, with the exception that the
water power from falls originates in unidirectional streamflow.

Tides are wave motions that occur regularly, although with varying
characteristics from time to time and place to place. Tidal flow consists of

8-15
*Passamaquoddy tidal power project, Eastport, Maine. Photograph looks from Pleasant
Point south along the line of the 1936 dams over Carlow Island toward Moose Island and
Eastport. (Photograph courtesy of Department of the Army, Corps of Engineers)*

the vertical motion of the rise and fall of the water level and a horizontal
movement in the tidal currents toward and away from shore. Such movements are oscillatory, having a frequency of about 12½ hours between
successive high tides. The movement of the tides depends on many factors,
of which the most important are the changing positions of the moon and
the sun with respect to the earth and also the tides' geographical locations.

Tidal current that moves toward the shore is called a *flood tide.* The
greatest velocity of the moving water occurs at the approximate midpoint
of the flood. The flood tide then eases off, a *high tide* point is reached,
and then the tide begins to *ebb* as the water flows out. Finally, a *low-tide*
point is reached, and the cycle begins again. The vertical distance between
the high- and low-tide marks is called the *tidal range.* In some places in the
world, the tidal range is as high as 50 feet; however, on open seacoasts and
islands the tidal ranges are usually as low as two or three feet. They are
highest in areas where either the topography of the sea floor or the
coastline configuration of the surrounding land areas creates a pronounced
inertia effect. This usually occurs in bodies of water that are funnel-shaped

or contained at one end, as are sounds, estuaries, and bays. It is in such areas that the greatest tidal ranges and strongest tidal currents are found. The pronounced floods and ebbs of these restricted areas are classified as the reversing type of tidal current. The oscillatory sloshing of water in bays and estuaries is complex and differs from the movements of tides in the open oceans.

It is in the restricted basin environment that the force of moving water can sometimes be harnessed to provide energy. Dams are built to enclose a basin in order to create a difference between the water level in the basin and that of the open ocean. As oscillation of the tides occurs, causing the basins to fill and empty, the flow of the water is used to drive turbines which generate electrical power. To be effective, the tidal range should be at least 30 feet.

The amount of energy derived from the tides throughout the world is insignificant compared with other sources of energy. It has been calculated that the world's potential tidal power amounts to less than one percent of its potential waterpower. In 1969 the total world energy provided by waterpower was slightly more than one trillion kilowatt-hours. In the United States hydropower accounted for only 3.8 percent of the power generated in 1970. There are a limited number of sites available for development of such power, and the installation of generating facilities is expensive. However, tidal-energy projects may be feasible, and even useful, on a local basis, especially if industry can be built in conjunction with them. The first big tidal power plant was built at the estuary of La Rance, a small river on the coast of Brittany in France. Other tidal-power sites are located in France as well as in Russia and on the Bay of Fundy (Fig. 8–11).

OTHER ENERGY SOURCES

To complete our analysis, we should mention other sources of energy that have been suggested in the past few years—the generation of gas, primarily methane, from organic wastes; the production of oils by certain organisms through photosynthesis; the use of wind to run generators; the development of fuel cells for power generation; and the replacement of natural gas by hydrogen. Of these, only the last two are realistic alternatives for the significant and widespread production of energy.

Hydrogen   Hydrogen has recently been given much serious consideration as a possible synthetic fuel to replace the fossil fuels as these sources become exhausted or, at least, too scarce to use. Hydrogen is not a natural fuel, but it can be readily synthesized from the natural hydrocarbons (coal, oil, and gas) and by the electrolytic disassociation of water molecules.

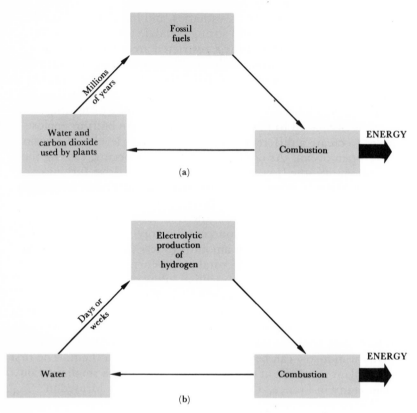

8-16
Comparison of regeneration times for: (a) fossil fuel, and (b) hydrogen fuel cycles.

There are many favorable reasons for using hydrogen as a fuel. First, it is nonpolluting; when it burns, the only product of combustion is water. Second, it can be transmitted and distributed by pipeline in essentially the same way as natural gas is handled today. Third, the energy content per unit mass of liquid hydrogen is about 2.75 times greater than that of fossil fuels, making it an ideal fuel for aircraft and rockets. Fourth, hydrogen is an extremely suitable fuel for fuel cells, making them at least 55 percent more efficient than do other fuels. Fifth, present technology is capable of producing hydrogen.

It has been suggested by proponents of the "hydrogen economy" that nuclear plants can generate the electricity necessary to synthesize the hydrogen. This in itself presents a problem. The amount of hydrogen necessary to provide the United States with a BTU value equivalent to that of a year's supply of natural gas would require three to four times as much electricity as is presently used in the United States. To handle this load, more than 1000 new nuclear power stations, each designed to produce 1000 megawatts, would have to be constructed. This, of course, is exclusive of those power stations which would have to be built just to fulfill the increasing demands for electricity.

Another negative feature of hydrogen is that it is hazardous material. Only one-tenth the energy required to ignite either gasoline or methane will ignite hydrogen. Since it has no odor or taste, leaks are an additional hazard. It is also dangerous to work with, as it burns with an almost invisible flame. These, however, are not considered to be significant problems to those who advocate a hydrogen economy because they feel that if strict codes were applied to handling and use, hydrogen would be no more dangerous than natural gas or gasoline. Odorants are added to natural gas today so that leaks can be readily detected, and these proponents say that if both an odorant and an illuminant were added to hydrogen, it could be smelled and its flame would be readily visible. Apparently, hydrogen may have some realistic potential as a significant source of energy in subsequent decades.

One more factor which makes hydrogen an attractive fuel is shown in Fig. 8-16. In addition to requiring a restricted set of conditions, the fossil-fuel cycle has a renewal time of millions of years, whereas hydrogen can be recycled and reused within a period of days or weeks.

Fuel cells   A fuel cell is an electrochemical device that converts the chemical energy contained in a fuel directly into electrical energy. Figure 8-17 shows the operation of a single fuel cell. In this model, hydrogen is fed into the unit at one electrode (the anode) where ionization takes place. The freed electrons travel to the other electrode (the cathode) via an external circuit, which is the means by which the energy of the cell is tapped. If an electric load, such as a light bulb, is placed in this circuit, the energy can be utilized. The hydrogen ions formed at the anode migrates through the electrolyte—which conducts current between the two electrodes, can be liquid or solid, and may be composed of many substances—to the cathode. At the cathode it combines with an oxidizer, oxygen in this illustration, to form a compound, which in this model is water.

Fuels other than hydrogen can be used in fuels cells. Among these are: carbon monoxide and air, methane (natural gas) and air, ammonia and air, zinc and oxygen, and aluminum and oxygen. Not all of these have proved practical to date, but research is being conducted on these and others. Because of the greater efficiency of fuel cells in comparison with conventional power plants, the advocates of this system suggest that natural gas would be more productive if used to feed fuel cells for the direct conversion of electricity than when used to generate electricity by conventional methods.

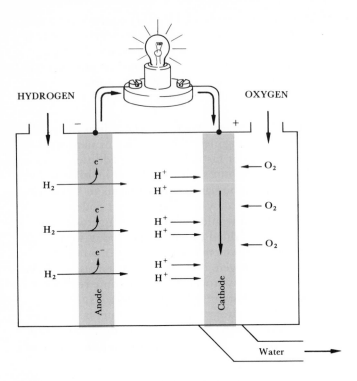

8-17
The generation of electricity from a simple hydrogen/oxygen
fuel cell. The reactions which take place are:
cathode: $O_2 + 4H + 4e^- \rightarrow 2H_2O$
anode: $H_2 \rightarrow 2H^+ + 2e^-$
overall: $2H_2 + O_2 \rightarrow 2H_2O$

Fuel cells are very versatile, and there is no appreciable change in efficiency relative to the number of cells joined together. Because of their ability to be joined into larger and larger units, they have been envisioned as running the full range of power-plant sizes, from those capable of running a single automobile or home to those generating thousands of kilowatts. Future research and development may make this a truly significant energy source.

Gas from organic wastes   The shortage of energy and the increasing amount of waste being generated annually would seem to make the idea of producing methane from organic waste an inspired solution to two very tough problems. When the gross figures of the amount of organic waste generated yearly (over two billion tons in 1970) are considered, the potential to be

8–18
Windmill, stock tanks, and irrigation on a Nebraska ranch. (Photograph courtesy USDA–Soil Conservation Service)

derived from this much waste is, indeed, rather large. There are, however, some problems involved in utilizing waste. First, a recent study done for the Bureau of Mines indicates that more than half of this weight is actually water. Second, more than 80 percent of the organic waste generated is so widely dispersed that it is, in effect, unavailable; only 15 percent of the net methane potential in total organic wastes is readily collectable and therefore available for use. This amounts to only about one-tenth of one percent of the total energy now used. Thus, it seems that this is not a viable option for easing the energy crisis.

Wind It is estimated that winds within 80 meters of ground level possess five times the total amount of energy we now use. Wind has been used to generate small amounts of electricity, usually for individual farm homes, in the West and Middle West for many years. However, these wind-generating electric plants are usually small, ranging from 100 to 2000 watts. Work is under way to study the feasibility of building larger plants. However, this source of electrical power would be localized and limited to areas where the wind blows steadily and continuously, and therefore it should not be considered a panacea for energy shortages in most areas.

Photosynthesis Suggestions have been made that algae that would over-produce one of the fatty acids of its cell membrane and then would decarboxylate it to oil be bred. This oil could then be collected and processed in a method similar to petroleum processing. However, it has been estimated that under present conditions less than one-quarter of one percent of today's energy needs could be filled by this method.

ENERGY AND THE COMING DECADES

Energy is indispensable to a developed, industrialized country, as is shown in Fig. 8–19. Conversely, if an underdeveloped country wishes to develop, it must avail itself of large amounts of energy. When we consider that at present approximately 75 percent of the world's population lives in developing countries which are making every effort to become industrialized, we begin to appreciate the demand that will be made on our energy resources in the next few decades, if these countries succeed. Between 1948 to 1968, the demand for energy (in all forms) in the United States grew at a rate of slightly more than 3 percent annually, while the demand for energy in the rest of the world grew at an annual rate of 6 to 7 percent (depending on what portion of that period one looks at). These rates indicate that the world must double its production of energy about every ten years. Since this growth is at least three times greater than the population growth rate,

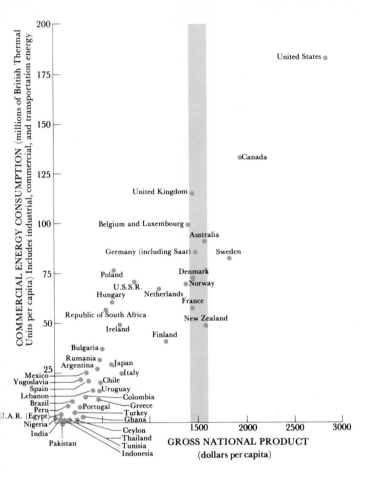

8-19
Per capita energy use and per capita gross national product for select-
ed countries, 1961. (Adapted from A. B. Cambel, et al., Energy Research and Development and National Progress, *Washington, D. C.:*
U. S. Government Printing Office, 1964.)

it is obvious that per capita demand, as well as demand related to population growth, is increasing. If current rates persist until the end of this century, we would expect that in the year 2000 the annual demand for all forms of energy in the United States will be almost twice its demand in 1970, and the energy demand for the entire world will be about three times as great as its 1970 demand. We have to ask ourselves how we can meet this demand.

Table 8–5 Estimated total original coal resources of the world, by continents[a] (in billions of short tons)

Continent	Resources determined by mapping and exploration	Probable additional resources in unmapped and unexplored areas	Estimated total resources
Asia and European U. S. S. R.	7,000[b]	4,000	11,000[c]
North America	1,720	2,880	4,600
Europe	620	210	830
Africa	80	160	240
Oceania	60	70	130
South and Central Americas	20	10	30
Total	9,500[b]	7,330	16,830[c]

Source: Paul Averitt, 1969, Table 8, p. 82.

[a] Original resources in the ground in beds 12 inches thick or more and generally less than 4000 feet below the surface, but includes small amounts between 4000 and 6000 feet.

[b] Includes about 6500 billion short tons on the U. S. S. R.

[c] Includes about 9500 billion short tons in the U. S. S. R. (Hodgkins, 1961, p. 6)

(M. K. Hubbert, "Energy Resources," in *Resources and Man*, Publication 1703, Committee on Resources and Man, National Academy of Sciences, National Research Council, W. H. Freeman and Co., San Francisco, 1969, p. 202. Reprinted by permission.)

The two chapters on energy have surveyed the probable energy sources available to us in the next two or three decades. Perhaps we should now assess their probable effectiveness in meeting the predicted demand. Of course, there exists the possibility that we will curtail our energy demands and therefore will have no problem to solve. However, if present trends continue, what sort of situation might we face? What resources are available to us to meet this demand? We should look at these resources from two points of view: (1) What is now known to be available and in what quantities? and (2) What is the full extent of our resources, including those which have not yet been discovered?

Coal Paul Averitt's study done for the United States Geological Survey gives us perhaps our best overview of coal as an energy resource for the future. Table 8-5 gives the estimates, compiled in this study, of the total amount of coal originally available to us. This total has been subdivided into coal which is known, through mapping and exploration, to exist and that which probably exists in areas not yet mapped or studied. However, not all this coal is available to us. Currently, only 50 percent of the total coal resources are capable of being mined. Using these figures and assuming 3.6 percent is the growth rate for coal usage for the period in which these resources were, and will still be, available, Fig. 8-20 depicts the total production of coal as a major source of energy. The area under the curve corresponds to the amount of coal initially present when man first discovered its usefulness. The amount of coal represented by this curve is approximately 50 percent of the known plus probable reserves. This curve has been plotted on the basis that the maximum rate of use will be eight times greater than the present rate. Assuming this, the peak rate will occur between 170 and 200 years in the future. If the peak rate actually turns out to be higher, the crest will come sooner; if it is lower, it will come later. A corresponding graph using only United States figures for combined known and probable reserves and a maximum rate of use eight times the present rate shows that the maximum rate will peak around the year 2220.

Petroleum   M. K. Hubbert estimates that the ultimate quantity of crude oil to be produced in the United States will total about 190 billion barrels. Others have presented estimates as high as 590 billion barrels, but we feel that Hubbert's estimate is the most realistic. The United States has already used nearly half the amount Hubbert estimated to be the total available. By the year 2000, the United States will have produced 90 percent of its available oil, and 50 years later the oil will be virtually gone.

For the world as a whole, the estimated production is considerably larger. According to Hubbert, figures by W. P. Ryman of the Standard Oil Company of New Jersey show that about 2090 billion barrels of crude oil can ultimately be produced, although the accuracy of this figure is not known at present. If current trends continue, the peak consumption of 37 billion barrels per year would occur between the years 1990 and 2000 (Fig. 8-19), and the production rate thereafter will decline rapidly. If this estimate proves to be correct, 80 percent of the total cumulative production will occur during the period from 1968 to 2032. One hundred years from now the world's supply of petroleum will have been depleted.

Natural gas   It is difficult to estimate the ultimate amount of natural gas and natural-gas liquids. Much of the natural gas associated with petroleum production has been burned off as waste at the refinery. Because there has been no market for it, not much has been done in the past to find and evaluate natural gas reserves. In the United States the ratio of natural gas to crude oil is about 6400 cubic feet per barrel, and the ratio of natural-gas liquids to crude oil is about 0.2 cubic feet per barrel. If we assume that these ratios are applicable to the rest of the world, we could expect to produce 420 billion barrels of natural-gas liquids and 12,000 trillion cubic feet of

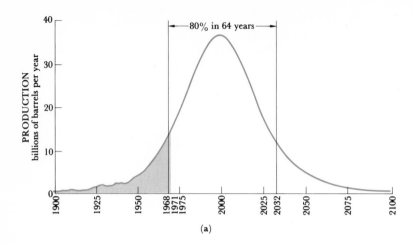

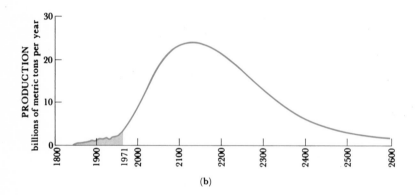

8-20
Production and projected production of oil and coal: (a) world oil production (total area under curve is equal to estimate of 2100 billion barrels ultimate production; shaded area is production to 1971); (b) world coal production (total area under curve is equal to estimate of 7.6 trillion metric tons ultimate production, i.e., recoverable coal; shaded area is production to 1970). (M. K. Hubbert, "Energy Re-sources," in Resources and Man, *Publication 1703, Committee on Re-sources and Man, National Academy of Sciences — National Research Council, W. H. Freeman and Co., San Francisco, 1969, pp. 196, 204. Adapted by permission.)*

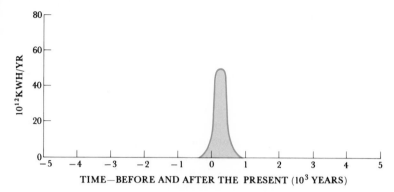

8-21
*The era of the fossil fuels, placed in historical perspective from plus to
minus 5000 years from the present. (M. K. Hubbert, Energy Resources,
Publication 1000–D, Committee on Natural Resources, National Academy of Sciences — National Research Council, Washington, D. C., 1969,
p. 91. Reprinted by permission.)*

natural gas for the 2100 billion barrels of crude oil estimated to be
ultimately available.

Oil shale and tar sand In 1965 the United States Geological Survey
estimated that as much as 305 trillion barrels of in-place oil (10 gallons or
more per ton) was locked within the organic-rich shales of the world. However, with 1965 technology, only less than 0.0001 percent could be recovered, and this has not significantly changed since. The little that is known
about tar sand is based almost exclusively on the deposits in Alberta, Canada,
which are thought to contain as much as 300 billion barrels of in-place oil.
Yet, like the oil shale, very little of this can be tapped with current technology.

Fossil fuels summary Figure 8–21 is perhaps the most succinct statement
of the significance, both past and future, of fossil fuels in the affairs of
man. In a total of approximately 1500 years, these fuels will have completed
the discovery-use-exhaustion cycle.

Uranium-thorium Two factors complicate the process of making sound
estimates of the ultimate, world-wide availability of these elements. One is
that the communist countries do not make public information concerning
their reserves. The other major factor is that for many years the small
market and consequent oversupply of thorium has not been conducive to

furthering the exploration necessary to make valid determinations of reserves.

It has been estimated that the free-world reserves of U_3O_8, which can be produced for less than $10 per pound, equal slightly over 1.5 million tons, while the total for thorium is about 1.7 million tons. If we continue to use burner reactors, it has been estimated that 0.7 million tons of U_3O_8 will be needed in the United States alone. Since there are many indications that the rest of the world will match our pace in the use of nuclear power, this could possibly account for the consumption of the remainder. Estimates of the electrical needs which nuclear power will have to satisfy keep being revised upward, and it is conceivable that the demands will be even greater. However, these predictions will be revised significantly if breeder reactors become commercially available and will become academic if fusion is achieved.

Other sources   If our technology could be improved, solar energy, of course, could potentially provide thousands of times the energy being used in the world today. The wind could also potentially produce around five times the energy consumed presently, yet in many cases it would not be in usable form and its distribution would be inequitable. If we combined hydropower, photosynthesis, organic wastes, tides, and geothermal energy as energy sources they would potentially supply less than 2 percent of the energy needed in the world.

SUMMARY

There are obviously many corollaries to the discussion presented in these two chapters on energy. First, and perhaps most important, is the fact that fossil fuels will become exhausted rapidly if they continue to be used as our major energy source. It is clear that we will need alternative sources of energy within the next generation, or sooner, in some areas.

A second point, which seems self-evident, is that nuclear power will replace the fossil fuels in furnishing power to a demanding world. Barring a completely unforeseen and spectacular development in some other area of energy, nuclear energy is the only alternative which has this potential and of which present technology is capable of realizing that potential. Yet, without breeder reactors or a feasible fusion process, even this source has its limits and could become only a short-term alternative.

Undoubtedly, in coming decades the technology needed to efficiently use solar energy will be developed, or possibly a presently untapped, unthought of source of energy will become a reality. We are confident that we are not witnessing the end of large-scale energy generation on this planet;

rather, we are seeing the beginning of a transition from the use of the fossil fuels as a prime energy source to the use of some other resource. This does not imply, however, that we no longer have to be concerned with energy. Indeed, with respect to energy, we may be entering one of the most critical periods of our history. Consider the consequences of exhausting the fossil fuels (or nearly so) before an alternative source of energy has been fully developed. We must have workable solutions for our energy problems before we can afford to relax. It is still possible that we will be subjected to rationing of energy, blackouts, and severe conservation measures before a new source of energy is completely capable of fulfilling our needs. Therefore, we must be fully aware of the problem and the possible answers so that we can use energy rationally.

REFERENCES

Aaronson, A. "The black box," *Environment* 13, 10 (1971): 10–18.

Barnes, J. "Geothermal power," *Scientific American* 226, 1 (1972): 70–77.

Bohn, H. L. "A clean new gas," *Environment* 13, 10 (1971): 4–9.

"Breeding power," *Newsweek*, June 14, 1971.

Cambel, A. B., *et al. Energy Research and Development and National Progress*, Washington, D. C., U. S. Government Printing Office, 1964.

"Chance for solar energy conversion," *Chemical and Engineering News* 49, (Dec. 20, 1971): 39.

Daniels, F. "Direct use of the sun's energy," *American Scientist* 55, 1 (1967): 15–47.

Daniels, F. and J. A. Duffie, eds. *Solar Energy Research*, Madison: University of Wisconsin Press, 1961.

Dutcher, L. C., W. F. Hardt, and W. R. Moyle, Jr. *Preliminary Appraisal of Ground Water in Storage with Reference to Geothermal Resources in the Imperial Valley Area, California*, U. S. Geological Survey Circular 649, 1972.

"The earth's heat: a new power source," *Science News* 98 (1970): 415–416.

Ehricke, K. A. "Extraterrestrial imperative," *Bulletin of the Atomic Scientists* XXVII, 9 (1971): 18–26.

Federal Power Commission. *World Power Data—Capacity of Electric Generating Plants and Production of Electric Energy—1969*, Washington: Government Printing Office, 1972.

Fenner, D. and J. Klarmann. "Power from the earth," *Environment* 13, 10 (1971): 19–34.

Ford N. C. and J. W. Kane. "Solar power," *Bulletin of the Atomic Scientists* XXVII, 8 (1971): 27–31.

Gillette, R. "Energy: President asks $3 billion for breeder reactor, fuel studies," *Science* 172 (1971): 1114–1116.

Gray, T. J. and O. K. Gashus, eds. *Tidal Power*, New York: Plenum Press, 1972.

Gregory, D. P. "The hydrogen economy," *Scientific American* 228, 1 (1973): 13–21.

Hammond A. L. "Solar energy: a feasible source of power?" *Science* 172 (1971): 660.

————. "Breeder reactors: power for the future," *Science* 174 (1971): 807–810.

————. "Energy options; challenge for the future," *Science* 177 (1972): 875–876.

————. "Solar energy: the largest resource," *Science* 177 (1972): 1088–1090.

————. "Fission: the pro's and con's of nuclear power," *Science* 178 (1972): 147–149.

Hoke, J. *Solar Energy*, National Academy of Sciences, National Research Council,: Franklin Watts, 1968.

Hubbert, M. K. "Energy resources," in *Resources and Man*, National Academy of Sciences, National Research Council, San Francisco: Freeman, 1969.

————. *Energy Resources: A Report to the Committee on Natural Resources*, Washington, D. C.: National Academy of Sciences — National Research Council, Publication 1000-D, 1962.

————. "The energy resources of the earth," *Scientific American* 224, 3 (1971): 60–70.

"Hydrogen fuel economy: wide-ranging changes," *Chemical and Engineering News* 50 (July 10, 1972): 27–30.

"Hydrogen fueld use calls for new sources," *Chemical and Engineering News* 50 (July 3, 1972): 16–19.

"Hydrogen: likely fuel of the future," *Chemical and Engineering News* 50 (June 26, 1972): 14–17.

"Imperial Valley - geothermal steam looks better," *Science News* 95(1969): 113–114.

Koenig, J. B. "Geothermal development," *Geotimes* 16, 3 (March 1971): 10–12.

Maugh, T. H. II. "Fuel from wastes: a minor energy source," *Science* 178 (1972): 599–602.

————. "Hydrogen: synthetic fuel of the future," *Science* 178(1972): 849–852.

————. "Fuel cells: dispersed generation of electricity," *Science* 178(1972): 1273–1274B.

Meinell, A. B. and J. P. Meinell. "Is it time for a new look at solar energy?" *Bulletin of the Atomic Scientists* XXVII, 8 (1971): 32–37.

"Methane from organic wastes," *Chemical and Engineering News* 50 (April 17, 1972): 23.

Metz, W. D. "Laser fusion: a new approach to thermonuclear power," *Science* 177 (1972): 1180–1182.

————. "Magnetic containment fusion: what are the prospects?" *Science* 178 (1972): 291–293

National Academy of Sciences, National Research Council. *Resources and Man*, San Francisco: Freeman, 1969.

Rappaport, Paul (chairman). *Solar Cells - Outlook for Improved Efficiency*, Ad Hoc Panel on Solar Cell Efficiency, National Research Council, National Academy of Sciences, 1972.

Rex, R. W. "Geothermal energy - the neglected energy option," *Bulletin of the Atomic Scientists* XXVII, 8 (1971): 52–56.

Solar Energy in Developing Countries: Perspectives and Prospects. Report of the Ad Hoc Advisory Panel of the Board on Science and Technology for International Development, National Academy of Sciences, 1972.

Sowers, A. E. "A novel proposal on energy," *Science News* **101** (1972): 386.

United States Department of the Interior. *United States Energy—A Summary Review,* Washington: Government Printing Office, 1972.

White, D. E. *Geothermal Energy*, U. S. Geological Survey Circular 519, 1965.

Yellott, John I. "Solar energy progress—a world picture," *Mechanical Engineering* (July 1970): 28–34.

Man must understand the surface processes if he is to live in harmony with his geologic environment. (Photograph by George Sheng)

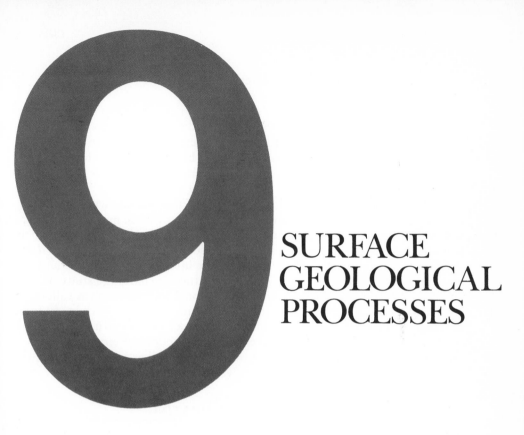

9

SURFACE
GEOLOGICAL
PROCESSES

INTRODUCTION

Much of the geological activity of the earth takes place at or very near the surface—the boundary between the lithosphere and atmosphere and between the lithosphere and hydrosphere. Living as he does at these boundaries, man is caught up in the midst of the complex interactions of many natural processes. It is necessary for man, with his great numbers and complex society, to understand these natural events and processes in order to live compatibly with them — the uplifting and downsinking of the land, weathering, erosion, transportation and deposition of rock material, rainfall, evaporation, and atmospheric and oceanic circulation that continually change the character of the land. Rocks and soil respond to changing environmental conditions through alterations in form or composition or by movement from one place to another. Sometimes the response is sudden and violent, as in the case of landslides; other times the response is slow, as was the case in the retreat of Ice Age glaciers.

One of the most important responses of rocks and soil to changes is *movement*. Earth materials are distributed and redistributed by agents that remove materials from one place, transport them, and deposit them elsewhere. Occasionally, the materials are stable, but this stability does not last forever, and sooner or later, the agents resume the processes of movement.

The agents of *erosion* (removal) and *deposition* include wind, glaciers, streams, ground water, waves, and currents. Each of these may act as either destructive agents, which wear away the rocks, or constructive agents, which build up depositional forms. Destructive actions work in opposition to *diastrophism*, the building up of the earth by the internal forces of heat and pressure. The study of the earth's surface features includes the study of diastrophic features, such as mountains; erosional features, such as valleys; and depositional features, such as river deltas and sandbars. Such study forms the basis of the science of *geomorphology* (Greek: *geo* — earth, *morphos* — form, and *logos* — study). We shall examine the actions of erosional and depositional agents on the surface of the earth, as well as some of the circumstances under which man becomes involved in these actions.

THE ROLES OF GRAVITY AND SOLAR HEATING

Surface processes involving the downward flow of streams and glaciers, the movement of the tides, and the downward mass movement of earth in landslides or creep depend on gravity. The settling of particles as the velocity of wind or water currents is reduced is also governed by gravity. Finally, the convective motion of both wind and water is caused by density differences resulting from inequalities of solar heating; this convection is a gravitational phenomenon. Were it not for gravity, then, glaciers and water would not flow downhill, and cold air and water would not sink. We would not have the action of streams, wind, currents, or waves. Were it not for gravity, there obviously would have been no earth in the first place. Were it not for the sun, the movement of wind and ocean currents would be negligible, weather would be essentially nonexistent, and the earth would be a frozen, lifeless planet—a strange and inhospitable place indeed.

WIND

The wind's environmental significance is dramatized by the destructiveness of tornadoes and hurricanes, but the greatest effects of wind take place daily with velocities of only a few miles per hour. The wind exerts a profound control on the hydrologic cycle by distributing water vapor over the earth and promoting evaporation. Wind is the driving force behind

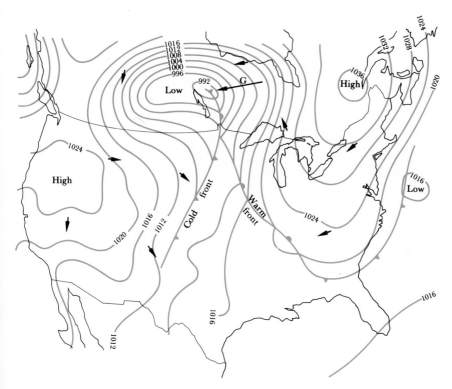

9-1

Weather map for September 20, 1972, showing high- and low-pressure cells. Pressure gradient exists from highs to center of low (G). Contours are lines of equal pressure called isobars, values in units of pressure called millibars. Direction of wind at various localities is shown by small arrows.

waves, which affect erosion and deposition along shorelines. Great quantities of surface material are eroded from the surface of the earth by the wind, which then carries this material for great distances — sometimes all the way around the globe. For example, after the gigantic 1883 eruption of Krakatoa, a volcano near Java, the winds distributed the dust from the explosion worldwide.

Atmospheric Circulation

Wind is the movement of air from an area of high pressure to one of low pressure. A low-pressure *cell* forms when an area of the earth's surface is warmed by the sun. The warm air expands, becomes less dense, and begins to rise, creating a lower pressure over the area than that in surrounding

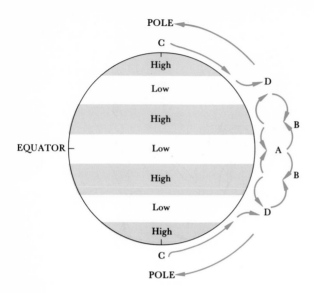

9-2
Latitudinal pressure cells of the earth.

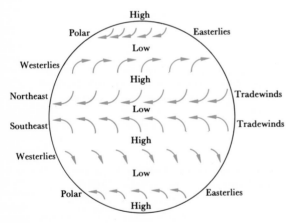

9-3
Prevailing wind systems of the earth.

areas. Cold, heavy air, on the other hand, sinks, creating a high-pressure cell (Fig. 9–1). A *pressure gradient* exists between a high- and a low-pressure cell, as shown by "G" in Fig. 9–1. The air moves in the direction of the arrow. The greater the difference in pressure between the two cells, the greater the pressure gradient and the velocity of the wind. The greatest wind velocities are found in tornadoes, which are small, extremely low-pressure cells around which air may blow at velocities of hundreds of miles per hour. Normal wind velocities, however, measure only a few tens of miles per hour.

Global effects caused by differences in heating between the equator and the poles, as well as local and seasonal variations in the heating of the earth's surface, produce winds. Warm air rises over the equator and creates a fairly permanent equatorial *low*-pressure belt (point "A" in Fig. 9-2). As the air moves north or south, it cools and sinks, forming *subtropical high-pressure belts* parallel to the equatorial low, (point "B" in Fig. 9-2). Cold air over the poles sinks, forming *polar highs* (point "C" in Fig. 9-2). As this air moves away from the poles, it warms and then rises, forming *subpolar low-pressure belts* (point "D" in Fig. 9-2). Winds blow toward the lows, forming three major latitudinal belts of *prevailing winds* (Fig. 9-3). As the earth rotates on its axis, however, the freely moving masses of air pass into parts of the earth which are moving either faster or slower than the parts in which the air originated. This apparently causes the winds to be deflected from their straight courses, westward if they blow toward the equator and eastward if they blow away from it in the Northern Hemisphere. If the land is hotter, air over the land rises, forming a low-pressure cell; the wind then blows from the water area toward the land. If the land is cooler, the opposite occurs. Such air movements occur seasonally and diurnally (day-night-day). An example of the ultimate effects of these air movements is the monsoon. In the winter, cold, heavy, dry air blows outward from central Asia. In the summer, however, winds from the Indian and western Pacific oceans blow toward the hot, low-pressure area of central Asia, drenching south and southeast Asia with rains.

Superimposed on global and other large-scale movements of wind are the movements of small pressure cells governed by local conditions. At any one place in the world, the nature of the winds is determined by the many complex factors of land, water, latitude, and altitude, thereby causing considerable variation in the geological effects of wind from one locality to another.

Wind Erosion

Wind operates as an agent of erosion in two general ways: deflation and abrasion. In the first case, particles are lifted from the ground and are carried aloft by the force of the wind. In the second case, the wind drives particles against other particles and solid rock, chipping and abrading them in a sandblasting action.

Because it lacks the density of ice or water, wind cannot move particles as large as those moved by glaciers or streams; wind can, however, remove great quantities of fine material, including clay, silt, and sand par-

9-4
Desert pavement in Libya. Removal by wind of fine particles leaves coarse particles behind. (Photograph by G. H. Goudarzi, U. S. Geological Survey)

ticles ranging in size from 10^{-5} mm to 1mm. The movement of particles by wind occurs in three ways. Fine particles are removed and carried in *suspension,* i.e., they are lifted up and held in the moving air as long as the wind velocity is sufficient to keep them suspended. Most suspended material is less than 0.1 mm in diameter; anything larger is removed only by the strongest winds and tends to settle rapidly. Because they tend to stick together, the smallest particles, usually of clay, are more difficult for the wind to remove than are slightly larger particles. Once these smallest particles are airborne, however, they may remain in the air almost indefinitely.

The movement of particles of about 0.1 to 0.5 mm generally occurs by *saltation,* a process whereby the particles are lifted momentarily, blown forward slightly, then dropped. When this occurs, they strike other particles, and the cycle is repeated. This process accounts for the movement of great quantities of loose fragments which lie within one to two feet of the ground surface. In arid regions, where the process is common, telephone poles, fence posts, and other structures become damaged as their bases are steadily worn away by abrasion. Wind abrasion wears away paint, pits automobile windows, and does many other types of damage to man's works, in addition to continuing the geological process of wearing away rock.

By far the greatest losses caused by the wind are of valuable soil. As we noted in Chapter 4, soil requires many, many years to form. Such a long time is needed for its regeneration that once it has been eroded,

9–5
Wind erosion in this field has stripped away as much as six inches of soil since the corn was planted. (Photograph USDA—Soil Conservation Service)

9–6
Dust storm across cultivated field during 35 mile per hour wind. (Photograph courtesy USDA—Soil Conservation Service)

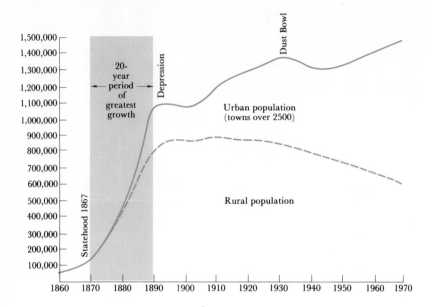

9-7
Growth of population in Nebraska, 1860–1970. Early growth on Great
Plains was mostly rural. Later growth has been urban, while rural population
has declined.

soil-dependent economies are destroyed. Entire farms or districts may be
destroyed in just a few windstorms.

When settlers migrated into the vast area of the United States between the Mississippi River and the Rockies, few of them understood the
significance of the ecological balance among wind, rainfall, soil, and
vegetation. The rapid settlement of a typical semiarid region is shown in
Fig. 9–7. Nebraska had a low population until well after the Civil War. By
1910, however, it had grown considerably, but then its growth leveled off.
The state was settled and most of its land cultivated in a quarter of a
century, which is, geologically, a very short time. Nebraska's population
growth since 1910 has occurred almost entirely in the cities and larger
towns rather than in the rural areas. In those few years of rapid agricultural development, the native grasses were plowed under, the soil was
exposed to the weather, and single-crop environments were established
over tens of thousands of acres. A number of years of unusually high rainfall deluded people into overestimating what the prairie land could reasonably support.

Figure 9–8 shows the 20-inch and 28-inch *isohyets* (lines of equal precipitation) in the central United States. Land that receives less than 28
inches of rainfall per year can support only a few crops. Land that re-

9–8
Twenty- and 28-inch rainfall isohyets.

ceives less than 20 inches is marginal at best, except where irrigation is practiced. The Great Plains, which lie approximately between the 98th and 105th meridians, were unduly exploited during the late 1800s and early 1900s. In the 1930s, disaster struck: much of the soil, including the topsoil, which is the most fertile, blew away. On May 12, 1934, howling winds stripped 300 million tons of soil from the Great Plains of western Kansas, Oklahoma, Texas, and adjacent parts of Colorado and New Mexico. The dust blotted out the sun over Washington, D.C., and ships hundreds of miles out in the Atlantic reported dust sifting onto their decks. The eastern half of the country received dust fallout of about 100 tons per square mile. In 1935 winds blew dust almost continuously from March to May. The amount of dust in the air near Wichita, Kansas, on March 20 was estimated to be five million tons over an area of 30 square miles and within a mile of the ground. At Lincoln, Nebraska, fallout from the dust storm, which lasted three days, was calculated to be 1536 tons per square mile. The total amount of dust that was deposited during the two-month period was many times this, especially to the west and south of Lincoln. Within a few years these storms and others destroyed the multimillion dollar product of thousands of years of rock weathering and soil formation. Coupled with the economic depression of the time, the agricultural disaster ruined the fortunes, and even the lives of many people.

Large areas of the Middle East and the Sahara were not deserts in past historical times. Forests and rich croplands abounded in Lebanon and Palestine during Biblical times. Because man had for centuries over-planted, cut trees, and engaged in other wasteful practices, the vegetative cover was destroyed, the ground was deprived of mineral nutrients, and the land became a desert. The Sahara is known to have had a humid climate during the last glacial stage of the Pleistocene epoch; the evidence is the fossil plants and animals found there. Stone Age artifacts indicate that man was present in areas of the Sahara where no nomads venture today. Probably about 3000 to 1000 B.C. the climate began to change, becoming increasingly drier.

In Roman times the city of Timgad in North Africa was prominent, having a library, a forum, mosaic baths, and rich fields surrounding it. Nomadic tribes attacked Timgad, however, and it finally fell. After the city was abandoned, the fields lay untended, and the desert winds swept over the dry, loose soil, destroying the farmland and burying the town. Only the Arch of Trajan remained, protruding through the cover of dirt.

Wind Deposition

Wind deposition is a problem in many areas. The huge tonnages of wind-eroded soil obviously will be dumped *somewhere*. Wind is the great disseminator among the geological agents of transportation. Since wind is not restricted to channels, valleys, or any other limits against move-ment, it can spread its deposits over wide areas. During the Dust Bowl era sheets of dust covered valuable land and buried farms and roads. As a result of large volcanic eruptions, great volumes of volcanic ash have been distributed afar by winds and have caved in roofs, covered fields, and even buried towns.

Although wind is indiscriminate about where it drops much of its load of fine particles, it is quite selective about where it deposits the less readily transported larger particles. Particularly in coastal areas and deserts, *dunes*, which are piles of very fine to fine-grained sand that form in locally restricted areas, accumulate. The wind loses velocity because of irregularities of topography or vegetative cover, which form barriers with sheltered places behind them in which the wind deposits its load of sed-iment. The size of dunes varies from heights and widths of a few feet to two or three thousand-foot high masses in some Saharan deposits. The shape, size, and relative stability of the dunes are governed by the wind velocity, the supply of available sand, and the extent of vegetative cover.

9-9
Stable dunes in the Sand Hills of western Nebraska. (Photograph by N. H. Darton, U. S. Geological Survey)

In many cases, loose soil removed by the wind from fields is trapped along fence rows at the edges of the fields. Eddying caused by such obstacles forms quiet places in which the entering, blowing sediment builds up a pile of dust along the fence. In some coastal areas, migrating dunes create problems as they move over highways, cultivated fields, or resort property. Removing the sand from roads is easily done with graders, but keeping sand out of some other areas can be troublesome and expensive.

Wind-deposited sediment called *loess* occurs in many places, including Argentina, central Europe, the USSR, China, and the United States. The major United States deposits are located in the Missouri and Mississippi river basins (Fig. 9-12). Loess is fine-grained material composed mostly of silt, with smaller variable amounts of clay and sand. Mineralogically, loess consists primarily of quartz, feldspar, clay minerals, calcite, and minor amounts of various other minerals. The materials of loess are too coarse to allow the formation of packed, sticky clods of clay, but are too fine to allow the rapid drainage of water, as occurs in sand dunes. Because of its minerals, its loose, workable condition, and its moisture content, loess is a primary agricultural resource.

9–10
Large barchane-type dune in the Libyan desert. (Photograph by G. H. Goudarzi, U. S. Geological Survey)

9–11
Sand dunes destroying oak and cedar forest along coast of North Carolina. (Photograph by J. A. Holmes, U. S. Geological Survey)

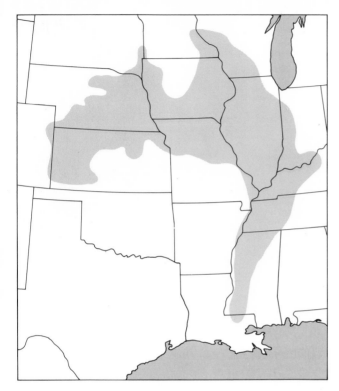

9-12
Loess deposits of the central United States.

Most loess is considered to be an eolian deposit, while much of it is thought to be derived from glacial-fluvial sediment formed during the Pleistocene Ice Age. Evidence also suggests that some loess may be of origins not related to glacial activity. The loesses of China, for example, may be derived from desert soils; those of Argentina, from volcanic products. Some of the loess in Kansas and Nebraska may have originated as a result of the weathering of Tertiary age sandstone and shale of the Ogallala Formation.

During the Pleistocene Epoch, meltwater from the continental ice sheets drained into the Missouri and the Mississippi river systems. These waters contained much sediment that had been carried in the ice. The outwash plains fronting the ice sheet and the river plains downstream contained countless braided channels which wove in and out as some became choked with sediment and others formed to bypass them. Since

9-13
Loess, overlying glacial till, exposed in road cut near Omaha, Nebraska. (Photograph by R. D. Miller, U. S. Geological Survey)

vegetation was insufficient to anchor the sediment on the plains, it was easily lifted and spread as a blanket over a wide area by the wind. The thickest deposits are near the major streams and are especially prevalent on the east sides of the valleys, where prevailing westerly winds dumped most of the sediment.

The importance of loess to agriculture and to the world food supply is not to be underestimated. Huge amounts of cereal grains and other crops are grown on loess-derived soils of the north-central United States, China, and Russia. In order to insure that loess will be preserved, it is necessary to invest wisely in proper farming and conservation methods.

Minimizing Wind Damage

Much of the world's future will depend on man's use of soil, vegetative cover, water, and plowing. A significant portion of the world's land is desert. Given the expected population growth and assuming that water will be available for irrigation, it will be necessary to cultivate at least

9-14
Farmstead and field windbreaks combined with wind strip cropping give protection from wind erosion. (Photograph courtesy USDA — Soil Conservation Service)

some of this land. Desert ecological balances will be destroyed, wild-life and plants will have difficulties surviving, with some species becoming extinct, and simple agricultural ecosystems will be established. By *ecosystem* we mean the total life of an area, together with its physical surroundings, including soil and water. Most ecosystems have a variety of life and maintain a number of checks and balances which insure the continuing stability of the system. Agricultural ecosystems, however, usually have one crop per field and include tillage of soil, burning of brush, application of herbicides and pesticides, and other practices that tend to exclude other organisms and simplify the system. The balance of such a situation is precarious because the variety of life and the associated checks and balances are not present in a one-crop ecosystem. Soil lies exposed between rows of the crop, and failure of the crop due to drought or a plague of insects can denude the ground. Wind and running water then remove the soil, and Nature's work of centuries is lost.

If land is to be stabilized and preserved, it is absolutely essential that wind and water erosion be reduced by decreasing the contact between loose soil and the wind and water. This can be accomplished in

9-15
Reclamation of desert wasteland by fixing of dune by vegetation. Trees will be planted next on the area. (Photograph by G. H. Goudarzi, U. S. Geological Survey)

many ways. Maintaining a good cover of grass and other vegetation helps to anchor the soil. Sand dunes along coasts can be stabilized by seeding them with various types of grass, by planting trees, and by constructing brushwood fences. Windbreaks and the more extensive shelterbelts, which consist of parallel rows of bushes and trees planted along edges of fields, along roads, or around farmsteads, serve to diminish the force of the wind near the ground. They are often established at right angles to prevailing winds.

Control of prairie and forest fires is essential to conserve vegetation and thus to protect the soil. Contour plowing, terracing of slopes, and maintaining a good vegetative cover not only retard water erosion by promoting increased infiltration, but also reduce the effects of wind as does any other means of conserving soil moisture. Irrigation is such a method and is especially important where fields are bare or have only thin row-crop cover.

STREAMS

Civilization has always been closely affected by streams. Primitive man was drawn to streams for his drinking water. Early civilizations developed in the Tigris, Euphrates, and Nile river valleys because of the availability of water and the presence of rich soil. Later, as civilization spread throughout the world, streams were the routes for explorers, who were followed by settlers. Today, stream valleys are among the most densely inhabited areas, and agriculture, industry, transportation, and recreation all vie for the resources and space of streams and their valleys.

A stream, its channel, and its valley comprise a highly complex system in which the hydrosphere, lithosphere, and atmosphere interact. Dynamic geological work which takes place in and around the civilization which has impressed itself on the stream system occurs here.

Erosion and Deposition

A stream's valley is formed by the force of running water, by dissolving, and by the grinding of sediment on the material over which the stream flows. The stream's energy determines its ability to do this and to carry away the debris. In turn, the energy possessed by a stream is greatly dependent on the velocity of the flowing water. Velocity increases with greater discharge and decreases with lessened discharge. In addition, velocity is proportional to the stream's *gradient*, or the steepness of the path of the stream, and is measured, for example, in feet per mile. From Sioux City, Iowa, to St. Louis, Missouri, the Missouri River has a gradient of slightly less than one foot per mile.

Sediment moved by streams is much more heterogeneous than that moved by wind. Fragmental sediment is moved both in suspension and by saltation. Particles may vary in size from the finest clay to large boulders, which may be washed down steep gullies in torrential rains. In addition to fragmental sediment, much dissolved material originating from the chemical weathering of rocks is borne by streams.

The dissolved material in streams is carried to the end of the stream, generally an ocean or lake. Fragmental sediment, however, is transported only if the velocity is sufficient to move it. The faster a stream flows, the larger the particles it can pick up and move. Silt can be transported by water which barely moves, but a velocity about ten times as fast is required to move coarse sand. Moving cobbles three inches wide requires a velocity of about ten feet per second (7 mph). Deposition occurs if the stream velocity drops below that needed to transport sediment. Clay can remain suspended almost indefinitely, but larger particles settle out more and more rapidly with increasing size. Deposition of chemical sediment is highly unusual and is limited to unique situations, e. g., deposition of carbonate or sulfate crusts in dry stream beds in arid regions.

The channel of a river is where the stream actually flows except during floods. Activity is greatest in the channel, not only because the water flows there, but also because sediment is continually affected by the flow, sometimes moving downstream, sometimes resting on the bottom.

The natural flow of a stream is not a perfectly straight path; instead, the stream winds. Generally, streams with steep gradients tend to be

9-16
Contrast between terraced farmland where erosion is minimized and unterraced land where cultivation is up- and down-hill and erosion is severe. (Photograph courtesy USDA — Soil Conservation Service)

9-17
Stream meanders near Rock Springs, Wyoming. (Photograph by J. R. Balsley, U. S. Geological Survey)

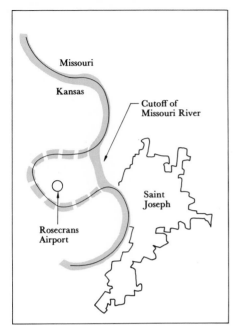

Missouri

Kansas

Cutoff of
Missouri River

Saint
Joseph

Rosecrans
Airport

9-18
Meanders and cutoff of Missouri River at St. Joseph, Missouri.

straighter than others, as most of their energy is spent cutting downward. Slower-moving streams with gentler gradients, however, are more easily deflected by irregularities in the channel. Ultimately, curves develop, some of them becoming broad, sweeping bends called *meanders* (Fig. 9-18). Although a stream may reach a stage where it lacks energy to cut further downward, it still may be capable of cutting laterally. Water flowing around a meander has its greatest velocity at the outside of the curve. Erosion continues there, and the meanders grow, both outward to widen the valley and toward each other to such an extent that they may be cut off in times of flood.

Cut-offs have resulted in various problems. Boundaries, for example, which had been established along rivers have been stranded along abandoned meanders as the rivers change course. A flood in 1952 cut off a meander of the Missouri River at St. Joseph, Missouri (Fig. 9-17). The St. Joseph airport, which had been built inside the meander, is now separated from the city by the cut-off, and it is necessary to cross the river into Kansas to reach the airport. The city of Omaha is more or less separated from its airport by the town of Carter Lake, Iowa, which was left on the Nebraska side of the river by a cut-off (Fig. 9-19). The lake, Carter Lake, is an *oxbow lake*, which is the name for lakes that occupy part of the meander abandoned by the stream. As might be expected,

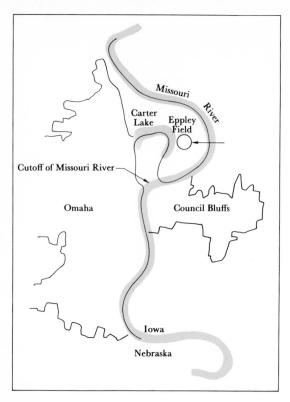

9–19
Carter Lake, an oxbow lake at Omaha, Nebraska.

9–20
Oxbow lakes and White River, Prairie County, Arkansas. (Photograph by J. R. Balsley, U. S. Geological Survey)

legal problems of residency, taxation, and ownership may arise if stream shifts are not accompanied by shifts in political boundaries. When political boundaries are established along streams, they must be carefully defined in order to avoid these problems.

As a stream carves out an increasingly broad path, a *flood plain* forms. Most of the time the stream flows within its channel. When there is a very high runoff, however, the discharge may be more than the channel can accommodate, and the stream overflows its banks onto the flood plain (Fig. 9-21). Since erosion has removed the high places and deposition of sediment has filled in the low areas, this flood plain has little relief. Obviously, such topography makes the construction of railroads, highways, airports, and cities easier.

The spread of flood waters over a flood plain is both destructive and beneficial. Water and sediment furnish organic and mineral nutrients to the flood plain, increasing the fertility of the soil. Agriculture therefore thrives, and in spite of the possibility of sudden and wholesale destruction of crops during the floods, it benefits by continued episodic flooding that renews the fertility of the soil. In addition, because ground water tables are usually nearer the surface in flood plains, water supplies from both the ground and the stream are available to agriculture.

9-21
Floodwaters of the Feather River at Nicolaus, California, with break in levee in the foreground. Flood of December 24, 1955. (Photograph by W. Hofman, U. S. Geological Survey)

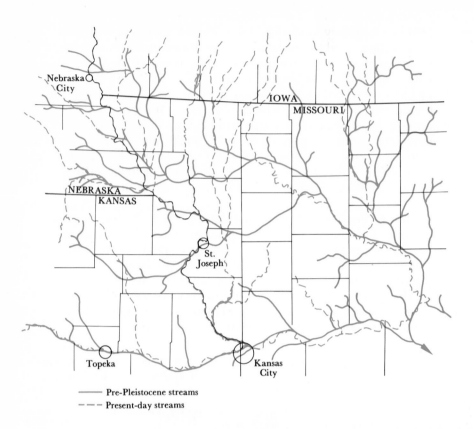

9-22
Pre-Pleistocene drainage system of northwest Missouri. (Adapted by permission from V. H. Dreeszen and R. R. Burchett, "Buried Valleys in the Lower Part of the Missouri River Basin," in Pleistocene Stratigraphy of the Missouri River Valley Along Kansas-Missouri Border, *Kansas Geological Survey, Special Publication 53, 1971, pp. 24–25.)*

Legend on figure:
———— Pre-Pleistocene streams
– – – – Present-day streams

Labels on figure: Nebraska City, IOWA, MISSOURI, NEBRASKA, KANSAS, St. Joseph, Topeka, Kansas City

   Channels generally contain coarser sediment than do the adjacent flood plains. The finer material is carried downstream or is deposited out on the flood plain as the flood waters lose velocity. Sand and gravel may form *bars* within the channel. These sandbars are dynamic, continually shifting position, size, or shape, depending on the discharge of the stream. The coarsest material remains and moves downstream only when high discharge furnishes enough energy to move it.
   Streams often change their courses rapidly. The abandoned channels then become buried under newer flood plain sediments. Saturated with water, the buried channel sands may furnish ground water supplies. During the Pleistocene Ice Age, the ancient drainage systems were almost com-

pletely rearranged by glaciation. The Missouri River, for instance, now flows southeast past Omaha to Kansas City, then turns eastward for the remainder of its course. The modern course of the Missouri was established primarily by the glaciation; prior to the glaciation the main river flowed eastward, considerably north of Kansas City. The buried channel sands of this river and its tributaries today furnish valuable supplies of ground water (Fig. 9-22).

Floods

Floods may possibly be the most costly of all natural phenomena in terms of loss of life, property, and land. Floods occur in almost all inhabited areas of the earth and range in size from short-lived spillage over banks of small creeks to huge inundations that kill many people and destroy millions of dollars' worth of property. Man has struggled with the problem of floods throughout history; although he has attempted to prevent or contain them, he has not yet fully solved the problem. In many cases, however, because he has cut forests and plowed grasslands, he has actually been the ultimate cause of floods. Stripping the land in these ways allows faster surface runoff and exacerbates flooding. Because of the variability in frequency and amount of rainfall, there will probably never be a way to stop flooding entirely. Yet it is still foolish to promote flooding, on the one hand, while battling it on the other.

  Precisely why a flood occurs at any given time and place depends on many things. Basically, a flood depends on the meteorological conditions and on the condition of the ground on which the precipitation falls or moves across. Flooding relates not only to the amount of rainfall, but also to the duration, intensity (amount per unit of time), and areal extent of the rain. Thus, if it falls in a very short time, — the"cloudburst" type of rain — almost any amount of rain can cause flooding. A lot of rain, if falling over a prolonged time, may not cause flooding if it can be absorbed or carried away efficiently.

  If the slope of the ground is not steep enough to allow efficient runoff, a flood may occur, but too steep a slope may also cause flooding. Obviously, the steepness of the slope required to result in a flood cannot be stated unequivocally because of the influence of other factors. If the ground is loose with open pores, as are sandy soils, infiltration is enhanced, but packed soil with clogged pores may lead to flooding. The pores may be inherently small, as in clay soils, or may become clogged if intense rain washes clay into them. Prolonged rain may saturate the soil to the extent that it is too waterlogged for further infiltration to occur.

  The presence of vegetative cover relates closely to floods. Forested areas, especially, are less prone to floods because trees intercept a huge amount of rain. Other vegetative cover also retards floods. The more ground covering there is, the better the flood control will be. In the event of prolonged or intense rain, plants may finally be unable to absorb any more water, thereby causing greater runoff.

An important aspect of floods is found in urban situations, where a significant amount of surface is covered by buildings and paved areas such as streets and parking lots. This reduces the ground surface that can be infiltrated by rainfall by a considerable amount, ranging from 15 to 20 percent in residential areas to nearly 100 percent in downtown business districts. However, there may be much bare ground in construction zones and areas under development. For example, in one area of Montgomery County, Maryland (suburban Washington, D. C.), storm runoff increased by 30 percent, and sediment yield increased 14 times because of urban construction on about 15 percent of a drainage basin which had previously been forest and pastureland. It was found that the sediment yield from active construction sites was *90 times* greater than that expected from the area under its original conditions of forest and pasture.

If the rainfall cannot soak in, it flows overland, and storm sewers and ditches are expected to carry away the unwanted water. Very often, however, debris clogs the entrances to storm sewers, so that they cannot function properly. The streets may fill with water, and low-lying areas may become inundated. The damage due to flooding in urban areas is compounded because of the density of housing and businesses.

The floods of January and February, 1969, in the Los Angeles area illustrate what can happen when urban sprawl interferes with the natural runoff in times of excessive rainfall. The normal geological situation used to be relatively stable. When heavy rains occurred, forested slopes in the mountains surrounding the coastal area held some water, while the remaining water flowed down canyons out of the mountains. When the streams decreased their velocity at the entrances to these canyons, the coarsest sediment would settle out, eventually forming an *alluvial fan* at those points. The remaining flood waters moved downward to the ocean and occasionally spread over the adjacent flood plains. Later, flood waters could be absorbed, to a large extent, in the porous deposits of the alluvial fan.

Now, Los Angeles and its satellites are situated on top of the entire coastal plain, including natural channels, flood plains, and alluvial fans. The rains in January and February, 1969, were heavy and intense, and the runoff water simply did not have the many available outlets it had had previously. The natural infiltration being greatly diminished and the natural routes to the ocean being constricted, the water had to flow overland through the city when the channels overflowed. Fortunately, manmade debris basins and reservoirs were able to catch much of the water and debris. Actual flood control had begun in the year 1915 with the organization of the Los Angeles County Flood Control District. By 1969 many reservoirs and debris basins had been built, over 350 miles of stream channel had been improved for efficient movement of flood water, and 950 miles of storm drains had been built. Yet in spite of the many controls which had been established, the city could not adequately handle the 1969 floods. The basic reason was that too many people had built too much on too wide an area, relative to the amount of flood control. Even the alluvial fans were covered with urban sprawl. ("Get out of the smog — homes with a view — buy now and save.") The total cost was 92 lives and 62 million dollars in January alone.*

One approach to the problem of urban flooding is *flood hazard mapping*. Chicago, Illinois, is one place where this has been done. There, the first flood we know of took place March 29, 1674, when the explorer Marquette was driven from camp by high water. Since then, various floods have occurred, and a complete set of hydrologic atlases has been prepared to show the flood hazards throughout the metropolitan area. These show areas likely to be flooded, drainage systems, flood-measuring stations, the heights to which water rose at different places in a number of historical floods, and the frequency expectation of floods of various magnitudes. The information is used as a guide to building and developing on flood plain areas, preventing hazardous uses and pollution from sources located on flood plains, and controlling and maintaining stream channels.

The anatomy of a flood: floods of the South Platte Basin, Colorado, 1965 A flood, actually a series of floods closely related to a single set of meteorological conditions, occurred on June 14 through June 17, 1965, in the basin of the South Platte River, located in northern Colorado. Since the character of floods varies greatly, there is no such thing as a "typical" flood, but the South Platte Basin flood was similar to many others. We shall examine this flood and its effects on man and will begin by emphasizing its profound effect on *people*. Eight persons died, and the hundreds who were driven from their homes suffered the immediate personal loss of homes, automobiles, crops, businesses, and other possessions. Thousands more were inconvenienced and affected by loss of income and other disruptions. Much of the tourist trade was cancelled. Every person in Colorado, in addition to many others, paid for the effects of the disaster in many ways. For example, the gasoline tax was raised by one cent to pay for damaged roads and bridges. The total cost of this disaster was one-half billion dollars.

---

*By comparison, the 1971 San Fernando earthquake killed 64 and cost about $500 million.

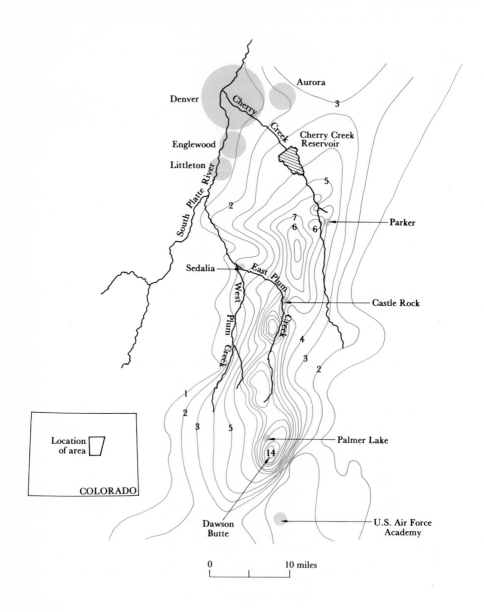

9–23
*Rainfall in the storms of June 16, 1965, at Plum Creek and Cherry Creek basins,
Colorado. Isohyets show inches of rain.*

One of the authors was entertaining friends at his home in Littleton, Colorado, on the evening of June 16. It began to rain, and during
dinner the storm intensified. The guests noted that they hoped the
weather would not get worse, because they had to drive to Boulder
that evening. Then, during the middle of the dessert course, the lights
went out. It was finally decided that the visitors should leave early so
that they would not be too inconvenienced in traveling.

Eighteen miles to the south, Dawson Butte had been inundated by
14 inches of rain in a period of about three hours late that same afternoon (Fig. 9–23). The entire area from a point five miles south of the
author's home south to Monument, Colorado, received more than five
inches of rain. East Plum Creek, along with others, exploded in a fury
of water, and water carrying logs, remnants of bridges and buildings,
and other debris surged northward into the South Platte River. When
it reached Littleton, the water picked up house trailers and headed for
the bridges downstream in Littleton, Englewood, and Denver.

As dessert had been served and the lights were out, the visitors left
and headed for the bridge past Centennial Race Track in Littleton. They
came to an abrupt halt: Centennial Race Track was under water, and
only the grandstand was above water. Using a circuitous route, they made
their way north through Denver and finally crossed the South Platte
ahead of the crest of the flood. Every bridge from Littleton southward
was gone.

Prior to the flood, rain had fallen periodically for about three
weeks. During the period of June 14 to 18, a large amount of moist air
had been blown by fairly strong winds toward Colorado from the Gulf of
Mexico. A large, low-pressure mass lay to the west, and Plum Creek was
located precisely in the area where very unstable air conditions were
created by the meeting of the low-pressure cell and the moist, rapidly
moving Gulf air.

On the afternoon of June 16, the unstable air erupted. Tornadoes
struck the area, one of them damaging 30 homes in the town of Palmer
Lake. Showers and thunderstorms followed. Because of the direction
of the upper-level winds, the thunderstorms remained just east of the
mountains, while the rain continued to fall in the same spot for about
an hour. Finally, the rains moved toward the northeast, although remaining in the Plum Creek and Cherry Creek drainage basins. Figure
9–23 shows the distribution of rainfall over the area on June 16, 1965.

Unstable atmospheric conditions continued through the next day,
June 17, which was cloudy, gray, and grim. Residents of the South
Platte Valley in Littleton, Englewood, and Denver were numbed by the
events of the previous night. The author was at home, since the electricity was off in his office in downtown Denver and the streets were

blocked or hazardous. Because the winds had shifted somewhat, the deluge, when it began again, took place to the southeast. This time, thunderstorms and showers dumped up to 12 inches of rain in a 20-mile wide band extending from Colorado Springs to the little settlement of Agate, 70 miles to the northeast. Kiowa and Bijou Creeks, dry washes much of the time, flooded severely. In contrast to the previous day, the path of the flood did not cross a metropolitan area of one million persons. An extensive network of roads, railroads, and bridges was destroyed, however, along with much private property.

That region of Colorado that had been destructively flooded had been settled for approximately 100 years. Stream-gauging stations had been in operation, and records of floods had been kept for various lengths of time up to a century. Nowhere, in all that time, had there been any indication of a flood of the magnitude of the 1965 flood.

The Multiple Use of Streams

According to an old cliche, "you can't have your cake and eat it too." Yet with respect to streams, we attempt to do that very thing. We would like to maintain our creeks and rivers in a pristine condition and to preserve forever all their beauty, clean and unpolluted water, and wildlife habitats. On the other hand, we need to use the water, to dispose of waste, to control erosion and floods, and to use the rivers for recreation and transportation. To attempt to combine these desires and needs is paradoxical: we must, but we cannot; we cannot, but we must. If there were so few people in the world that there was no competition for recreational space; if there were so few people that there was no need to maintain additional fertile land along flood plains; if there were so few people that their wastes could be entirely diluted before reaching the next settlement downstream, then the problem would resolve itself.

At the beginning of the nineteenth century, Lewis and Clark were sent by President Jefferson to see what he had bought from the French. They journeyed up the Missouri and reported extensively on one of the great rivers of the world. Since that time, the Missouri River has become the most important geographical feature in the extensive area reaching from St. Louis to the Rockies to the Canadian Border. The river has served as a water supply, as a transportation route, as a sewer, but occasionally has killed people and destroyed their establishments by flooding.

Disastrous floods which caused severe erosion, great damage to property, and loss of life were all too common along the Missouri, year after year. The Missouri is a big river and has had big problems. It was recognized that conditions had to be ameliorated, and beginning in 1912 a project was begun to develop a navigation channel from Kansas City to

9–24
Fort Randall Dam, South Dakota, a part of the Missouri Basin Project. (Photograph courtesy U. S. Department of the Army, Corps of Engineers)

St. Louis. Improvements were made subsequently, and in 1944 Congress passed the Flood Control Act, which provided a comprehensive plan for building dams to store water and generate electric power, for irrigation of new land, for flood protection, for river stabilization and navigation improvement, for recreation and wildlife propagation, and for the conservation of land, forests, and water. This plan was one of the most extensive ever conceived for the development of a river resource.

Presently, five major dams impound reservoirs in the upper two-thirds of the river, while the lower third is devoted to commercial transportation. The reservoirs intercept flood waters and have eliminated flooding in their portion of the river, while greatly reducing it downstream. Some of the stored water is used to irrigate about 3.5 million acres of land. Water is released in controlled amounts so that the downstream channel will be properly maintained for navigation. Recreation, including swimming, boating, and fishing, has developed into a major business on the reservoirs. A large quantity of hydroelectric power is

9-25
Stone-filled dike, used for channel control on the Missouri River. (Photograph courtesy U. S. Department of the Army, Corps of Engineers)

produced by the power installations at the dams. Their total capacity is 1,648,000 kilowatts, to be increased up to 3,000,000 kilowatts within a few years.

Yet, with the filling of the five reservoirs, other problems developed. People were displaced from their homes, some of them from places where they had lived all their lives; for many of them the psychological and economic shock was severe. In addition to the personal problems, there were those of a different nature. Water stored in the reservoirs is subject to significant evapotranspiration loss, especially in this arid to semiarid region. This results in great loads of sediment settling in the reservoirs. The Missouri and its tributaries contain much sediment, again because this system has flowed through dry country. Because of silting in, the life of the reservoirs is shortened to the extent that they will cease to be useful in a few decades.

Downstream, the Missouri has had its meandering channel stabilized. Pile dikes are built out into the river, causing sediment to deposit, form a bar, and then a low bank behind the pile dikes (Fig. 9-25). As accretion

9-26
The Missouri River at its function with the Big Tarkio River. The right bank has been sta-bilized, and silt is being deposited behind the dike system along the left bank. (Photo-graph courtesy U. S. Department of the Army, Corps of Engineers)

occurs in these areas, the river moves into a deeper, narrower channel be-yond the dikes. As the river stabilizes, stone dikes, which are more perma-nent, are built as a substitute for the pile dikes. Instead of a straight channel, a series of broad bends is constructed by proper placement of dikes along the river. Straightening the river would increase the gradient because the river would flow a shorter distance to drop the same height. This increased gradient would increase the velocity, bringing about accel-erated erosion. Because water flowing around bends is deepest and fast-est along the outside of the curve, which is where the erosion occurs, it is necessary to construct revetments, usually made of stone, to prevent the erosion.

   With the sediment load greatly decreased below the dams, and with channel control in the downstream portion, navigation is greatly enhanced. Commercial barge traffic has increased significantly, and it is estimated that there will be over five million tons of haulage annually along the Missouri River. Recreational facilities have been expanded, and much land formerly occupied by the sprawling river is now arable. Spawning grounds

9-27
Towboat on the Missouri River at Omaha. (Photograph courtesy U. S. Department of the Army, Corps of Engineers)

of some varieties of fish have been destroyed, reducing the numbers of those fish, but other fish that are more adaptable to the new regimen of the river have been introduced. Pollution is still a problem.

Aswan – a case study   One of the most interesting and important illustrations of man's interaction with a stream regimen is the construction of the Aswan High Dam and Lake Nasser in Egypt. The Nile River is the most vital artery of northeastern Africa. Since before the times of the pharaohs, it has nurtured man by serving as a transportation route, by furnishing water, and by spreading layers of fertile sediment over its flood plain. In its annual floods, the river replenished the fertility of the land by depositing a new layer of sediment. Farmers then brought water from the river to the fields and raised their crops. Over the course of millennia the Nile formed its delta with sediment that had not been dumped farther upstream. Nutrients which sustained schools of sardines and other fish flowed into the Mediterranean through the mouths of the Nile.

The High Dam at Aswan was built during the 1960s. Although a low dam already existed there, it did not affect the river regimen as profoundly as does the High Dam now. The High Dam impounded the Nile to form Lake Nasser, which is one of the largest lakes in the world and holds about 80 billion cubic meters of water (15 times the storage capacity of the lake behind the old dam). People who had lived in areas now flooded by Lake Nasser had to be relocated. The wildlife habitat was radically altered, and the balance of the fauna and flora along the lake was realigned. It is hoped that fish production from the lake will be greatly increased. While the lake holds a vast amount of water, large quantities of water are continually lost because of the high evaporation rate in this arid region. There has also been some question about possible leakage of lake water into permeable sandstone formations, but the Ministry of Power of the United Arab Republic suggests that no discernible problem has yet arisen.

Electric power produced at the High Dam equaled 7 billion kilowatt hours by 1971, thus helping to conserve the diesel oil previously used for power generation. Electrification of villages and new factories has commenced. Additional water and electricity have made the construction of facilities for the fishing industry and the manufacture of fertilizers feasible, with further construction still planned.

Although the annual floods of the Nile maintained the fertility of the soil, they could also be destructive. In some years droughts occurred, and sufficient water supplies for crops were not available. The presence of the High Dam has stopped the floods and has assured a continuous supply of irrigation water. Serious flooding was averted in 1964, and for the next four years droughts due to low floods were alleviated. Perennial irrigation has been applied to the increasing amounts of land which have been reclaimed. The value of the land has increased, and two or more crops, instead of one, can be grown each year.

Unquestionably, the High Dam at Aswan has produced great changes in the agriculture, fisheries, industry, and economy of Egypt. Many of these changes have been very beneficial. At the same time, however, we must consider the problems that have resulted from interfering with the Nile system. The evaporation and possible seepage from Lake Nasser result in the consumption of water that otherwise might have been used. Silt which is being deposited in the lake causes the storage capacity to be reduced. According to the Ministry of Power, the section of the reservoir designed for silt accumulation should last for at least 500 years.

Other changes occur downstream. The water that leaves the dam is mostly free of sediment. Besides the obvious fact that there is no longer an annual flood and consequently no silt deposit on the flood plain, there is also the problem of increased erosion. Until the High Dam was constructed, the Nile had been essentially a graded stream, with the exception of brief periods of very high or very low water. Without its sediment load, however, the Nile is no longer in equilibrium and must therefore adjust itself in order to become a graded stream once again. In doing so, it will erode its channel and banks. In addition to the increased erosion along the banks of the Nile, there is increased erosion due to wave action in certain parts of the delta. Because of the reduced supply of sediment, there is not enough sediment to replace that which is eroded by the waves along some parts of the shoreline.

Without the continual addition of nutrient matter to the soil, the ever-decreasing fertility along the Nile flood plain will become a problem. Farmers will be forced to apply manufactured fertilizer to the land to compensate for the loss of natural nutrients, especially if more than one crop is to be raised every year. The reduction of nutrients flowing into the sea will create an additional problem. The eastern Mediterranean sardine industry is nearly defunct because there is no food for the sardine. Marine scientists are now attempting to track down the sardine population, assuming it has moved elsewhere and not died out.

9–28
The Athabaska Glacier, outlet glacier of the Columbia Ice Fields, Alberta. (Photograph by F. O. Jones, U. S. Geological Survey)

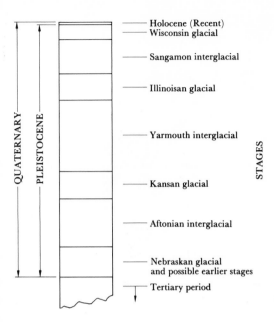

Holocene (Recent)
Wisconsin glacial

Sangamon interglacial

Illinoisan glacial

Yarmouth interglacial

Kansan glacial

Aftonian interglacial

Nebraskan glacial
and possible earlier stages

Tertiary period

9-29
*The Quaternary period and its subdivisions in
North America.*

GLACIERS

Although having less of a direct effect on man than wind or streams,
glaciers are, nevertheless, of considerable environmental importance because of their past activity. At various times during the earth's long
history, sheets of glacial ice have spread over large expanses of the earth's
surface. Why this happened is still not known, although many theories
attempting to explain it have been advanced. The glaciation that affected
us the most occurred during the past two to three million years — essentially during the time known as the Pleistocene Epoch. The Quaternary
Period is usually divided into the Pleistocene and Holocene epochs
(Fig. 9-29); the Pleistocene is usually considered to include all of the
Quaternary up to about 10,000 years ago, the Holocene thereby being
the shortest geological time of all.

It has frequently been said that the Quaternary Period was atypical
of the earth's geological and climatological history — with the continents
standing high above sea level, with few epicontinental seas, and with wide
variations in climate including severe glacial conditions. As a corollary to
this observation, it has also been suggested that the evolution of man, and

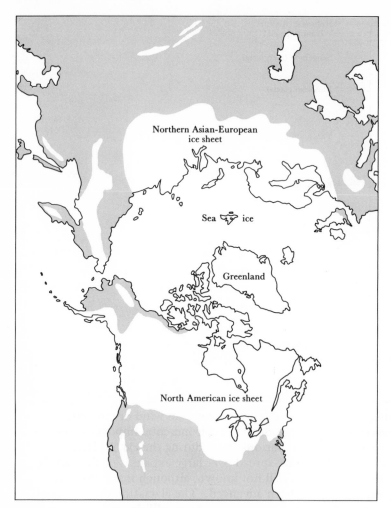

9–30
*Extent of ice advance during Pleistocene glaciation in the Northern
Hemisphere.*

even of civilization, has been fostered by the unique environment of the
Pleistocene. The truth of this statement is debatable. Certainly we know
that some of man's close ancestors predate the Pleistocene. Work in east
Africa provides evidence that man evolved as a recognizable genus during
the Pleistocene in an area having suitably temperate conditions. Undoubtedly, his evolution would have proceeded without interruption, in spite
of the Ice Age. Thus, man was not born of the glacial epoch; he has, how-

ever, been known to have lived in periglacial regions during that time. The specific course civilization took, for hundreds of millions of people, has been influenced to a great extent by the climate, landforms, soils, hydrology, and life of the glacial epoch.

There were four distinct advances of the ice sheets over North America during the Pleistocene. Figure 9-30 shows the maximum extent of the ice. The duration of these advances totaled much less than half of the Pleistocene. Warm climates prevailed between individual advances and since the last advance.

There were a number of centers of low temperatures, precipitation, and topography that favored accumulation of snowfall. As snow piled up, year by year, faster than it could all melt, the layers became packed and formed ice. Ice, although a crystalline solid, has weakly bonded molecules; therefore, it was easy for the ice to flow laterally as the weight of layers totaling a thickness of perhaps 15,000 feet caused the bonds to break, reform, and break again. The ice flowed out from the centers of accumulation, coalesced, and spread as a continental icecap over northern North America. Simultaneously, glaciers formed at high elevations in the mountains from Alaska to New Mexico and flowed downward, carving out much of the mountain scenery visible today.

The people of North America, Europe, and Asia have been left a surprisingly important legacy by the glaciers of the Ice Age. Because the glaciers retreated so recently, the land bears very fresh imprints of glacial activity. Many of the characteristics of the hydrosphere on the continents were established by glaciation. Many soils are of glacial derivation. Man himself most probably entered the New World from Asia, crossing over dry land where the Bering Straits now lie. Climatic changes resulted in great transformations of fauna and flora. For various unknown, or dimly known, reasons many species became extinct, so that present populations are diminished in number and variety, compared to previous geologic time.

Glacial Deposits

Various types of deposits were laid down by the glaciers, by streams that issued from the glaciers, and in lakes. Some of the older deposits exhibit a relatively high degree of weathering and have formed mature soils. Others that have been deposited more recently, particularly those of the latest glacial stage (Wisconsin), are considerably fresher. Since it takes about 50,000 years for a soil to develop to maturity, there has been little chance for soil formation to progress in these latest deposits.

The unconsolidated debris deposited by glaciers is called *till* and consists of an unsorted mixture of clay, silt, sand, and other fragments

9–31
Channel of Cranberry River in glacial till, showing bouldery bed material of glacial debris, Ontonagon County, Michigan. (Photograph by J. J. Hack, U. S. Geological Survey)

which can reach boulder size. Till may be composed of almost any mineral or rock materials, depending on where the glacier picked up its load. Till, deposited as a sheet under the moving glacier or in piles as the glacier melts, has been deposited in thicknesses of a few feet over much of the glaciated area of the United States. There are accumulations of 200 to 300 feet in some places. Some deposits of sand, silt, and clay

9-32
*Map of part of the western United States, showing
known or inferred locations of Pleistocene lakes.
(Adapted from J. H. Feth,* Short Papers in Geologic and Hydrologic Sciences, *U. S. Geological Survey Professional Paper 424–B, 1961.)*

were laid down by meltwater streams issuing from the edges of icecaps
or were left in the lakes associated with Pleistocene glaciation. In contrast to till, which is nonstratified, these deposits are stratified and, in
most instances, better sorted. They form stream beds and channel sandbars, flood plains, outwash plains, and lake beds.

Large lakes occupied basins in many areas during the Pleistocene
(Fig. 9-32). The climate was more humid, and a large supply of water
was available. Later, when conditions became arid, most of the lakes
dried up or shrank to the size of their present remnants, which include
the Great Lakes, Lake Winnipeg, Lake Manitoba, and Great Salt Lake.
Richly fertile soils developed from many of the sediments deposited
in the lake basins and today constitute rich croplands. Eastern North
Dakota, for example, possesses rich plains consisting of lake deposits
dating from the Pleistocene.

9-33
Morainal topography in Michigan, showing irregular, hilly character and rocky soil. (Photograph courtesy USDA — Soil Conservation Service)

Glacial Landforms

Geomorphologists recognize a wide variety of landforms caused by glacial activity. Some of these are erosional, some are depositional. Much of the spectacular mountain scenery of the Alps and Rockies, especially the large glacial valleys which are typically U-shaped in cross-section and the sharply pointed peaks ("horns") and jagged ridges, is the result of glacial erosion. Many lakes occupy basins eroded by glaciers, and many waterfalls occur where streams in small tributary glacial valleys ("hanging valleys") plunge into larger, deeper main valleys.

Other glacial landforms are depositional; for example, some important landforms consist of till. These are classified on the basis of shape and origin, chief among which are the depositional features known as *moraines. Ground moraines* are bodies spread out as sheets beneath the icecaps; these may cover extensive areas. *End moraines,* or *terminal moraines*, are features formed of till heaped at the end of a valley glacier or an icecap. As the glaciers receded spasmodically, the pauses which occurred during the retreat led to the creation of a series of *recessional moraines.* Many moraines cover large areas of the Middle West of the United States. The unconsolidated, gravelly till found in them very often houses an important supply of ground water.

Cause of the Ice Age

We have already admitted that we do not know the cause of the large-scale glaciations that have taken place at various times in the earth's history. In one sense, investigating the cause is an academic exercise; on the other hand, some hypotheses describe changes in the atmosphere which may relate to modern atmospheric changes being made by man. There are literally dozens of hypotheses for the cause of the Ice Age. Many of them have serious flaws, many rely on unique happenings on the earth or in the solar system, and many require special physical or chemical conditions. As yet, we lack the knowledge to prove *any* hypothesis.

The various theories can be grouped into (1) astronomical and (2) terrestrial hypotheses. The first group includes those which involve changes in the relationship of the earth to the sun or other bodies. For example, it has been variously suggested that glaciation has been brought about by : (1) a decrease in radiation given off by the sun, (2) an increase in distance between the earth and sun resulting in colder temperatures on earth, and (3) a decrease in the tilt of the earth's axis, causing a decrease in the sun's energy received at higher latitudes in the winter. It has also been suggested that the earth, traveling through the galaxy with the rest of the solar system, has passed through concentrations of dust. The dust would provide extra nuclei around which drops of precipitation (snow) could form.

The second group of hypotheses include those which refer to changes taking place as the result of processes occurring on earth. These involve possible changes in the land masses, in the oceans, in the atmosphere, or in combinations of these. For example, a time of increased volcanism might have added dust to the air which served as condensation nuclei. Uplift of a land mass could have diverted or blocked oceanic currents, thereby changing the climate, especially if polar areas were affected. Conversely, the submergence or splitting of a land mass could have changed the oceanic circulation. It has been suggested that polar wandering and continental drift operated in such a way that the continents and Arctic Ocean were placed in precisely the right positions for the intricate mechanisms of evaporation, condensation, and precipitation to cooperate in a sequence which produced an icecap over places now far removed from the poles.

Hypotheses which indicate the atmosphere as the cause of glaciation mostly involve changes in the amount of carbon dioxide (CO_2), water vapor, or dust present. Because CO_2 absorbs long heat waves given off by the earth, an increase of atmospheric CO_2 might cause a warming of the earth's surface. Increased evaporation might then occur, adding moisture to the air which, in high latitudes, would fall as snow. On the other hand, the increased moisture content could produce an increased cloud cover, and the clouds could cause the solar energy to be reflected back into space. Eventually, therefore, the earth might cool due to the lack of solar energy, and because melting of the snow would be slowed, a glacial age would ensue.

As yet, the available data are insufficient to allow predictions based on a change of the variables of the system to be made. This explains the reasons for the recent great debate about the consequences of moisture which would have been added to the atmosphere by the proposed supersonic air transport (SST). We are at least aware of the fact that the lithosphere, hydrosphere, and atmosphere comprise a highly complex system that includes a great number of physical and chemical variables. Taking the time dimension into account, we would clearly realize that the reason it is so difficult to understand the causes of the Ice Age is that we are so removed from the time in which these events occurred.

Why dwell on this? This becomes understandable if one considers the following: first, the earth is dynamic, not static. Who is to say that we have seen the last of the Ice Ages, when we do not even know what caused them? Was the Wisconsin stage the last, or will there also be a Minnesotan or a Kentuckian stage? Second, the onset of the stages of the Pleistocene occurred rapidly in geological terms, as well as in terms of man's history. Civilization could be affected drastically if another glacial stage were to occur. Third, if adding carbon dioxide to the atmosphere is indeed a cause of glaciation, we might be capable of bringing about another Ice Age.

We know that since the beginning of the industrial age we have added roughly another 10 percent to the carbon dioxide content of the atmosphere. This has been entirely a result of the burning of fossil fuels. If you note the graphs (Chapter 8, Fig. 8–20) which indicate that we have, so far, burned less than one-tenth of the world's coal and petroleum, it should be obvious that the continued burning of fossil fuels will contribute an enormous amount of carbon dioxide to our atmosphere.

What might happen if the ice in Greenland and Antarctica should melt or even begin to increase? If all the ice in the icecaps were to melt, enough water would be added to the ocean to raise the sea level about 200 feet. Consider the results: coastal areas would be inundated, places such as Holland and Florida would be almost entirely under water, and all of the world's seaports would have to be abandoned. In the United States, many of the major cities would be included: New York, Boston, Philadelphia, Baltimore, Washington, Miami, New Orleans, Los Angeles, San Francisco, San Diego, Seattle, and hundreds of others would be completely, or almost completely, under water. Rich and important agricultural lands would disappear, and hordes of the world's populace would jam onto the shrunken continents. The ecology of fisheries and wildlife habitats would alter radically as the coastline configurations were changed.

On the other hand, what if the icecaps grew and another glacial age occurred? Since water would then be withdrawn from the oceans and incorporated into icecaps, the sea level would be lowered. By mapping

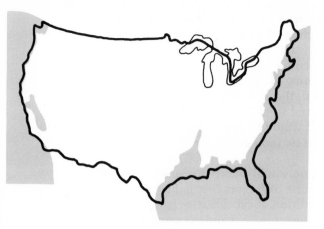

(a) # IF ALL THE ICE MELTED

Scientists estimate the sea level of the
earth's oceans would rise 200 to 250
feet if all the ice in the world melted.
The map above shows how a rise of
250 feet would change the coastlines
of the continental United States.

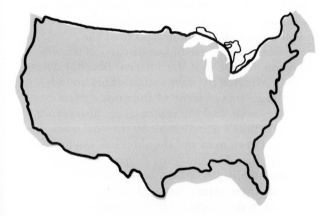

(b) # OR THE EARTH COOLED

During the Ice Age, a heat loss of 3 to
4 degrees in the earth's temperature
created gigantic glaciers, which lowered the sea level 300 to 400 feet. As a
result, lands now under the oceans were
exposed. Over 2.5 million square miles
of land were added to the coastlines.

9–34
Possible effects of: (a) melting; or (b) readvancing of the icecaps. (Source: River of Life,
vol. 6, Conservation Yearbook Series, U. S. Department of the Interior, 1970.)

features that form above water and which are now submerged, we know that the sea level was as much as 300 feet lower during the Pleistocene glacial stages than it is today. Large areas of the continental shelves — those portions of the continents which lie under the sea — were exposed at that time. Were this lowering of sea level to occur again, our seaports would be located on dry land, the plant and animal ecology would be greatly changed, and more land would be available for living space or other use. Whether this land would be suitable for agriculture is another problem, as conditions might be too cold or too dry to permit cultivation of the land.

In either case, the climate would change because of the advance or retreat of the icecaps or because of factors that induce such an advance or retreat (Fig. 9–34). In addition, local climatic changes would occur here and there because of changes in land and sea geography and the shifting of ocean currents. All of these changes would involve temperature and precipitation changes, with concomitant changes in water supply, wildlife, agriculture, and the lifestyle of man.

SHORELINE PROCESSES

We have already noted that much geological activity takes place at the boundaries between the great spheres of the earth — the lithosphere, hydrosphere, and atmosphere. We will now examine some aspects of the boundary common to all three. For the purpose of our discussion we will call this boundary the *shoreline* and include in it the adjacent beach, tidal flats, river deltas, and the sea floor affected by wave action. This boundary where the three spheres meet is an area of some of the most dynamic geological activity anywhere on the earth, and the seacoasts are also scenes of bustling human activity. Because man's activities along the coasts are many and varied, the interaction between man and the shoreline environment is intense. Many of these activities demand some degree of geological stability; efforts to achieve this stability may be nominal or highly complex and expensive. Sometimes these efforts succeed, sometimes they fail, and sometimes they result in detrimental effects elsewhere along the shore. The usual reason for problems that occur is lack of understanding that the shoreline environment obeys certain rules within its own dynamic system. Clearly, man's works along a coast must function as a part of this system. Any control, alteration, or use of the shoreline zone must take place within the natural rules of the system.

The processes along the shoreline are the same as elsewhere: erosion and deposition, with varying intervals of relative and temporary stability between times of active erosion or deposition. Waves and winds are the agents which carry out these processes. We have previously described

9-35
Marine terrace near Arecibo, Puerto Rico, formed by planing off of rock by wave action. (Photograph by C. A. Kaye, U. S. Geological Survey)

the action of the wind in moving sediment and noted that it is the driving force behind water waves, whose action we will now describe.

As winds blow across the surface of a body of water, the frictional drag creates a wave in the water. The energy that is transferred from the moving air to the water moves through the water in the form of a wave. Water waves may also be caused by seismic (earthquake) disturbances. These waves, called *tsunamis*, may be of great importance and will be described in detail in Chapter 10. Still other water waves are generated by tides, but the importance of true tidal waves is negligible.

The height, distance between, and frequency of waves depend on a number of factors. (For a thorough understanding of the complexities of wave generation and motion, you are referred to texts on oceanography. Here, of course, we are most interested in the effects of waves on the shoreline in terms of erosion and deposition.) As waves approach the shore, they enter water shallower than that in which they originated. Because of the effect of drag on the sea floor, the waves are transformed into local *currents* and *surges*. These movements of water cause much of the erosion and movement of the rock material on the shore.

Erosion along Shorelines

Probably the most important way in which waves erode is by hurling and churning rock fragments about in the surf. This abrasive action (*milling*) wears away the materials of the shore by chipping and grinding. As finer and finer fragments are produced, they are carried away; the coarser ones remain to continue the milling. The energy available to erode greatly increases with the height of the wave (kinetic energy is proportional to the square of wave height). Therefore, a relatively large share of erosional damage takes place during storms.

In addition to the milling action of rock particles in turbulent water, there are other ways of eroding shorelines. One way is through the pressure created by the weight of moving water (*hydraulic action*), which in storms may be thousands of pounds per square foot. Such pressures, themselves, may be sufficient to break rock; an even more important effect, though, is that this pressure can move very large rocks and increase the milling action. Another way in which water may erode the shore is by *solution*. The chemical breakup of rock may be relatively pronounced, depending on the composition of the rock and the surface exposed.

All degrees of erosion occur along shorelines. The factors that regulate erosion include the composition of shoreline rocks, the degree of induration of the rocks (jointing, cementation, and so forth), humidity and rainfall, strength and direction of waves, frequency and duration of storms, and topographic influences (elevation, exposure to open sea, bottom configuration, etc.).

Longshore Currents and the Movement of Beach Sand

In most instances, whatever erosion or deposition is taking place at any one locality on the shoreline is influenced by events taking place further along the shore, especially those in the direction from which the currents come. In many cases, the prevailing winds and the topographic configuration of the coast constrain the waters near the shore to move obliquely toward the shore. Waves move the water toward the shore *en masse*, and the only direction in which it can reach the shore is downshore. The waves almost always wash up on the beach at an acute angle, but their return flow is directly perpendicular to the beach. Subsequent waves carry the water onto the beach at distances farther along the beach. Sediment particles are transported by such *longshore currents* in a direction generally parallel to the shoreline, causing some movements of beach sand to cover many miles. Thus, what might appear to be stability to a casual observer is in reality part of a dynamic system.

There is considerable variation in the amount of sediment supplied to the shoreline at various spots. The Pacific and Gulf coasts of the United States receive new sediment in fairly large amounts. The Mississippi River basin drains a third of the country, and the sediment supplied to the Mississippi Delta forms a major portion of the total sediment deposited along the United States coastlines. Rivers that empty into the

9-36

Waves approaching shore obliquely at Atlantic City, N. J., setting up southward longshore drift of beach sand. (Photograph courtesy of National Ocean Survey — NOAA)

Atlantic, on the other hand, contribute a relatively insignificant amount of sediment to the entire shore from the tip of Florida to Long Island, New York.

Destruction of the natural dunes and beaches caused by man's construction may result in permanent damage to the shoreline. The conditions of dynamic stability created by geological processes are altered by the building of resorts, industries, and other facilities. Erosion may be speeded up, and in the case of shorelines such as the Atlantic coast with its dense population and minor supply of sediment, the loss of beaches and dunes will become permanent.

Measures taken to stabilize the shoreline or to protect shore facilities from damaging waves usually create problems at the same time that they solve others; at best they are partially effective. Structures built along shores include, for example, groins and breakwaters. Groins are long structures built of timber, rock, steel, or concrete that project out into the water at right angles to the shore. Sometimes just one may be built along an entire stretch of beach, while at other times they are built in a series. Their purpose is to act as a barrier to longshore currents in order to break the longshore drift of sediment.

Breakwaters are built outside many harbors to diminish the force of storm waves or to interrupt prevailing longshore currents which may be eroding or depositing sand in the harbor. Just what may happen when this is done can be seen at Santa Barbara, California, where a breakwater, detached from the land, was built offshore. Interruption of the prevailing waves decreased the energy behind the breakwater. When the waves no longer had sufficient energy to move sand down the coast, the sand was dumped behind the breakwater in the harbor area. In just one year the harbor had filled with sand to the extent that it had to be partly closed off by building a connection from the breakwater to the shore. After this was done, sediment accumulated upcurrent, leaving a deficit downcurrent. It next became necessary to pump sand from where it had accumulated to the other side of the harbor. Thus, where man blocked the natural flow of sand, he had to assist Nature in keeping his harbor clear and the sand well distributed.

This situation also occurs at Santa Monica, California. Deposition of sand has occurred in the low-energy zone behind a detached offshore breakwater. The beach there is very wide, while the beach downcurrent is very narrow as a result of the erosion which has occurred due to the movement of high-energy waves around the end of the breakwater.

Deltas and Estuaries

Deltas and estuaries are some of the most delicate and ephemeral features to be found along shorelines. Deltas are extensions of the land and are composed of sediment deposited by streams as they enter the ocean or lakes. They are low-lying and readily subject to attack by waves. They are also fertile and constitute the natural habitat of a great number and

9-37
Delta of Nelchina River into Tazlina Lake, Alaska. (Photograph by J. R. Williams, U. S. Geological Survey)

variety of living things. When man moves in to cultivate the deltas, his activities actually promote their destruction.

The Mississippi River delta is a large, complex system that is actually a group of merged deltas. The coastline has changed significantly even in modern times, partly because it has been affected by man. Because levees for channel and flood control have been built, the river now must carry its load of sediment farther out into the Gulf. As the sediment is no longer spread out as a sheet over areas adjacent to the river, in times of flood no new land is formed, and no new deposits are available to replace the land which has been eroded. As a result, delta land is being lost to the sea.

The entire Mississippi delta area is slowly tilting Gulfward. The St. Bernard portion of the delta, east of New Orleans, has sunk since the flow of the Mississippi shifted to the Plaquemines delta to the southeast. Indian artifacts and mounds are found underwater there, indicating that submergence took place within the past few hundred years.

Erosion of the Nile delta has increased considerably since the construction of the Aswan High Dam. The Nile no longer leaves Aswan carrying a load of sediment as it did for centuries before the dam was built. There is not enough sediment to replace the deltaic material eroded by the wave action of the Mediterranean Sea. While we do not yet know the accurate rate of erosion, we are aware that much valuable land is in jeopardy. A glance at the dark area of the delta (Fig. 9–38) contrasted with the lighter tone of the adjacent desert vividly emphasizes the importance of the delta as an agricultural region.

D. W. Pritchard of the Chesapeake Bay Institute has defined an estuary as "a semi-enclosed coastal body of water which has a free connection with the open sea and within which sea water is measurably diluted with fresh water derived from land drainage" (Pritchard, 1967). Estuaries include glacial fiords, submerged river valleys, and down-faulted blocks. Chesapeake Bay, San Francisco Bay, and the estuaries of the Seine and Thames rivers are among the better-known estuaries. The environmental character of the estuary is complex. Typically, estuaries have a rich diversity of life and physical characteristics. In addition, they are areas of rapid change, with the rates of natural processes such as sedimentation being more rapid than in almost any other enironment. Man has concentrated much of his population and activity in estuaries; because estuaries are highly sensitive environments, he can effect rapid and drastic changes in their condition.

The alteration of river flows can be achieved by removing vegetation cover through cultivation or logging, by constructing dams and reservoirs, and by straightening channels. When the amount of flow reaching an estuary changes, there are alterations in salinity, nutrient supply, sedimentation, water circulation, in oxygen content of the water, and ultimately in the very life of the estuary, as well as in its navigational and recreational potentials. Man can not only alter the flow of entering rivers but can make other changes as well. Pollution from sewage, chemicals, and heat are critical problems that affect the living organisms and water supply of estuaries.

The filling that has been done around the margins of San Francisco Bay has reduced the area of the Bay from over 600 square miles to 435 square miles. This reduction has lessened the amount of water carried in and out of the Golden Gate by the tides, and there consequently is less water available to flush out pollutants. An additional problem is that the filled

9-38
The Nile Delta (dark area) as seen from the Gemini IV space craft. (Photograph courtesy NASA)

land is composed of unconsolidated materials which can be highly dangerous in earthquakes (see Chapter 10).

L. E. Cronin has pointed out (1967) that seven of the ten largest metropolitan areas in the world (New York, Los Angeles, London, Tokyo, Osaka, Buenos Aires, and Shanghai) are situated next to estuaries and that, in the United States alone, one-third of the population lives near them. It is evident from these observations that man has had qualitative as well as quantitative effects on estuaries.

Man and the Coast — The Netherlands

An outstanding example of the drastic alteration of the estuarine environment is found in The Netherlands. Throughout their history, the Dutch have lived precariously at the triple interface of land, air, and sea, but have become wise and adept at achieving a harmony with their environment. In order to both protect themselves from the sea and increase the extent and capacity of their land, the Dutch have undertaken two mammoth projects: the conversion of the Zuider Zee from an arm of the sea partly to a land area and partly to a fresh water body, and the construction of a system of dams, flood and storm gates, and other waterway controls in the estuaries of the southwest part of the country where a number of Europe's large rivers enter the sea.

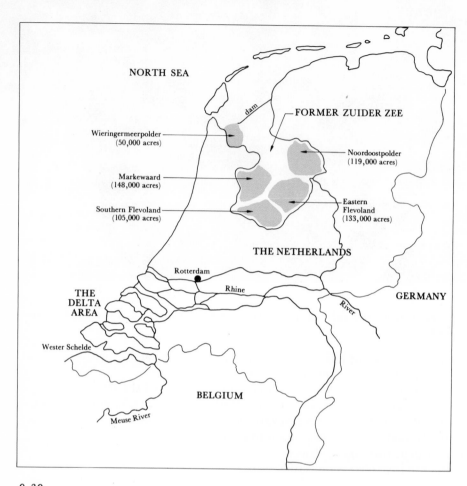

9–39
Map of The Netherlands, showing location of the Delta and Zuider Zee projects.

The Zuider Zee (Fig. 9–39) was essentially an estuary, until a dike was built across its mouth in 1932. With only fresh water flowing into it, the body gradually changed into a fresh-water lake. The Dutch are gradually converting the shallow lake into land by building successive dikes around parts of it and then pumping out the water. In this way, about two-thirds of the original 2700 sq km of estuary have been converted to new land.

The Delta Plan is concerned primarily with flood control rather than with making new land. North Sea storms have caused trouble for centuries. In February 1953, an unusually severe storm broke the existing dikes and allowed the sea to flood about 375,000 acres. Eighteen hundred persons died, and agriculture was crippled by the spread of silt and salt over the flooded land. Although work had already begun in the late 1940s, the 1953 storm gave impetus to the plan. Dams have been built across the mouths of all but the Wester Schelde (Fig. 9–39). Auxiliary bridges, locks, canals, dikes, and flood gates have been built as part of the highly complex system. Some of the basins will be segregated from the sea and will eventually contain tideless fresh-water lakes, while the northernmost basins will still be connected to the sea.

As is the case with any large-scale tampering with a dynamic natural system, there are mixed results in the Zuider Zee and Delta projects. Favorable effects are that the destructive floods will be stopped, lives and property will be preserved, and new land will be added to the tiny, crowded nation. The problem of sea salt encroaching landward into valuable farm land will be solved as the fresh water collecting in the basins flushes back the salt. Certain types of food production will increase. Finally, there will be increased knowledge of the biological and hydrological conditions of estuaries and of their response to change. On the other hand, the unfavorable effects include wholesale destruction of the fauna and flora of the estuaries. The mussel and oyster industry and fishing will be greatly altered, and in the case of certain species will be wiped out altogether. Erosion and deposition may not be a serious problem in the Dutch projects but can be in other estuary projects. The cost of the projects is enormous, especially for a small, albeit wealthy, nation. The Dutch, however, believe the results to be worth the cost.

MASS MOVEMENTS OF ROCK

Surface materials of the earth are frequently subject to downward movement *en masse*. Destabilizing influences, such as water, earthquakes, or the digging away of supporting material by man, act to weaken rock and soil masses to the point where they slide, slump, flow, or fall. In many cases, this downward mass movement leads to the acceleration of erosion and to the destruction of lives and property. The type of construction and its positioning on slopes subject to mass movement are therefore of considerable importance. For example, F. Beach Leighton suggests that by 1990, as many as 25,000 homes near Los Angeles, California, may be located in hills prone to landslides.

Classifying mass movements according to type helps us to understand them, as each type of movement presents its own peculiar problems. Very slow, imperceptible movement of loose material is known

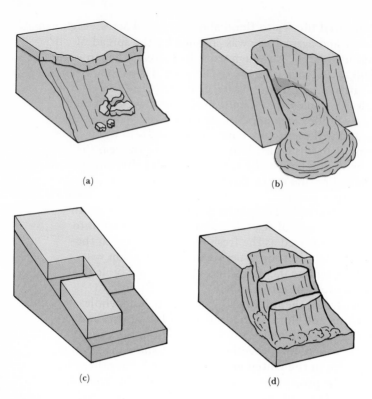

9-40
Types of rapid, mass earth movements: (a) rockfall; (b) debris slide, mudflow; (c) rock slide; (d) slump blocks.

as *creep*. Hillsides may bulge downward as this occurs, causing fence lines to bend or telephone poles to tilt after a few months or even years, but the net effects of creep are not violent or greatly destructive.

Other movements are more rapid than creep. These are known by a variety of names, depending on the rapidity of movement and the nature of the material being moved. Thus, we have slumps, mudflows, landslides, rock falls, debris avalanches, and others (Fig. 9-40).

Most mass movements are caused by certain existing conditions in conjunction with some event that, because of the conditions, can easily start the rock mass in motion. Steep slopes are more responsive to gravity than are less steep ones, level land, obviously, not being subject to downhill movement. Another factor is the nature of the rock itself. Rocks composed of clay and certain other minerals are less "competent" than others; in other words, they are more likely to slide. The major

9-41
Earthflow developing in weak, unconsolidated material. (Photograph by K. D. Gilbert, U. S. Geological Survey)

9-42
Creep in vertically bedded shale, Washington County, Maryland. (Photograph by G. W. Stose, U. S. Geological Survey)

9-43
Large landslide which occurred in southern California, blocking the highway for a number of hours until it was cleared. (Photograph courtesy of the Department of County Engineer, County of Los Angeles)

incompetent rocks are shale, siltstone, mudstone (sedimentary), tuff and basaltic volcanic rocks (igneous), and schist and serpentinite (metamorphic). The presence of water, especially in clays, reduces the cohesion of rocks. In addition to wetting the clay, water provides increased buoyancy. It may also freeze and thaw cyclically, breaking up the rock (see *weathering*, Chapter 4). Both the buoyancy effect and the freeze-thaw cycle lessen the resistance of the rock mass to downward movement. Another important condition is the structure of the rock mass. The presence, number, and direction of faults or joints in the rock affect the propensity of the mass to move.

What event could trigger the movement of a rock body whose composition, water content, position, and structure make it a ready

9-44
Rock slide along rock cleavage planes into roadway in North Carolina. (Photograph by J. C. Reed, Jr., U. S. Geological Survey)

9-45
Aerial view of Madison Canyon slide and Earthquake Lake, Montana, 1959. (Photograph by J. R. Stacy, U. S. Geological Survey)

subject for sliding? It is evident that the shaking produced by an earthquake might do it. This has indeed happened many times, and an examination of many landslides shows them to have been caused by earthquakes. Less commonly, some landslides have been caused by high winds, the crashing of waves on the shore, or flood waters. It is likely, however, that most mass movements of earth have been caused by the addition of water to the ground to such an extent that the material is weakened beyond its ability to withstand the pull of gravity. Heavy rains are usually the source of the added water, but on occasion intensive irrigation of lawns or fields and the creation of reservoirs have added enough water to cause a slide. Man, of course, is responsible for both of these additions.

Man contributes to the landslide process in other ways, too. For example, he may dig away the base of a hill. This both steepens the slope and removes supportive material, the net effect being the lessening of the force necessary to cause movement. In other instances, hills are denuded by bulldozers or graders, destroying the vegetation which retards erosion. In southern California, severe mudflows have occurred after bulldozing, grading, and brush fires (some of which have been set by man) have destroyed the protective vegetation. In fact, many mudflows occurred around Los Angeles in 1969, when unusually heavy rains beat down on hills that had been burned during the previous summer. In addition to the floods caused by these rains, which we described earlier, the devastating mudflows destroyed millions of dollars' worth of property and caused some loss of life.

Landslides

Landslides, slumps, and more or less fluid mudflows are often loosely referred to as landslides. These features occur in many places in the world and collectively form a significant geological hazard. Man is often a causative agent by virtue of his own activities. Let us look at a few examples of landslides to illustrate some of the different movements and their causes.

Sometimes unstable masses of material slide as a result of earthquakes. Just below Hebgen Lake, Montana, a huge mass of dolomite situated high on a mountainside was shaken loose by the 1959 Hebgen Lake earthquake. The rocks, estimated to weigh a total of 37 million tons, slid along the contact between the dolomite and the underlying rock, across the Madison River Valley, and over 400 feet up the mountain on the opposite side of the valley. About 19 campers were buried alive in a campground in the valley. The landslide also dammed the Madison River and caused the formation of Earthquake Lake as the impounded river swiftly began to fill the valley behind the slide. Because of the high

9-46
Landslide scars at edge of Franklin D. Roosevelt Lake, Washington. The slides cut into an ancient stream terrace composed of silt and clay.

probability that the rising waters would breach the newly formed dam, earth-moving equipment was quickly brought in to clear away the top part of the slide. A potentially destructive, or even disastrous, flood downstream was thus avoided.

The Alaska earthquake of 1964 triggered a number of slumps at Turnagain Heights in Anchorage. Homes in a nice residential area were located on top of a high bluff commanding a splendid view of the ocean and the mountains. The top of the bluff consisted of a layer of glacially deposited gravels, while underneath the gravel the bluff was formed of a thick, water-saturated clay. We have already noted that clay is weak and is very subject to sliding, especially if waterlogged. As the shaking initiated the movement, slump blocks moved downward, slipping along curved surfaces. The result was an enormous jumble of broken ground and

9-47
*Homes devastated by Turnagain Heights slide, Alaska earthquake, 1964. (Photograph by
W. R. Hansen, U. S. Geological Survey)*

destroyed houses (Fig. 9-47.) The houses had been built in spite of the
fact that the material composition of the bluff had been well known, and
published reports had warned against construction on it.

A slide of rock debris down a steep mountainside into Vaiont
Reservoir, Italy, in October 1963, resulted in the death of almost 3000
persons. Nearly a quarter of a billion cubic meters of material fell into
the reservoir, displacing the water with such force that the water sloshed
over the dam and thundered down the valley, obliterating towns in its
path. The soundly constructed dam stood firm due to its remarkable
design and construction; however, the disaster was a grim reminder of
the fact that the dam should never have been built in the first place.

While the dam was being built, it was discovered that the rocks along
the walls of the canyon were unstable. At one time in the earth's history,
they had been deeply buried and were therefore under great pressure.
More recently, stream erosion had rapidly cut the canyon of the Vaiont
River, exposing the heretofore deeply buried rock. The release of pressure,
or unloading (Chapter 4), causes cracks to form; in the case of the rocks in
the Vaiont area, unloading is still continuing. Thus, when the dam was

completed and the reservoir began to fill, water entered the cracks and other openings in the rocks of the reservoir walls, creating buoyancy and reducing cohesion in the rock. With the reservoir continuing to fill, a landslide was almost inevitable. Sliding began, continued for three years, and about a week before the disaster, greatly increased its rate. The rate of sliding had been measured, and when it became evident that the rate was increasing significantly, an attempt was made to release water from the reservoir. Unfortunately, it could not be released fast enough, and the reservoir was still nearly full when the mountainside gave way.

Debris Avalanches and Pyroclastic Flows

One of the world's worst disasters took place in Peru on May 31, 1970. A strong earthquake triggered the fall of a slab of ice and rock from a point about 17,000 to 18,000 feet high on the 21,860-foot north peak of Nevados Huascaran, the highest mountain of the Cordillera Blanca of the Peruvian Andes. The slab, one-half mile across, crashed down the slope in an orgy of destruction, completely destroying a number of towns, including Yungay and Ranrahirca, in the Rio Santa Valley, and causing damage as far away as the Pacific Ocean. Over 15,000 persons of a population of 19,000 in Yungay simply disappeared. Only 50 out of a total of 1850 persons were left alive in Ranrahirca. Blocks of rock, some weighing more than three tons, were hurled as much as a half mile through the air, and the rock torrent killed many persons and livestock. One block weighing 14,000 tons was dumped near Ranrahirca. The slide, well lubricated by the ice and water flowing down a steep slope, flowed downstream with unbelievable speed. It traveled the ten miles to Yungay in two to four minutes (150 to 300 mph) and reached the town of Tablones, about 80 miles downstream, 3½ hours after the earthquake. Highways were buried, the airfield at Caraz was swamped, the Choquehaca bridge across the Rio Santa was destroyed, the diversion dam and railroad bridge at the hydroelectric plant near Huallanca were demolished, and the Rio Santa itself momentarily flowed backward as the avalanche splashed into it. This total mayhem was the result of the movement of perhaps 50 million tons of solid material.

Most of the material was mud and boulders. Much of the ice was melted by friction, but blocks of ice, along with trees, manmade objects, and other debris were incorporated in the slide. The avalanche moved with a deafening noise and was accompanied by turbulent air. Astonishingly, it moved over ground that was later found to be relatively

9–48
Overturned bus caught in debris avalanche in Yungay. The depth of the material here is about 15 feet. (Photograph by U. S. Geological Survey)

undisturbed; however, this happened only in the upper reaches of its course, where it moved with such great speed that it apparently overrode the air, which cushioned the flow.

The Huascaran slide is of the type known as a *debris avalanche*. Debris avalanches are rapidly flowing masses of usually well-lubricated mud and rock, often including ice and vegetation. They occur where materials are unstably situated on extremely steep slopes and can be triggered by a number of occurrences, including heavy rains and earthquakes. It should be noted that many debris avalanches have occurred in the Nevados Huascaran area. One in 1962 destroyed most of Ranrahirca, the same town that was destroyed in 1970, but the 1962 avalanche was not set off by an earthquake. Obviously, settlements should not be built in the potential paths of debris avalanches. To decide which are the safe areas, however, requires extremely careful and thorough mapping of the topography, hydrologic conditions, weather, and geological features such as rock and soil types, breaks in the rocks, and possible planes of slippage such as bedding planes. Statistical records of where and how often avalanches occur must also be available. Even then, it may be extremely difficult to perceive the likelihood of a disaster such as the one at Yungay, which was destroyed only after the avalanche rolled up and over a ridge 300 to 600 feet high.

Another type of avalanche is the *pyroclastic flow*, associated with certain types of volcanic eruptions. A classic example was the eruption of hot gases (mostly air) and ash from Mount Pelée on the island of Martinique in 1902. The volcano had been rumbling for some days, and

the citizens of the city of St. Pierre were jittery. An election was sched-
uled for May 10, and the ignorance of the populace plus coercion by the
politically motivated governor kept most of the people in town. The
population, augmented by nervous people from the countryside, had
swelled to about 30,000 by May 8. On the morning of May 8, Mount
Pelée erupted. A great cloud rolled down the mountainside and overwhelmed the city. The election was never held; only two persons sur-
vived. The force of the avalanche knocked down stone walls three feet
thick. These avalanches, also called *nuées ardentes* (French: "glowing
clouds"), constitute a prime hazard in volcanic areas, where the rocks and
magmas contain a high percentage of SiO_2 and are highly viscous and explosive; the Indonesian and Caribbean volcanoes have such characteristics.
In 1971 the volcano Soufriere, which behaves much like Mount Pelée,
threatened the islanders on St. Vincent. Many people departed, recalling
that pyroclastic flows had previously caused death and destruction there.

The greatest dangers of pyroclastic avalanches are probably heat
and suffocation. Although some of the gases may be poisonous, superheated air probably is a greater danger. Observers have referred to the
avalanches as "fiery clouds" (*nuées ardentes*). The clouds also have a
great amount of incandescent ash in them, which alone may kill many, as
it is known to have done. Because pyroclastic flows are sufficiently dense
to maintain contact with the ground surface, they may be considered to
be avalanches. The pyroclastic avalanche probably changes gradually into
other eruptive phenomena; however, these others, including intense ash
falls with perhaps less hot gas, such as may have occurred at Pompeii, are
not, in strict terms, flows or avalanches.

Control of Mass Movements

When we consider the relationship of man and his works to the hazard
from mass movement of rock and soil, we must really consider two aspects
of dealing with this hazard. First, we must understand that there are some
movements that cannot be controlled, e.g., most debris avalanches, such
as the Huascaran avalanche, and pyroclastic flows. Ultimately, these are
caused by unrestrainable forces within the earth. The only means to protect ourselves from such phenomena is to recognize the hazardous places,
map them adequately, and ensure that they are avoided, or if they cannot
be, to monitor them continually and install a warning system. For
example, we pointed out that the dam and reservoir at Vaiont should not
have been built. They were constructed but were also monitored. Yet so
little attention was given to adequate mapping and monitoring, that when
it became evident that trouble was to be expected, the frantic attempt to
lower the water level came too late.

The second aspect of landslide control involves treating the ground in order to prevent the occurrence of a slide or to stabilize an old slide. Such treatment includes terracing and grading of hills to maintain proper slopes, filling and compacting some areas, building retaining walls to provide support, and constructing drainage systems to ensure removal of water. In some cases there will have to be regulation of the irrigation of fields or lawns or a prohibition, during the wet season, of certain building practices such as grading, which lays bare the soil. Selective planting of trees, bushes, or grass may help to stabilize slopes. The location of roads and other installations can be planned to avoid the excavation of the "toe" of a potential slide. Finally, laws designed to ensure adequate study of rock and soil conditions and proper building practices can be passed. Land-use zoning and planning by governmental agencies on all levels can help to ensure that people will live safely and in harmony with their geological environment.

REFERENCES

Glaciers

Eardley, A. J. "Glaciers and climates of the Pleistocene Epoch," in *General College Geology*, New York: Harper & Row, 1965.

Fehrenbacher, J. B., B. W. Ray, and J. D. Alexander. "Illinois soils and factors in their development," in *Quaternary of Illinois*, ed. R. E. Bergstrom, Urbana: University of Illinois College of Agriculture Special Publication No. 14, 1968, pp. 165-175.

Hackett, James E. "Quaternary studies in urban and regional development," in *Quaternary of Illinois, op. cit.*, p. 176.

Longwell, C. R., R. F. Flint, and J. E. Sanders. "Glaciers and glaciation," in *Physical Geology*, New York: Wiley, 1969.

Peck, Ralph B. "Problems and opportunities — technology's legacy from the Quaternary," *Quaternary of Illinois, op. cit.*, pp. 138-144.

Strahler, A. N. "Glacier landscapes and the Ice Age," in *The Earth Sciences*, New York: Harper and Row, 1963, pp. 517-542.

Thorp, James. "The soil—a reflection of Quaternary environments in Illinois," *Quaternary of Illinois, op. cit.*, pp. 48-55.

Tricart, Jean. *Geomorphology of Cold Environments.* New York: St. Martin's, 1970.

Mass Movement

Erickson, G. E., *et al. Preliminary Report on the Geologic Events Associated with the May 31, 1970 Peru Earthquake*, U. S. Geological Survey Circular 639, 1970.

Gilluly, James, A. C. Waters, and A. O. Woodford. *Principles of Geology*, 3rd ed., San Francisco: Freeman, 1968, pp. 182-206.

Kenney, Nathaniel T. "Southern California's trial by mud and water," *National Geographic*, 136, 4 (October 1969): 552-573.

Kiersch, G. A. "Vaiont reservoir disaster," *Geotimes* 9, 9 (1965): 9-12.

Leighton, F. B. "Landslides," in *Geologic Hazards and Public Problems*, Conference Proceedings, Office of Emergency Preparedness, Region Seven, Santa Rosa, Calif., 1969, pp. 97-132.

————. "Origin and control of landslides in the urban environment of California," *Engineering Geology*, 13, Proceedings of the 24th International Geological Congress, Montreal, 1972, pp. 89-96.

Longwell, C. R., R. F. Flint, and J. E. Sanders. *Physical Geology*, New York: Wiley, 1969, pp. 159-181.

Morton, D. M. and Robert Streitz. *Landslides: Man and his Physical Environment*, Minneapolis: Burgess, 1972, pp. 64-73.

Putnam, W. C. *Geology*, New York: Oxford University Press, 2d ed., rev. by A. B. Bassett, 1971, pp. 180-195.

Terzaghi, Karl. "Mechanics of landslides," *Application of Geology to Engineering Practice*, Berkey Volume, Geological Society of America, 1950.

Williams, G. P. and H. P. Guy. "Debris avalanches—a geomorphic hazard, in *Environmental Geomorphology*, ed. D. R. Coates, Binghamton: State University of New York, 1970, pp. 25-46.

Zaruba, Q. and V. Mencl. *Landslides and their control*, New York and Amsterdam: Elsevier, 1967.

Shorelines

Cronin, L. E. "The role of man in estuarine processes," in *Estuaries*, ed. G. H. Lauff, American Association for the Advancement of Science, Publication 83, 1967, pp. 667-689.

Davis, J. H. "Influence of man upon coast lines," in *Man's Impact on Environment*, comp. Thomas R. Detwyler, New York: McGraw-Hill, 1971, pp. 332-347.

Johnson, J. W. and P. S. Eagleson. "Sediment problems at coastal structures," in *Estuary and Coastline Hydrodynamics*, ed. A. T. Ippen, New York: McGraw-Hill, 1966, pp. 462-492.

Klimm, Lester E. "Man's ports and channels," in *Man's Role in Changing the Face of the Earth*, Vol. 2, ed. W. L. Thomas, Chicago: University of Chicago Press, 1956, pp. 522-541.

Krumbein, W. C. "Geological aspects of beach engineering," in *Application of Geology to Engineering Practice, op. cit.*

Martinez, J. D. "Environmental geology at the coastal margin,"*24th Int. Geol. Congress Symposium on Earth Sciences and the Quality of Life*, Montreal, 1972, pp. 45-58.

McDougall, Harry. "The Delta Plan,"*Canadian Geographical Journal* LXXIX (1969): 64-75.

Pestrong, Raymond. "San Francisco Bay tidelands," *California Geology*, 25 (1972): 27-40.

Pritchard, D. W. "What is an estuary—physical viewpoint," in *Estuaries*, *op. cit.*, pp. 3-5.

Russell, R. J. "The coast of Louisiana," *Applied Coastal Geomorphology*, ed. J. A. Steers, Cambridge, Mass.: M. I. T. Press, 1971, pp. 84–97.

Shepard, F. P. and H. R. Wanless. *Our Changing Coastlines*, New York: McGraw-Hill, 1971.

Streams

Emerson, J. W. "Channelization: a case study," *Science* 173 (1971): 325-326.

Gillette, Robert. "Stream channelization: conflict between ditchers, conservationists," *Science* 176 (1972): 890-894.

Leopold, L. B., M. G. Wolman, and J. P. Miller. *Fluvial Processes in Geomorphology*, San Francisco: Freeman, 1964.

Matthai, H. F. *Floods of June 1965 in South Platte River Basin, Colorado*, U. S. Geological Survey, Water-Supply Paper 1850-B, 1969.

Morisawa, M. *Streams: Their Dynamics and Morphology*, New York: McGraw-Hill, 1968.

Rantz, S. E. *Urban Sprawl and Flooding in Southern California*, U. S. Geological Survey Circular 601-B, 1970.

Sheaffer, J. R., D. W. Ellis, and A. M. Spieker. *Flood-Hazard Mapping in Metropolitan Chicago*, U. S. Geological Survey Circular 601-C, 1970.

Stall, J. B. "Man's role in affecting sedimentation of streams and reservoirs," in *Proceeding of the 2nd Annual American Waters Resources Congress*, ed. K. L. Bowden, 1966.

Sterling, Claire. "Aswan Dam looses a flood of problems," *Life*, February 12, 1971.

Strahler, Arthur N. "The nature of induced erosion and aggradation," in *Man's Role in Changing the Face of the Earth*, Vol. 2, ed. W. L. Thomas, Chicago: University of Chicago Press, 1956, pp. 621-638.

Thornbury, W. D. *Principles of Geomorphology*, 2d ed., New York: Wiley, 1969, pp. 99-177.

Wind

Bennett, H. H. *Elements of Soil Conservation*, 2d ed., New York: McGraw-Hill, 1955.

Dorr, J. A., Jr. and D. F. Eschman. *Geology of Michigan*, Ann Arbor: University of Michigan Press, 1970.

Eardley, A. J. "Work of the wind," Chapter 11 in *General College Geology*, New York: Harper & Row, 1965, pp.189-207.

Easterbrook, D. J. *Principles of Geomorphology*, New York: McGraw-Hill, 1969, pp. 289-303.

Jacks, G. V. and R. O. Whyte. *Vanishing Lands: A World Survey of Soil Erosion*, New York: Doubleday, Doran, 1939.

Krinitzsky, E. L. and W. J. Turnbull. *Loess Deposits of Mississippi*, Geological Society of America Special Paper 94, 1967.

Schwab, G. O., R. K. Frevert, K. K. Barnes, and T. W. Edminster. "Soil erosion by wind and its control," in *Elementary Soil and Water Engineering*, 2d ed., New York: Wiley, 1971.

Strahler, A. N. "Wind as a geologic agent," in *The Earth Sciences*, 2d ed., New York: Harper & Row, 1971, pp. 694–705.

Thornbury, W. D. *Principles of Geomorphology*, 2d ed., New York: Wiley, 1969, pp. 263–302.

Dynamic forces within the earth may wreak terror and destruction, but man continues to crowd into areas of seismic and volcanic activity. (Photograph courtesy Sigurdur Thorarinsson)

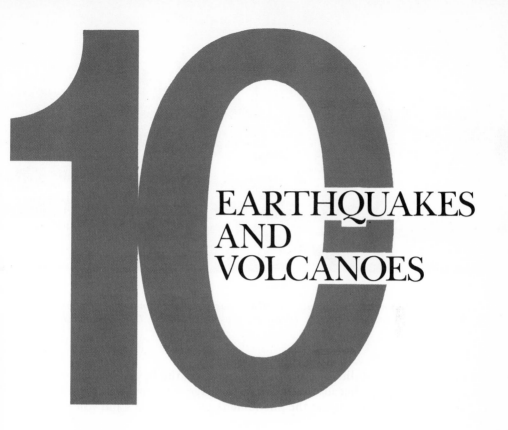

EARTHQUAKES
AND
VOLCANOES

INTRODUCTION

Few events demonstrate the dynamic nature of the earth as vividly as do earthquakes and volcanic eruptions. Both have terrified man for centuries, and even today we see man and his works being shaken to the ground, buried, set afire, or washed away by these inexorable phenomena of nature. The disasters or potential disasters posed by earthquakes and volcanoes loom even larger today as an increasing population crowds into seismic and volcanic regions. For instance, California now has a population of about 20 million, which is ten times greater than it was when the great San Francisco earthquake occurred in 1906. If an earthquake the magnitude of the 1906 earthquake (or that of the 1964 Alaska earthquake) were to strike today in California or some other heavily populated area such as Japan, its enormity would be overwhelming. If Mount Vesuvius were to

10-1
Buckled curbstone and pavement, with tilted houses and fire damage visible in background, San Francisco earthquake, 1906. (Photograph by G. K. Gilbert, U. S. Geological Survey)

behave today as it did in A.D. 79, devastation in the Bay of Naples area could be terrible. Because they constitute large-scale environmental hazards, we should investigate the phenomena of earthquakes and volcanoes.

THE THEORY OF PLATE TECTONICS

Lisbon, 1755; San Francisco, 1906; Anchorage, 1964; Managua, 1972 — some of the worst earthquakes — and Vesuvius, A.D. 79; Tambora, 1815; Krakatoa, 1883; Pelée, 1902 — some of the greatest volcanic eruptions — all remind us that the earth we live on is alive, restless, and dynamic. Not only are we reminded that the earth is not a perfectly secure body, "solid as rock," but we are also made to wonder what, indeed, does take place below its surface.

In the past two or three decades we have witnessed a vast increase in knowledge about the interior of the earth. Establishment of earthquake recording stations on a much broader scale, satellite measurements of gravity, drilling into rocks of the ocean floors, and many other sophisticated technologies have given us a better knowledge and at long last are enabling geoscientists to put together a coherent, unified picture of the earth. The movements — up, down, and sideways — of the outer few miles of the earth can now be explained by the *theory of plate tectonics.* (Tectonics is the study of the crust of the earth, its structure, and the forces producing the structures.) Although the theory of plate tectonics is not universally accepted, and many of its details are yet to be explained satisfactorily, more and more evidence which reinforces the credibility of the theory is being accumulated daily. Simply stated, the theory holds that the outer 70 miles or so of the earth consist of a dozen large plates, plus a number of smaller ones, that continually move about (Fig. 10-2). The motion of these plates helps to explain the existence, as well as the distribution of, earthquakes and volcanoes, which are generally concentrated along the boundaries of the great plates.

The theory of plate tectonics originated in part at least one hundred years ago, when a number of people suggested that the Western Hemisphere had once been joined to Europe and Africa, with no Atlantic Ocean in between. This idea evolved from observations that the general outlines, especially of South America and Africa, corresponded and that rock formations and their fossils would tend to match if the continents were moved back together. At the turn of the century, Eduard Suess of Austria theorized that all the continents had originally been one supercontinent, which he called Gondwanaland, named after an area in India where geological evidence supported this theory. Later, in 1912, Alfred Wegener, a German meteorologist, presented a detailed and definitive theory which came to be known as the *theory of continental drift*. It stated that the supercontinent broke up about 200 million years ago and that the fragments have since drifted apart into their present positions.

For many years the theory of continental drift was unacceptable to most geologists and physicists. How could continents float through a sea of solid rock lying beneath them? What kind of driving force could have possibly initiated this split? Then, bit by bit, information was gathered. The Mid-Atlantic Ridge, a vast chain of mountains almost totally under water, was discovered. More about the earth's gravity and magnetism was learned and mapped. Rock formations were studied; fossils were discovered; samples of the ocean floor were drilled; and nearly all the evidence favored the concept of drifting continents.

Convection cells in the mantle of the earth have been postulated to be the driving force that moves the plates (Fig. 10-3). It is still not known why such cells exist, but an assumption of their existence provides an explanation of the major features of the earth — the oceanic ridges and trenches, mountain ranges, and large faults. In the Atlantic area, hot rock material, rising through the surrounding rocks of the mantle because of its lower density, reaches the surface and bows it upward at the Mid-Atlantic Ridge. Extra flow of heat has been measured along the ridge. Shallow earthquakes also occur along the ridge. These phenomena reinforce the suggestion that convection cells are located here, while further reinforcement is provided by the volcanic rocks of the sea floor. All islands along the ridge are composed of volcanic rocks, so that the 1973 volcanic eruption in Iceland was typical of activity along this area. A central valley,

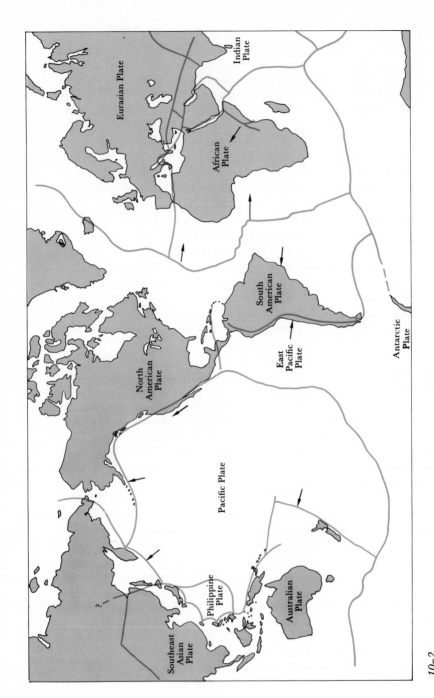

10-2
Some of the major lithospheric plates. Arrows show direction of movement.

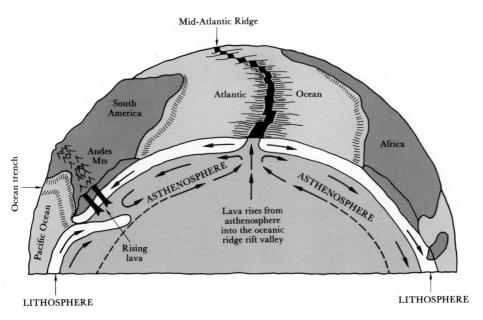

10-3
Movement of lithospheric plates by means of postulated convection currents in the mantle
of the earth. (Peter J. Wyllie, "Earthquakes and Continental Drift," University of
Chicago Magazine, Jan./Feb. 1972, p. 16. Reprinted by permission.)

called a *rift* valley, has also been discovered along the ridge. The dropped
central block in a rift valley is the result of the pulling apart (*tension*) of
rocks on either side (Fig. 10-2).

One of the most important pieces of evidence was the discovery of the
magnetic patterns of the rocks parallel to the axis of the ridge. As molten
rock rises to the surface along the center of the ridge, it cools and solidifies.
Crystals of iron-bearing minerals align themselves, like compass needles,
with the earth's magnetic field, and as the entire rock solidifies, these tell-
tale compasses are frozen into position. As it is easy to distinguish the
"north" end of the minerals from the "south," we have been able to
learn that the iron minerals indicate that the earth's magnetic field has reversed itself many times. We have also discovered that strips of rock lying
parallel to the ridge are alternately polarized; therefore, it is apparent that
as rock rises and forms during one epoch of magnetic polarity, it shoves
aside rocks formed in the previous epoch of reversed polarity.

The continual movement of the crust and upper mantle rocks (collectively known as the lithosphere) away from the oceanic ridges has been

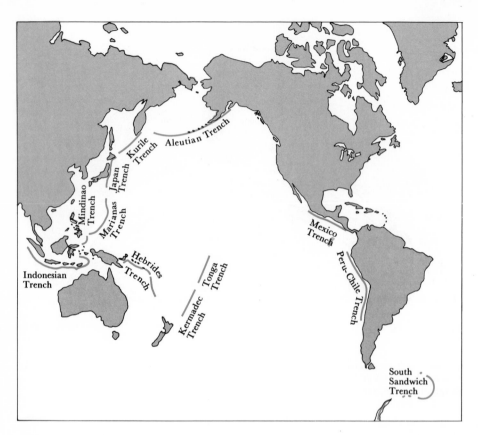

10-4
Trenches of the Pacific.

called *sea floor spreading*. Thus, it appears that the continents are indeed drifting, but not as separate and independent entities. They are, instead, parts of larger plates which also include parts of the ocean floors. Consequently, continental drift and sea floor spreading are integral parts of overall plate tectonics.

According to the theory, new lithosphere is created along the oceanic ridges. However, if the surface area of the earth remains constant, the plates cannot grow perpetually; somewhere, material must be subtracted from the plates and returned to the depths of the mantle in order to replace the material which has risen from below. We find the probable answer in the oceanic trenches of the Pacific (Fig. 10-4). If a supercontinent existed at one time and broke into fragments which moved apart, the expansion in area caused by the creation of new lithosphere (essentially Atlantic lithosphere) must have occurred at the expense of the ocean surrounding

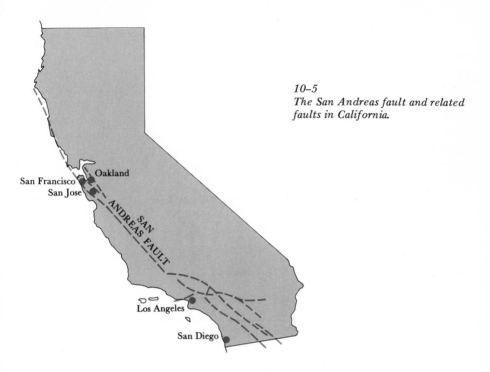

10–5
*The San Andreas fault and related
faults in California.*

the continent. The Pacific Ocean is descended from that primordial ocean, for North America, South America, and other plates are crowding into the Pacific. As this takes place, material is carried downward at the trenches and back under the in-crowding plates. Earthquakes indicate that slippage is occurring along *subduction zones,* along which rock material from the lithosphere is pulled down and under the in-crowding plates. The fact that the deepest earthquakes of all occur here suggests that the lithospheric rock remains cooler and more brittle at a greater depth. Buckling occurs, and melting takes place as the rock reaches warmer depths or is heated by friction. As a result, mountain ranges form, and volcanism becomes common as the magma rises through conduits from below.

Although we have already described the boundary between plates which are moving apart (the oceanic ridges) and the boundary between colliding plates (the trenches and subduction zones), the boundary between plates which are sliding past one another must still be described. An outstanding example occurs where southwestern California is moving northwestward as part of a plate which is laterally sliding past the North American plate. The giant tear in the lithosphere along which the plates move is the San Andreas fault, a strike-slip fault (Fig. 10–5) so well known that it has been designated a National Natural Landmark of the United States.

What is known about the continents and ocean basins before the supercontinent broke up? If the earth is 4600 million years old, and break-up began 200 million years ago, then the presently existing plates have been in motion about 1/25 of all the time of the earth's life. Since the earth is dynamic, it is very possible that there were prior cycles of movement, whose nature remains unknown and obscured by the events of the last 200 million years. Can anything be predicted about the probability of future cycles? Studies suggest that plate movement will continue. Movements have been projected for another few million years: North America is to move west and north, South America will move west, Australia is to move north, Asia will pivot to the southeast, and so on.

Knowing something about plate tectonics, we are now ready to explain the phenomena of earthquakes and volcanoes. We know that earthquakes occur most frequently along plate boundaries, that the interiors of plates are more stable but, nevertheless, subject to some degree of stress, and that volcanoes are found along plate boundaries where rock may be moving up or down between the crust and mantle.

EARTHQUAKES

What are earthquakes, what causes them, and what makes them such an environmental problem? Briefly, an earthquake is a shaking of the ground. The shaking ranges from unnoticeably gentle to incredibly violent. Fortunately, the really violent earthquakes do not occur frequently. Although the cause of earthquakes is not fully understood, we know that the earth, as a dynamic body, builds up strain; when this strain exceeds the breaking point of the rocks involved, something gives. The shaking that is experienced is thought to be the result of rocks rebounding after the break has occurred. The slippage, or movement, of the material occurs on either side of faults (Fig. 10-6). The largest shakings are almost incomprehensible in scale. The U.S. Geological Survey's study of the 1964 Alaska earthquake stated that "the entire earth vibrated like a tuning fork."

The energy released during an earthquake usually causes a series of tremors which, in most cases, consist of one large tremor followed by lesser tremors called *aftershocks*. Occasionally, *foreshocks* occur prior to the main tremor. An earthquake lasts only a few seconds, sometimes having short intervals between individual jolts. For instance, the 1906 San Francisco earthquake attained its maximum intensity in about 40 seconds, then ceased abruptly for about 10 seconds. It then started again, more violent than ever, and continued for about 25 seconds. This 75-second interval constituted the main shock and was followed by numerous, less intense aftershocks.

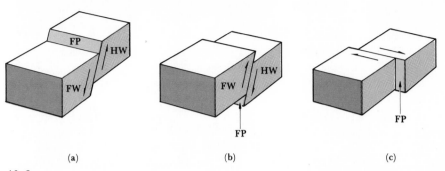

10-6
Basic fault movements: (a) normal fault; (b) reverse fault; (c) strike-slip fault. HW represents a hanging wall, FW is a footwall, and FP is the fault plane.

Occurrence of Earthquakes

It is likely that some motion of materials occurs throughout the entire earth; but as for earthquakes, we need only be concerned with the upper 700 km of the earth, as no earthquakes have been recorded below that depth. This would indicate that only above 700 km are rocks brittle enough to fracture and undergo sudden slippage. Below that depth, any yielding of rock takes place entirely as plastic flow.

Within the 0 to 700 km range, earthquakes may be grouped into three depth zones: shallow, intermediate, and deep. The shallow quakes are those which occur essentially in the crust (depth 0 to 33 km). The others take place in the upper mantle, the intermediate range extending from 33 to 300 km and deep quakes occurring at depths of 300 km to 700 km. The greatest number of earthquakes, including most of the large ones, occur in the crust.

We have noted that earthquakes take place along faults, which are large fractures in the earth's crust and upper mantle. Faults vary widely in size, in terms of both their length and the amount of displacement on them. California's San Andreas fault is hundreds of miles long, and the slippage (one side moving relative to the other) accumulated throughout geologic history is thought to be as much as 300 miles. The vertical displacement on some of the faults bordering mountain ranges may be well over two miles. On the other hand, the lengths and displacements of some faults are only a few feet, and the movements along them may never have caused noticeable trembling.

Earthquakes may occur anywhere at any time, although they are more prevalent in certain areas. Perhaps this fact is overemphasized: many persons living in the Middle West or eastern United States, for example, seem to feel that they do not live in an earthquake area and that only

10-7
Main fault between Point Reyes Station and Olema, 1906 California earthquake. (Photograph by G. K. Gilbert, U. S. Geological Survey)

10-8
Fault scarp at Blarneystone Ranch, Hebegen Lake area, Montana. The picture was taken on the downthrown block looking toward the scarp, which is about ten feet high. (Photograph by I. J. Witkind, U. S. Geological Survey)

residents of California and Alaska should be concerned with the problem. Actually, some of the largest earthquakes of recent history occurred near New Madrid, Missouri, in 1811–1812. Although the area was sparsely settled at the time and the earthquake occurred prior to the development of modern seismological measuring abilities, reports indicate that some of the New Madrid major shocks (shocks abounded over a period of a few months) were very severe. They may, in fact, have been the most severe ever to have occurred in the United States, at least until the Alaska earthquake of 1964. Thus, in spite of the infrequency of earthquake activity in the "stable" regions, major earthquakes *can* and do occur there.

Earthquakes are most common in a few well-defined zones of the world (Fig. 10-9). The largest of these is the circum-Pacific belt, which suggests that a belt of major weakness exists in that part of the earth. A great majority of the world's earthquakes occur there, including virtually all of those classified as "deep" shocks. Most of the remaining earthquakes occur in the Tethyan belt, which includes the Mediterranean area and Asia from Turkey to southwestern China. There are other minor belts, including the various oceanic ridges such as the Mid-Atlantic Ridge, which are areas where rock material is thought to be welling upward from the mantle, causing the sea floors to be spreading apart from the ridges in either direction, thus making them extremely active areas.

Earthquake Waves

When an earthquake occurs, energy is released and is dissipated as shock waves which move on a more or less spherical front expanding outward from the place where the slippage occurred (Fig. 10-11). This place, called the *focus,* is usually not a point, but rather a zone along the fault.

Two basic types of waves are associated with the transmission of the shock energy: *body* waves and *surface* waves. Body waves pass through the earth and consist of two types. One type, called a *P* wave, is compressional. The particles of rock in a *P* wave vibrate parallel to the direction of wave propagation through the rock. *P* waves can occur in any kind of material — solid, liquid, or gaseous. The other type of body wave is the *S* wave, or shear or transverse waves, which are called that because the particles of rock vibrate in a direction which is transverse to the direction of propagation of the wave. Body waves are of interest because they both enable us to locate the site of the earthquake and provide us with most of our knowledge of the earth's interior.

Surface waves are long waves that spread over the surface of the earth after the energy from the focus reaches the surface. In one type of surface wave the motion is orbital in a vertical plane in the direction of wave propagation; in the other type, motion is transverse to the direction of wave propagation.

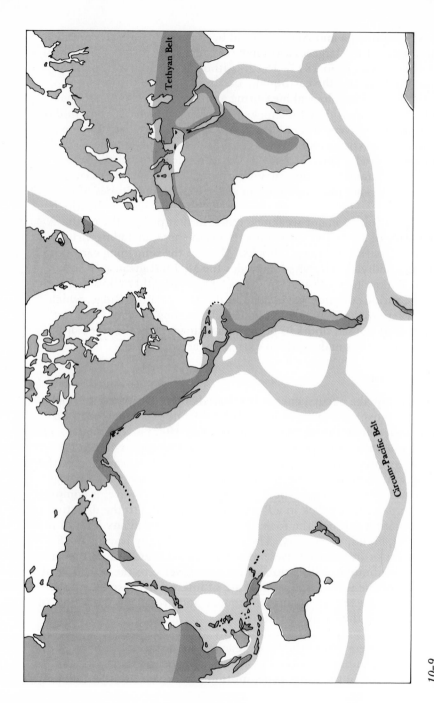

10-9
Earthquake belts of the world.

10-10
Reelfoot Lake, Tennessee, formed by depression of the land in the New Madrid earthquakes of 1811–1812. (Photograph by F. E. Matthis, U. S. Geological Survey)

Wave velocities depend on the density of rocks through which they pass. Near the surface, *P* waves travel from 5.5 to 7.0 km per second, which is about 1.73 times faster than the velocity of *S* waves. Surface waves are the slowest, traveling about 3.2 km per second.

Recording Earthquakes

Earthquakes are recorded by sensitive instruments called seismographs (*seismos*, from a Greek word meaning *to shake*, = earthquake). In principle, a seismograph operates by means of a heavy weight attached by a spring or hinge device to a frame. Because of the way in which it is attached to the frame, the weight is essentially independent and because of its inertia tends to remain stationary as the earth and frame tremble. *Inertia* is the tendency of a body to resist being moved, or if in motion, to being stopped. The heavier the object, the greater its inertia. This device, which has a heavy weight attached to the earth by a spring, hinge, or other device in order to reduce friction, constitutes a *seismometer*. A *seismograph* also has a continuous recording device attached to it. This device may either have a source of light that casts a thin beam of light on a clock-driven, rotating drum covered with photographic paper, or may transmit an elec-

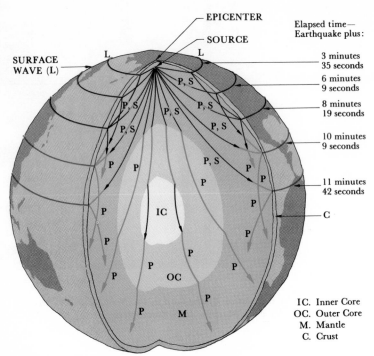

10–11
*Cross-section of the earth, showing ray paths of P, S, and surface waves.
(Source:* **Earthquakes,** *U. S. Department of Commerce, Environmental
Science Services Administration, Washington, D. C.: U. S. Government
Printing Office, 1969.)*

10–12
*Simplified seismogram of earthquake of magnitude 6.5 in the Aleutian
Islands, July 13, 1959.*

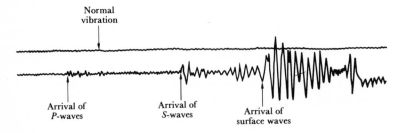

trical impulse to a stylus on the drum. As the drum rotates, a line is traced that shows the earth's vibrations and the times they occur. Such a record is called a *seismogram,* a typical example of which is shown in Fig. 10-12. Seismographs can be placed so that they record east-west, north-south, or vertical components of shaking.

Intensity and Magnitude

When attempting to learn more about the nature of earthquakes, it is useful to describe them in terms of destruction caused and energy released. A scale in which certain observed effects are assigned a number is used to describe intensity, which is a relative measure of the strength of the earthquake. The various effects described by this scale include sound, visible and felt vibration, type and severity of damage, and others. Scales of this type are the Modified Mercalli Scale, summarized in Table 10-1, which is in general use in the United States; the Rossi-Forel Scale, much used in Europe, and from which the Modified Mercalli Scale was developed; and the JMA (Japan Meteorological Agency) Scale.

The intensity of an earthquake, naturally, depends on a number of factors, most important of which, perhaps, is the distance between the observer and the epicenter. The *epicenter* is the point on the earth's surface which is directly above the point of actual rock failure (focus). It is often described as the surface expression of the earthquake and is the point at which the damage would be at a maximum. Another factor is the kind of rocks present and their degree of consolidation. In general, the intensity decreases with increasing distance from the epicenter. For example, the powerful Alaska earthquake of 1964 was most intense in Alaska and became progressively less so in areas farther away. People in the mid-western United States did not feel it, but water in a well in South Dakota fluctuated as much as 23 feet, attesting to the severity of the shock. Other wells were affected as far away as South Africa, half way around the globe.

Soon after an earthquake, cards are mailed to many people in the vicinity of the quake in order to gain important information. With this information, intensity values are assigned and plotted geographically on a map. Thereafter, it is easy to contour the map, showing the area of greatest intensity, areas progressively less affected, and finally areas where the earthquake was not felt. Such a map is called an *isoseismal* map. These maps may prove to be of great value to geologists, geophysicists, and engineers who are studying earthquakes in an attempt to reduce the hazards in quake areas. An isoseismal map of the Charleston, South Carolina earthquake is shown in Fig. 10-13.

Table 10–1 The Modified Mercalli Scale of Seismic Intensity, abridged

I. Not felt except by a very few under especially favorable circumstances.
(I Rossi-Forel Scale.)

II. Felt only by a few persons at rest, especially on upper floors of buildings.
Delicately suspended objects may swing. (I to II Rossi-Forel Scale.)

III. Felt quite noticeably indoors, especially on upper floors of buildings, but
many people do not recognize it as an earthquake. Standing motorcars
may rock slightly. Vibration like passing truck. Duration estimated.
(III Rossi-Forel Scale.)

IV. During the day felt indoors by many, outdoors by few. At night some
awakened. Dishes, windows, and doors disturbed; walls make creaking
sound. Sensation like heavy truck striking building. Standing motorcars
rocked noticeably. (IV to V Rossi-Forel Scale.)

V. Felt by nearly everyone; many awakened. Some dishes, windows, etc.,
broken; a few instances of cracked plaster; unstable objects overturned.
Disturbances of tree, poles, and other tall objects sometimes noticed.
Pendulum clocks may stop. (V to VI Rossi-Forel Scale.)

VI. Felt by all; many frightened and run outdoors. Some heavy furniture
moved; a few instances of fallen plaster or damaged chimneys. Damage
slight. (VI to VII Rossi-Forel Scale.)

VII. Everybody runs outdoors. Damage negligible in buildings of good design
and construction; slight to moderate in well-built ordinary structures;
considerable in poorly built or badly designed structures. Some chimneys
broken. Noticed by persons driving motorcars. (VIII Rossi-Forel Scale.)

VIII. Damage slight in specially designed structures; considerable in ordinary
substantial buildings, with partial collapse; great in poorly built
structures. Panel walls thrown out of frame structures. Fall of chimneys,
factory stacks, columns, monuments, walls. Heavy furniture overturned.
Sand and mud ejected in small amounts. Changes in well water. Persons
driving motorcars disturbed. (VIII+ to IX Rossi-Forel Scale.)

IX. Damage considerable in specially designed structures; well-designed frame
structures thrown out of plumb; great in substantial buildings, with
partial collapse. Buildings shifted off foundations. Ground cracked
conspicuously. Underground pipes broken. (IX+ Rossi-Forel Scale.)

X. Some well-built wooden structures destroyed; most masonry and frame
structures destroyed with foundations; ground badly cracked. Rails bent.
Landslides considerable from river banks and steep slopes. Shifted sand
and mud. Water splashed (slopped) over banks. (X Rossi-Forel Scale.)

XI. Few, if any (masonry), structures remain standing. Bridges destroyed.
Broad fissures in ground. Underground pipelines completely out of
service. Earth slumps and land slips in soft ground. Rails bent greatly.

XII. Damage total. Waves seen on ground surfaces. Lines of sight and level
distorted. Objects thrown upward into the air.

(Source: USCGS, *Preliminary Determination of Epicenters*)

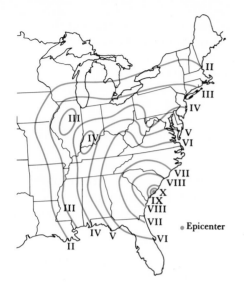

10-13
Isoseismal map of the Charleston, South Carolina earthquake, August 31, 1886. The Rossi-Forel Scale is used.

The *magnitude* of an earthquake is indicated by a scale known as the Richter scale, named after Dr. Charles F. Richter, professor of seismology at the California Institute of Technology and developer of the scale. The magnitude is an expression of the amount of energy released by an earthquake and is determined by instrumental measurements rather than by subjective observations. The farther away from an earthquake a wave is recorded, the smaller the amplitude of the wave. The total energy given off by an earthquake can be measured at any distance within recording range, since the amplitude of the wave is related to the distance. The magnitude, M, of an earthquake is a number derived from the amplitudes of the earthquake in relation to a reference earthquake.

Earthquakes having magnitudes slightly greater than zero and of more than 8 have been measured. Since the scale is logarithmic, an earthquake of magnitude 8 is 10^8 times as strong, in terms of energy released, as the smallest earthquake. No magnitude of 9 or over has ever been recorded, nor does such a value appear to be possible. It is probable that there is insufficient strength in the rocks of the earth to allow that much stored energy (somewhere on the order of 9×10^{24} ergs — a truly enormous amount!) to accumulate.

Locating an Earthquake

As noted earlier, the body waves generated by earthquakes are made up of *P* waves and *S* waves which, because they travel through the entire earth (Fig. 10-11), have enabled seismologists to learn much about the interior of the earth. Because *S* waves disappear upon reaching the core boundary, these scientists have learned, for example, that the outer portion of the earth's core is molten. Since fluids cannot transmit these shear, or S, waves, the outer core is assumed to be liquid. Of greater environmental importance, however, is the fact that P and S waves are used for locating epicenters. Since P waves travel faster than S waves, if the speeds of two waves and the difference in their times of arrival at a seismograph station are known, then the distance to the point from which the waves originated can be easily calculated. We know that distance equals velocity multiplied by time.

Although this method can be used to calculate the distance from the seismograph to the earthquake, it does not indicate the direction of the earthquake. However, the earthquake must be located somewhere on a circle drawn around the seismograph station having a radius equal to the distance. Suppose an earthquake has been determined to be located 500 miles from Pasadena, California. We could plot a circle on a map (Fig. 10-14a). If additional information received from Salt Lake City were plotted, we would have two intersecting circles (Fig. 10-14b) and would know that the earthquake was located at one of the two points of intersection. If a third circle were plotted, using information, for example, from the University of California at Berkeley, we would know that the approximate location of the epicenter was at the one point common to all three circles. In our illustration, the epicenter is somewhere in central Nevada (Fig. 10-14c). Because of heterogeneity of the rocks and other factors, it is rarely possible to have all three circles intersect exactly at one common point. However, the method works sufficiently well to locate earthquakes fairly accurately.

Destructive Agents Associated with Earthquakes

Tsunamis   An old Japanese folk tale tells of a wise and venerable grandfather who owns a rice field at the top of a hill. One day, as the harvest season nears, he senses that an earthquake has occurred. Suddenly, the water of the sea pulls back from the shore, exposing a wide expanse of tidal flats and attracting the attention of the villagers, who all rush down to the shore to see what is happening. The old man knows that there is great danger to the people and the village. What should he do? Finally, he thinks of something: he and his grandson light torches and dash through his rice fields, setting the grain on fire. Soon the villagers notice the ascending plumes of smoke and, forsaking the beach, hurry up the hill to the aid of their neighbor. As the people beat out the flames, they see the old man and the child scurrying ahead of them, lighting new fires. Of course they believe the pair to be daft, but in reality grandfather and the boy are ensuring that everyone will be drawn away from the danger area.

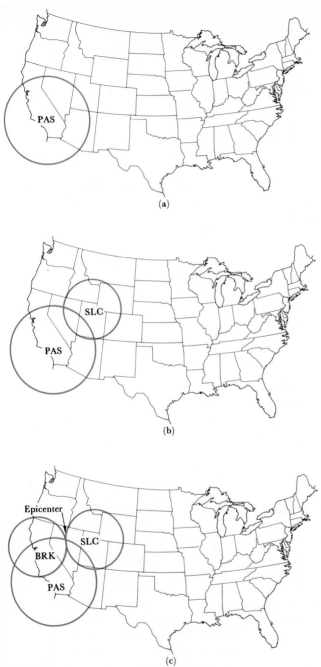

10-14
Determination of earthquake epicenter.

Suddenly, a tremendous wave appears on the horizon of the sea and moves toward the shore at great speed, overwhelming the previously exposed flats and washing up over the town, destroying all the homes. It continues and thunders up the mountain, destroying all the rice fields. It finally spends itself just below the grandfather's house. But in spite of the damage, everyone is saved, and now they all realize how grandfather has saved their lives. Grandfather's large home is more than adequate to shelter them all until they can rebuild their own, and his rich storehouses of grain are sufficient to feed everyone until the next season's harvest — which, by the way, was the greatest one ever.

When warning of a similar wave went out in Crescent City, California, in 1964, the part of the grandfather was played by the National Tsunami Warning Center. Once, then twice, the wave crashed over the public dock and the lumber boats. A number of persons, having withdrawn to higher ground when the warning was given, returned to the danger area, thinking that the danger was past. But the wave came back — *three* more times. The fourth one did the most damage, washing 11 persons out to sea.

Meanwhile, at Kodiak, Alaska, the fishing boat *Yukon,* along with others, was hurled into the center of town. At Seward, Alaska, railroad cars were dashed about like matchsticks in front of a firehose. At Chenega, long-time natives of the area urged everyone to move to higher ground, but even so, one-fourth of the inhabitants of that little village were swept out by the waves.

Waves such as those described above are called *tsunamis.* (*Tsunami* is a Japanese word which was not, by the way, originally used in this sense.) Tsunamis are sea waves generated by seismic activity. Although they are often mistakenly referred to as tidal waves, tsunamis have no genetic relationship to the tides; however, their effects may be more serious at high tide. Their exact cause is not well known, but most authorities attribute them to the up-and-down-movement of blocks of the ocean floor. They can also be caused by undersea mud slides. Lateral movements, such as those along the San Andreas fault, do not seem to produce tsunamis. Some tsunamis have been produced by large volcanic eruptions, such as that at Krakatoa, which capped a small island in the sea near Java. In 1883 this volcano decapitated itself in one of the most spectacular eruptions in history. As a few cubic miles of rock were blown out of the sea in the form of powdery dust, a tsunami was born which spread across the seas and ultimately drowned 30,000 people in Java and Sumatra. Nevertheless, tsunamis are usually associated only with earthquakes. When an earthquake occurs under the sea or on nearby land, the sea waves spread outward from the epicenter. Though the waves are concentric, they do not necessarily spread along circular wave fronts. They may, in fact, be highly

10-15
Tsunami damage in railroad yards at Seward, Alaska, 1964 Alaska earthquake. (U. S. Army photograph, from R. W. Lemke, U. S. Geological Survey)

10-16
Wrecked boats and other debris washed ashore by tsunami, 1964 Alaska earthquake. (U. S. Army photograph, from R. W. Lemke, U. S. Geological Survey)

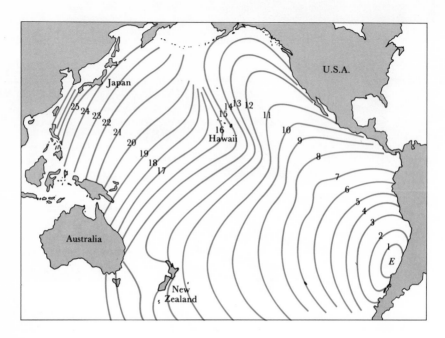

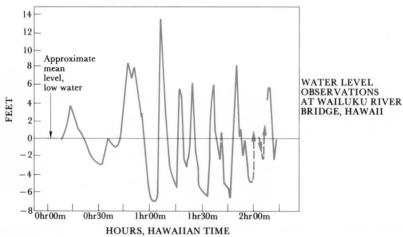

WATER LEVEL
OBSERVATIONS
AT WAILUKU RIVER
BRIDGE, HAWAII

10-17
*The tsunami of May 22-23, 1960, that occurred as a result of the large earthquake just off
the coast of Chile on May 22, 1960. The lines indicate the number of hours it took the
tsunami to arrive at various points across the Pacific after spreading from the epicenter
at E. The graph shows that the water level fluctuated as much as 20 feet at a station in
Hawaii. (Graph of the water level observations at Wailuku River Bridge, Hawaii, from T. H.
Eaton, et al., "Tsunami of May 23, 1960, on the Island of Hawaii," Bulletin of the Seismological Society of America 51, 2, 1961, p. 135. Reprinted by permission.)*

directional, as the Alaskan ones seem to have been. The tsunami sent out by the Chilean earthquake of 1960 is mapped in Fig. 10–17.

Tsunamis have great wavelengths of about 100 miles or more and move rapidly, sometimes as fast as 600 miles per hour. Their wave heights are not high in the open sea and are scarcely noticeable to ships; yet when they approach the shore, the waves build and sometimes reach heights of more than 50 feet. Such waves caused much of the damage, as well as 96 of the 115 deaths, in Alaska in 1964 and have been responsible for thousands of deaths throughout history. Their significance as an environmental hazard ranks with the shaking of the earthquake itself, and their destructiveness should not be underestimated. Japan is well acquainted with them, as illustrated by the folk tale. The older people at Chenega and other Alaskan villages were as wise about them as was the grandfather. Many areas around the Pacific have experienced them, yet even though they are most common in the Pacific, tsunamis occur in all the oceans. Probably the most infamous tsunami that occurred in the Atlantic was the one that overwhelmed the Lisbon seafront in 1755. Of an estimated 55,000 fatalities in the Lisbon earthquake, perhaps one-fourth to one-half were due to the wave.

Tsunamis are certainly highly destructive of life and property. As long as there are earthquakes and as long as man inhabits the coasts, he will have to deal with them. But unlike earthquakes, tsunamis *are predictable*, at least to the extent that warnings of them can be issued after an earthquake. The National Tsunami Warning Center issues warnings when tsunami-creating conditions occur. Although only a few moments warning may be given in the case of a nearby earthquake, it may be hours before a tsunami reaches a distant shore. Generally, tsunamis seem to occur in association with the larger earthquakes, especially those of shallower focal depth. Active investigations of earthquakes and tsunamis are being carried out continually, and as we learn more, we will be better able to defend ourselves.

Fire   Another destructive aspect of earthquakes is fire. Fire was responsible for much of the damage and loss of life in the Lisbon earthquake of 1755. In San Francisco in 1906, damage caused by the trembling itself is thought to have been about 20 to 25 percent of the whole, while the majority of the remaining damage was due to fire. In the Niigata, Japan, and Alaska earthquakes of 1964, fires were started when oil storage tanks were ruptured (by tsunamis, incidentally) and their flammable contents spread. Upset oil lamps and cooking fires have caused fires to spread, especially in the past or in underdeveloped countries where the use of these has been common. This happened in the great 1923 Tokyo-Yokohama earthquake, which occurred as the midday meal was being prepared. Aided by a breeze, fires raged through the area. Many thousands of people took refuge in the open area near the Military Clothing Depot in Tokyo when the coalescing fires raged out of control. However, a shift in the wind which brought the

10-18
The city of San Francisco burning two days after the earthquake, seen from the Oakland Ferry, 1906 San Francisco earthquake. (Photograph by W. C. Mendenhall, U. S. Geological Survey)

10-19
Fire damage, 1906 San Francisco earthquake. (Photograph by W. C. Mendenhall, U. S. Geological Survey)

flames their way resulted in perhaps 40,000 persons dying ghastly deaths as the oxygen around them was consumed or as they breathed superheated air.

Fire is obviously a potentially severe threat in wood or other non-fireproof construction, in concentrations of oil or chemicals, and in facilities in which gas mains may break. The danger is compounded where there is an inadequate or vulnerable water supply. Water mains were broken by the earthquake in San Francisco, and firefighters found themselves without water. The fires raged on, and only after three days were they contained by means of dynamiting and clearing lanes that could stop their spread. In the meantime, more than $300,000,000 worth of property was lost. Since that time auxiliary water systems have been installed to insure a continuing supply in the event of disaster. In any earthquake-prone area, people are advised to check their utilities and leave them turned off, to open windows to prevent gas buildup, and to await the "all clear" from the authorities.

Man Versus Earthquakes: Some Defensive Strategies

As long as earthquakes are part of man's environment, it is essential to find some satisfactory way of dealing with them. Two different approaches to this are to protect ourselves from them when they do occur and to prevent them in the first place. The actual prevention of an earthquake is a very remote possibility, so remote in fact that many earth scientists dismiss it entirely. Protecting ourselves from the effects of earthquakes, however, is a very real, and an entirely feasible, necessity.

In recent years there has been much talk about protecting ourselves from earthquakes, and some positive measures have been taken. One example is the aforementioned construction of an auxiliary water system in San Francisco. Many measures, indeed, have been taken, but the total size of the problem still outweighs them. After every earthquake, there is much talk about revising building codes and the like. But, in contrast to the seismograph, which operates because of inertia, man's *inaction* seems to stem from his own inertia. Thus, especially in this country, in spite of renewed talk after every earthquake about protecting ourselves, not much is accomplished. More action on the subject of earthquake safety is definitely needed.

Most people know that earthquakes occur frequently in Japan and California. What we specifically need to know, however, is the likelihood of a damaging earthquake occurring at any given time. The more we can refine our assessment of this, the better able we will be to defend ourselves. Detailed and accurate gathering and recording of earthquake infor-

10-20
Ground breakage in San Fernando, California, earthquake, 1971. Note compression as shown by bulging of open ground and compression of sidewalk and street. (Photograph courtesy U. S. Geological Survey)

mation is essential to making these refinements. Figure 10-21 shows the location of the seismically active areas of the United States. Obviously, earthquakes are concentrated in a few places, notably in the western United States. Using such data, a number of persons and agencies have prepared maps that show the damage risk. This information is shown along with the earthquakes and indicates that the high-risk zones merely outline the areas in which severe earthquakes have occurred.

However, there are problems in using this kind of information; for example, zone 3 appears in areas on both the east and the west coasts. South Carolina, which is rarely thought of as "earthquake country," is located in as high a risk zone as is most of California. Much of the basis for assigning such a high-risk value to South Carolina is simply the occurrence of a single severe earthquake in Charleston in 1886. What would the map look like if the Charleston earthquake had never occurred? Suppose the earthquake were to occur ten years from now: until that time the map would have been misleading, and Charleston might have been lulled into the belief that it was in a fairly "safe" area. Also consider Colorado. As far as is known, other than a now-famous swarm of tremors probably associated with the injection of fluid down a deep well near Denver, the state has had only one major earthquake in its settled history. But who can say that another major earthquake — or for that matter, a number of them — will not occur in Colorado? After all, there are scores of major faults in the area, some of which could become active again. In the final analysis we must continue to rely on seismic risk maps because they are still valuable guides for the planner and builder. We must also persist in mapping the seismicity of our country and the world in order to gain a better understanding of the geographic distribution of earthquakes.

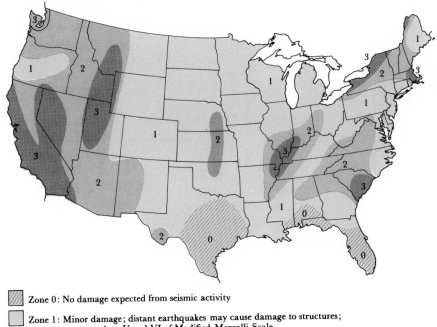

Zone 0: No damage expected from seismic activity

Zone 1: Minor damage; distant earthquakes may cause damage to structures; corresponds to V and VI of Modified Mercalli Scale

Zone 2: Moderate damage; corresponds to VII of Modified Mercalli Scale

Zone 3: Major damage; corresponds to VIII and higher of Modified Mercalli Scale

10-21
Zones of seismic risk in the United States. The zones are based on the distribution of historical, damaging earthquakes, their intensities, evidence of strain release, and distributions of geological structures related to earthquake activity. The frequency of possible earthquakes within the zones is not considered on this map.

In addition, it is essential to know about the *time* distribution of earthquakes, and again, cataloging helps us. Very small earthquakes occur continually, and those of M 3.0–3.9 take place about once every 11 minutes. This continual release of energy is further proof of the earth's dynamism. You should note that two or three earthquakes of M 5.0–5.9, shown to be damaging by the Modified Mercalli scale (intensity VI–VII), occur each day. If the 139 earthquakes with intensities greater than VII (magnitudes over 6) are added to these, it is obvious that there are many damaging earthquakes in the world annually or even monthly. Despite the fact that many of these are under the oceans or in uninhabited areas, it is still obvious that damaging earthquakes are an ever-present danger; this is

Table 10-2 Approximate relationship of magnitude and frequency of earthquakes

Magnitude	Energy-released[3] (ergs)	Expected annual incidence[1]	Felt area (sq. miles)	Distance felt (statute miles)[2]	Intensity (maximum expected Modified Mercalli)[3]
3.0 – 3.9	$9.5 \times 10^{15} - 4 \times 10^{17}$	49,000	750	15	II – III
4.0 – 4.9	$6 \times 10^{17} - 8.8 \times 10^{18}$	6,200	3,000	30	IV – V
5.0 – 5.9	$9.5 \times 10^{18} - 4 \times 10^{20}$	800	15,000	70	VI – VII
6.0 – 6.9	$6 \times 10^{20} - 8.8 \times 10^{21}$	120	50,000	125	VII – VIII
7.0 – 7.9	$9.5 \times 10^{22} - 4 \times 10^{23}$	18	200,000	250	IX – X
8.0 – 8.9	$6 \times 10^{23} - 8.8 \times 10^{24}$	1	800,000	450	XI – XII

[1] B. Gutenberg and C.F. Richter, *Seismicity of the Earth and Associated Phenomena*, Princeton, N.J.: Princeton University Press, 1954, p. 18.

[2] H. Benioff and B. Gutenberg, "General Introduction to Seismology," *Earthquakes in Kern County During 1952*, State of California, Division of Mines. Bulletin 171, San Francisco, 1955, p. 133.

[3] C.F. Richter, *Elementary Seismology*, San Francisco: Freeman, 1958, pp. 353, 366.

(Source: *Earthquakes*, ESSA, 1969, p. 11.)

10-22
Sag pond along the San Andreas fault, California. Such low spots help us to recognize and locate faults. (Photograph by E. F. Patterson, U. S. Geological Survey)

especially true in light of the huge amounts of energy involved. A single earthquake of magnitude 8 releases more energy than do all those of magnitude 3 in a whole year. If you examine Table 10-2 carefully, you will notice that most of the global energy release is accomplished by very few earthquakes.

Much of our knowledge of earth movements has come from measuring very subtle changes in the ground which have taken place over long periods of time. As strain accumulates along a fault, the earth on either side of the fault slowly bends. The bending can continue only to what the strength of the rock permits. When rupture of the rocks along the fault finally occurs, the rocks on either side rebound a bit.

Along faults, a slight movement, known as *creep*, often continues for an extended time. This movement may or may not be accompanied by noticeable earthquakes. Railroads, pipelines, or buildings that lie across the fault may be damaged as the creep continues. Evident creep has recently been observed along the Hayward fault, a large strike-slip fault which lies east of San Francisco Bay. Among other structures that lie on

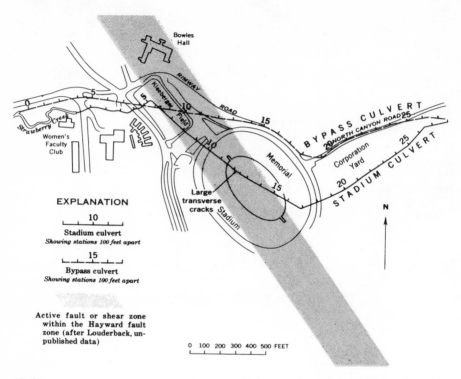

10–23
Location of the Hayward fault zone with respect to the football stadium, University of California, Berkley. (Source: Tectonic Creep in the Hayward Fault Zone, California, *U. S. Geological Survey Circular 515, 1966)*

the fault in this heavily populated area is Memorial Stadium on the Berkeley campus of the University of California (Fig. 10–23). The Hayward fault runs directly through the stadium and down the length of the football field. Creep along the fault has caused some damage to a large culvert that lies under the stadium.

Movement between points on opposite sides of faults is being carefully measured in more and more places in an effort to understand the behavior of faults. Once this can be done, it may considerably help us to protect ourselves and also to predict earthquakes. Strain seismographs are installed at two points about 100 feet apart. A rigid rod is extended between a solidly anchored pier at one point and a similar position at the other point. An electronic signal, the character of which depends on the velocity or displacement of the points, is produced and indicates movement between the points.

10-24
A strong-motion accelerograph. (Photograph courtesy Kinemetrics, Inc., San Gabriel, California)

Another aspect of seismology that is developing rapidly and will aid in the determination of the effect of seismic waves on structures is strong-motion seismology, which deals with motions of the ground that are large enough to affect structures. Instruments called *strong-motion accelerographs* measure the acceleration of the moving earth. *Velocity* is speed in a particular direction; if either the speed or the direction changes, the velocity changes. The *rate* at which this change takes place is called *acceleration.* Readings from strong-motion accelerographs are usually expressed in relation to the acceleration due to gravity, which equals 32 feet per second per second. The Long Beach, California, earthquake (1933) gave us some of our first strong-motion readings, and the Imperial Valley earthquake (1940) provided more. Readings of about 0.25 of the acceleration due to gravity, on

which some of the buildings codes were based, have been recorded. Unfortunately, these values of g are beginning to appear to be too low, but
until recently they formed a good share of the available information. Only
in Japan and on the West Coast of the United States are there effective
networks of accelerometers. The lack of information is underscored by
the fact that strong-motion data from all of the following major earthquakes are completely lacking: Mexico, 1957; Chile, 1960; Agadir,
Morocco, 1960; Iran, 1962; Skopje, Yugoslavia, 1963; Alaska, 1964; and
Turkey, 1966.

It is necessary to consider the effect of earthquakes on property. Property damage in the San Francisco earthquake of 1906 amounted to about
$400 million; that in the San Fernando earthquake of 1971 was roughly
comparable. The San Fernando earthquake, however, was not nearly as
strong ($M = 6.6$) as the San Francisco quake, and it therefore caused much
less property damage. This increased dollar value for less property damage
is, of course, indicative of the rise in prices and property values since 1906.
Each year about $10 billion worth of construction is done in California,
which suggests the size of the investments we need to protect. Yet, these
figures omit the cost in lives and injuries caused by the collapse of structures.

To a great extent, the degree to which the works of man are threatened depends on three things: the kind of structure, the design of the
structure and the quality of construction, and the placement of the structure. Let us look at each of these. First, experience has indicated that,
depending on the geologic conditions, certain structures are suitable for
some areas, whereas others are not. After the 1964 Alaska earthquake,
it was recommended that construction in certain areas in which slides had
occurred be limited to light buildings not over two stories high. Massive
and multistoried buildings were felt to be unsuitable for some areas.

From time to time questions arise as to the feasibility of building a
variety of structures in certain areas. Part of the opposition to the construction of a trans-Alaska oil pipeline, for example, stems from the knowledge that it will cross territory prone to earthquakes and that the rupture of such a line could cause damage as oil spilled over the countryside.
There was similar opposition to the construction of a nuclear power plant
in Colorado at about the time that the swarm of earthquakes was occurring
in the Denver area. Fear was expressed that an earthquake could damage the
plant and cause an ecological disaster as a result of escaping radioactive
materials.

The proper design of earthquake-resistant structures is, in large part,
an engineering problem. But at the same time, it is only with knowledge
of the seismicity of an area and an understanding of ground motion and
bedrock and soil characteristics that an architect or engineer can incor-

10-25
Collapsed overpass connecting Foothill Boulevard and Golden State Freeway, San Fernando, California, earthquake, 1971. (Photograph courtesy U. S. Geological Survey)

porate safety factors into his designs. Many cities have building codes that require protection, yet few codes are adequate. Surprisingly high values (some over 1 g) for ground acceleration were recorded in the 1971 San Fernando earthquake. Building codes here, as well as in other parts of the world, that have used older data must now be updated to include more stringent requirements. Even the San Fernando earthquake, the best monitored of all the earthquakes, which provided much data, still did not indicate enough about the behavior of the earth.

Some of the effects of the San Fernando earthquake on buildings are revealing. First, most of the buildings that had been constructed according to the more recent building codes, in which earthquake-resistant design was required, withstood the quake reasonably well. Schools built since the passage of the Field Act did not suffer great damage, whereas many older schools were severely damaged. The Field Act, passed after the 1933 Long Beach earthquake, requires earthquake-resistant construction in school buildings. The old Sylmar Veterans Hospital underwent much destruction:

10-26
Damage to Van Norman Lake Dam, San Fernando, California, earthquake, 1971. At the time the picture was taken, much water had already been released from the reservoir to prevent further disaster. (Photograph courtesy U. S. Geological Survey)

two of its buildings collapsed, and the deaths of 44 of its inhabitants accounted for two-thirds of the deaths in the San Fernando earthquake. On the other hand, parts of the new reinforced-concrete Olive View Hospital, which were supposedly earthquake-resistant, also collapsed. A number of new bridges and roadways on the freeway system, which were constructed to withstand the shock of earthquakes, failed to do so. The immediate area has few tall buildings, but instrument readings in tall buildings in Los Angeles, 25 miles away, suggested that a severe earthquake could cause them problems.

As the ground shakes, so, too, do the structures. The structures have different periods of vibrations, depending on such characteristics as height. (The period of vibration is the length of time between vibrations.) The "right" earthquake — a strong one — might establish a period of vibration within a building that could cause the building to fall. It is essential for us to determine the behavior of structures influenced by vibrations. Finally, it is mandatory that building codes be drafted accordingly and that their provisions be enforced. It is also apparent that some structures, built according to otherwise adequate codes, have nonetheless failed. Many times we see that the reason for failure is not necessarily an inadequate code, but rather the use of inferior materials or even shabby or dishonest workmanship. Variations exist in materials; concrete is an example. Most manufacturers run careful quality-control tests to be sure that their products meet the required standards. It is necessary that we do this in our construction too.

10-27
Damage due to sinking of building on weak foundation soil, Niigata, Japan, 1964. (Photograph courtesy T. Minakami, Earthquake Research Institute, University of Tokyo)

The Field Act regulated school construction in California with fairly strict standards. One legislative omission, however, was a provision defining *where* schools could be built; consequently, well-built schools could possibly end up being located on a fault. Fortunately, the law has been amended properly. Expensive homes at Turnagain Heights in Anchorage, Alaska, were carried away in a massive landslide in the 1964 earthquake. Well-constructed buildings at Niigata, Japan, simply toppled over during the Niigata earthquake. Although they were barely damaged, they were useless in that position.

Such problems are ones of *siting*. It is obvious that we must know the geology of the sites where we plan to construct our buildings. Three-dimensional mapping, which takes into account not only the surface, with its two dimensions, but also the subsurface, in which soils and rocks can be studied at depth, is essential. Excavation, core drilling, and the information obtained from wells help us to do this. Since faults often appear at the surface of the ground, we can easily avoid placing structures at these locations. Other problems to be considered in siting include the settling and sliding of the ground. Areas of bedrock are most likely to be free of these problems, but areas of loose, uncompacted ground have been troublesome. Figure 10-28 shows a direct relationship between the areas of artificial landfill and the area of greatest seismic intensity in the 1906 San Francisco earthquake. The Ferry Building dropped about two feet as the land under it settled. At Niigata, the problem was one of *liquefaction* of the sandy

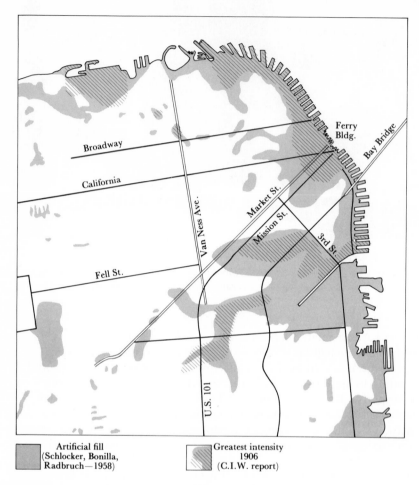

10–28
Relationship between intensity of earthquake and filled land, San Francisco, 1906. (G. O. Gates, "Earthquake Hazards," in Geologic Hazards and Public Problems, *Conference Proceedings, ed. Robert A. Olson and Mildred M. Wallace, Santa Rosa, Cal.: Office of Emergency Preparedness, 1969.)*

soil. Most people have heard of quicksand, a water-saturated, loose sand having negligible strength. Many areas consist of ground of this type; although they are firm enough to walk on and even to build on, they liquefy and fail (becoming "quick") when shaken in an earthquake. Severe damage is usually done to the structures as the ground settles or slides. Struc-

10-29
Wreckage of school, Anchorage, Alaska, due to sliding of weak, unconsolidated sediments. The earthquake occurred after school hours. (Photograph by W. R. Hansen, U. S. Geological Survey)

tures can sometimes be built on loose, uncompacted, wet ground, but this presents special engineering problems. Too often in the past, structures have been built on such ground due to either carelessness or a lack of understanding of the behavior of these materials.

Prediction of earthquakes In terms of practicality, the ability to predict earthquakes is urgent. The results of two recent earthquakes illustrate this point. One of these was the 1971 San Fernando, California, earthquake. The shocks took place at 6:01 A.M., local time, when there were very few people on the highways, no children yet in school, and only a few stores or businesses open for the day. As noted previously, the newer schools generally withstood the shocks well, but the question of what the casualty lists would have looked like in the older schools if the earthquake had taken place during school hours and had been a little more severe still

Table 10-3 Populations of inhabited places damaged in 1964 Alaska earthquake, with total populations of Alaska, California, and Japan

Afognak	190
Anchorage, including military	82,833
Cape St. Elias	4
Chenega	80
Chugiak	51
Cordova, including airport	1,168
Eagle River	130
Ellemar	1
Girdwood	63
Homer	1,247
Hope	44
Kodiak Fisheries Cannery	2
Kaguyak	36
Kodiak, including Naval Air Station	4,788
McCord	8
Old Harbor	193
Ouzinkie	214
Point Nowell	1
Portage	71
Port Nellie Juan	3
Seldovia	460
Seward	1,891
Valdez	1,000
Whittier	70
	94,548

	1960 Population	1970 Population
Alaska	226,167	294,607
California	15,717,204	19,696,840
Japan	93,418,501	104,700,000 (1971 est.)

haunts the people. Of possibly greater concern is the mayhem that would have occurred on the highways had the earthquake taken place at the height of the infamous California rush hour. The freeway system was heavily damaged, and although two persons died when an overpass collapsed on a truck, the numbers of deaths could have been compounded an hour and a half later.

The second example is the 1964 Alaska earthquake, which caused 130 deaths. The National Academy of Science has pointed out that casualties would certainly have been higher had the earthquake occurred during school or working hours, at night, in very cold weather, at high tide, at the height of the construction season three months later, or at the height of the tourist and commercial fishing seasons four months later.

Let us also consider the size of the population and the area involved. Table 10–3 gives the populations of the damaged communities in the part of Alaska that was affected. The total population of that area was about 100,000. Since this land area comprises 50,000 square miles, the population density was only two persons per square mile. Japan, however, had 655 persons per square mile in 1960, and California had 100.4. By 1970 the density in Japan had risen to about 695; in California, to about 124.1. If an earthquake of 8.4 magnitude were to strike Japan or California, the effects could be horrifying.

Obviously, a correlative to protection is prediction. Assuming adequate building codes and proper emergency planning are enacted, hundreds or thousands of injuries or deaths could be prevented by being able to announce an impending earthquake. Both the time and place of a predicted earthquake must be known as accurately as possible, and the populace must be presented with understandable evidence which will so convince them of the danger that they will react properly.

What is involved in the prediction of earthquakes? Our understanding of natural phenomena that relate to earthquakes must be greatly increased by means of a collection of an enormous amount of data about earthquakes. The most important development has probably been the establishment of a worldwide network of seismograph stations (Fig. 10–30). These stations report on seismic events and thus help to build a great storehouse of data which we can use to learn more about the earth.

The studies of earthquake prediction are proceeding along a number of lines. The various phenomena now being studied include accumulation of strain in rocks, creep along faults, micro-earthquakes, distribution patterns of earthquakes, the time distribution of seismic activity, magnetic and thermal effects in the earth, and the nature of earth motions, including rotation, revolution, and wobble.

The use of strainmeters to study subtle movements along faults has previously been described. Since much can be learned from the information provided by strainmeters, a sufficient number of strainmeter stations carrying out measurements for a relatively long time would be very useful in helping to predict earthquakes.

After one earthquake in Germany, an increase in the concentration of methane gas was noted in a research shaft in the ground. A scientist suggested that subtle earth movements preceding the earthquake allowed the gas to rise through a fault and enter the shaft. The concentration of gas returned to normal, but because it later increased again the scientist predicted an aftershock, which did occur shortly thereafter.

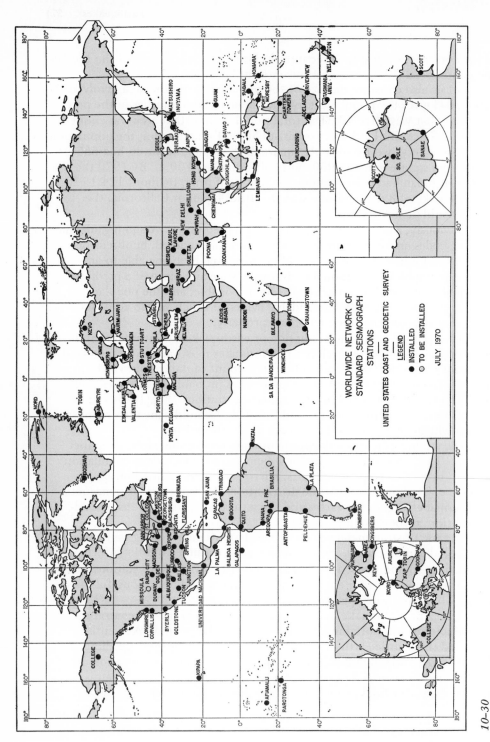

10-30
Worldwide network of standardized seismograph stations. (Map courtesy of J. Oliver and L. Murphy.)

Surveys in Sicily have been made from the air by using both infrared film to photograph hot spots where the heat flow from the earth is greater than average and conventional film to photograph fractures and damage patterns. If hot spots are found to occur along fractures and it is discovered that heat and earthquakes are related, continued monitoring of such areas could lead to accurate earthquake predictions.

It has been observed in a number of instances that changes have taken place in the earth's magnetic field just prior to an earthquake. This could possibly be the result of the effect of stress on the magnetic characteristics of the rocks just prior to an earthquake. Careful monitoring of the magnetic field with sensitive magnetometers may provide us with more information and another possible means of predicting earthquakes.

Through the use of tiltmeters, it has been discovered that differences in the amount and direction of tilt have occurred prior to some earthquakes. Tiltmeters measure the inclination of the ground and can detect very minute changes in the amount and direction of the tilt. Such changes were observed prior to earthquakes in Japan as long ago as 1943. In California, tiltmeters showed a change in the direction of regional tilt toward the town of Danville about 29 days before an earthquake occurred there. The amount of tilt increased until just ten hours before the earthquake and then ceased. After the earthquake, the direction of tilt changed back to approximately its original position. This change suggests that the epicentral area had sprung upward, relieving the accumulated strain. This evidence correlates with the elastic-rebound theory of earthquake action.

Interpretations of seismograms have indicated that strain release preceding large earthquakes often varies considerably from the usual minor seismic activity of the area. If enough seismic data are accumulated to accurately compare the "normal" activity with its variations, then this could possibly prove to be a useful tool in predicting earthquakes.

Scientists at the Arctic Institute of North America have been able to correlate increases in the frequency of earthquakes with the passage of the moon through its apparent point of greatest apparent latitude. During one period of study, 43 percent more than the average number of earthquakes occurred. Studies of this nature have continued and may be promising in predicting earthquakes, since various astronomical events occur regularly and are therefore themselves predictable.

Prevention of earthquakes The question of whether earthquakes can be prevented is often raised. Although the probable answer is no, the feasibility of preventing large, damaging ones has been seriously considered. Even though there apparently is no way to stop the forces which act within the earth, it might someday be possible to release the accumulating strain gradually rather than let it build to a point where a major earth-

10-31
A geologist checks a tiltmeter installation near the San Andreas fault. The tiltmeter situated on a concrete slab at the bottom of the vault measures changes in the attitude of the ground and sends the data to Sacramento for interpretation. (Photograph courtesy Autonetics Division, Rockwell International)

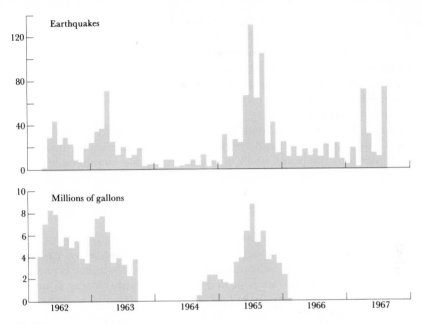

10-32
Relationship between amount of liquid waste pumped down injection well and occurrence of earthquakes near Denver, Colorado. (Reprinted from Colorado School of Mines Quarterly **63**, *1, p. 13, by permission of the Colorado School of Mines. Copyright © 1969 by the Colorado School of Mines.)*

quake occurs. In 1963 the Rocky Mountain Arsenal decided to dispose of noxious waste products by pumping them down a 12,000-foot well northeast of Denver, Colorado. Almost immediately after the pumping started, many small earthquakes began to occur in the Denver area. Denver geologist David Evans showed a startling correlation between the pumping of the waste water and the pattern of earthquake occurrences (Fig. 10–32). He suggested that the water, acting essentially as a lubricant, permeated fractures in the Precambrian rocks and allowed slippage to occur and strain to be released. Because of the strong evidence, pumping operations were halted. The earthquakes, however, continued for a while, causing some scientists to doubt whether injection of the waste water actually caused the earthquakes. But the evidence favoring injection as the cause was too convincing for it to be seriously questioned.

tions on the relationship between earthquakes and fluid removal or injection. One experiment was performed at the Rangely Oilfield in northwest Colorado. Here, too, it was observed that injection of water was accompanied by minor tremors. After a number of such observations, suggestions that water be pumped down wells along such great faults as the San Andreas naturally arose. "Greasing the skids" in this manner would allow gradual release of strain through a series of small and, hopefully, imperceptible earthquakes. However, our present knowledge offers absolutely no assurance that such a program of injecting fluid would be either effective or controllable.

Suggestions have also been made to the effect that underground detonation of nuclear devices would help to release the strain and thus prevent its buildup to the point where a major earthquake would occur. In addition to the questions of effectiveness or controllability, the one of radioactive pollution arises. The possibility of radioactive pollution of the ground would seem to make detonation of nuclear devices an unacceptable option.

There are many ramifications of the problem, including the issues of cost, liability, and technical aspects. If a damaging earthquake is caused by fluid injection, for example, who is liable if the wall of a house is cracked, if a valuable piece of glassware is jiggled off a shelf and shatters, if one person is injured, or if hundreds are killed? Dr. Robert Hamilton of the United States Geological Survey was quoted as saying, "it would be one thing if God caused it, and quite another if the Geological Survey caused it."

All in all, we must admit that we simply do not have adequate knowledge of the seismicity of the earth, the cause of earthquakes, the local or regional geology of most areas, or the dynamic behavior of earth materials. Earthquake prevention does not presently appear to be possible, let alone promising. Still, it is dangerous and foolish to say something can never be done, because never is a long, long time.

VOLCANOES

Volcanoes have always intrigued man and although they have dealt him some severe blows, they have provided him with fertile soil, various economic products, and a subject of beauty, romance, and mythology. Man has long been intimately associated with volcanoes, and many volcanoes are located in such heavily populated areas as the Mediterranean, Japan, and Indonesia. Figure 10-33 shows that volcanoes are located in the major earthquake belts, and among the active volcanoes, 62 percent are situated around the Pacific rim. A few more volcanoes are found in the central Pacific, while most of the remaining third of the world's volcanoes are located along the Mid-Atlantic Ridge, the Caribbean arc, the Mediterranean, in northern Asia Minor, and in the East African rift area.

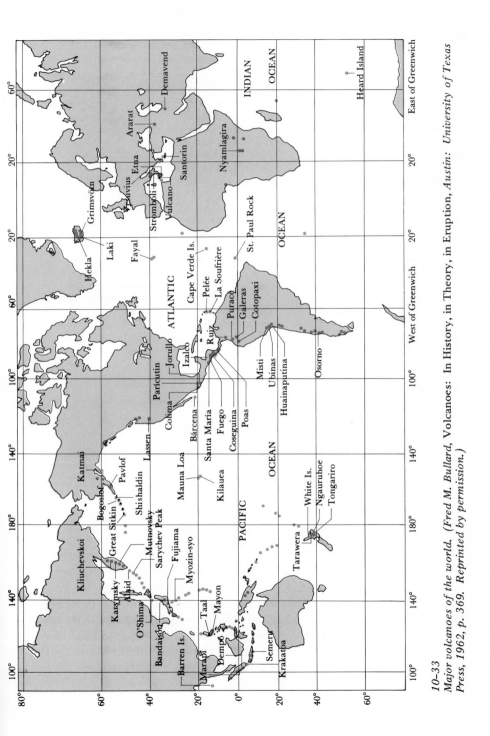

10-33
Major volcanoes of the world. (Fred M. Bullard, Volcanoes: In History, in Theory, in Eruption, Austin: University of Texas Press, 1962, p. 369. Reprinted by permission.)

Table 10–4 Products of volcanic eruptions

Gases	Lava-derived products, with typical rocks	Fragments from ejected lava	Ejected solid fragments
Important: steam, H_2O	basic lava (less than half SiO_2) — basalt	ashes — usually a mixture of glassy, cooled lava and pulverized rock from the walls of the vent	lapilli — 2 mm to 2 cm in size
hydrogen, H_2			blocks — up to many tons
hydrogen chloride, HCl	intermediate lava (one half to two thirds SiO_2)		. . . . The composition
hydrogen sulfide, H_2S	— andesite — trachyte	pumice — various sizes of porous, glassy, solidified	of lapilli and blocks is variable — may be old
carbon monoxide, CO	— dacite and others	lava fragments	lava rock, scoriae, nonmagma igneous, sedimentary, and
carbon dioxide CO_2	acidic lava (more than two thirds	cinders (scoriae) — moderately vesi-	metamorphic wall rocks
hydrogen fluoride, HF	SiO_2) — rhyolite	cular fragments	
. . . . Minor: methane, CH_4	— obsidian (glass)		
ammonia, NH_3	. . . .	bombs — large, solidified globs of	
hydrogen thiocyanate, HCNS	Rocks of the intermediate and acidic	lava	
nitrogen, N_2	groups form light-colored rocks, collec-	. . . . compacted rocks composed of: tuff (mostly ash)	
argon, Ar	tively referred to as *felsite*	breccia (various fragments, usually angular)	

Products of Volcanism

The products of volcanic activity consist mainly of the material that originates in magma and include steam and other gases which escape from the magma. Basalt, felsite, and other rocks are formed through the solidification of lava, which is the molten fraction of the magma that reaches the surface. Other products include cooled bits of lava that are explosively

10-34
Lava fountain and cinder cone, Kilouea Volcano, Hawaii, 1955. (Photograph by G. A. MacDonald, U. S. Geological Survey)

discharged, ranging from fine (ash) to large (bombs) particles. Fragments of other rocks that are present in the vent, including old volcanic or any other kind of rock, may also be hurled out. Blocks of limestone from strata situated at depth, for example, were among the products ejected by Mt. Vesuvius. A summary of volcanic products is shown in Table 10–4.

It is worth noting some characteristics of the products of volcanic eruption. Until the eruptive activity ceases and the products become stabilized, we find that these products are, variously:

> hot, sometimes incandescent
> flammable
> poisonous
> wet
> voluminous, bulky, heavy
> mobile—flowing, floating, or drifting.

The hot materials may set fires or kill people, as noted earlier (Chapter 9) in regard to the incandescent cloud that destroyed the city of St. Pierre, Martinque, as Mt. Pelée erupted in 1903. Flammable gases, such as hydrogen and carbon monoxide, are usually consumed when released into the air. However, this does not constitute a particularly significant environmental hazard, since this occurs in the crater or otherwise very near the site of activity. Poisonous gases, however, may be distributed farther away from the crater by the wind and can be dangerous to nearby populaces. Most of the gases emitted and most of the products of

Table 10–5 Examples of volumes of pyroclastic flows and areas covered by pyroclastic flows

Volumes

New Zealand	8300 km^3
San Juan Mountains, Colorado	9500 km^3
Mt. Katmai, Alaska	26 km^3 in a 60-hour eruption

Areas Covered

Valles Mountains, New Mexico	900 km^2
Lake Toba, Sumatra	$25,000 \text{ km}^2$
New Zealand	$26,000 \text{ km}^2$

(Data from Ollier, 1969)

chemical activity after emission are poisonous. Therefore, the atmosphere around a volcano is certainly more hazardous than the air of downtown Tokyo or Los Angeles during rush hour.

Let us examine in some detail the volcanic avalanches briefly looked at in Chapter 9. Volcanoes give off enormous quantities of steam which condenses in the air; forms clouds; and, converted to water, falls as rain during and after the eruption. This adds to the misery of refugees and, more importantly, creates a first-order environmental hazard by causing mudflows.

Great quantities of fine ash, lapilli, or other fragments accumulate during the eruption and pile up on unstable slopes. When the rains, often heavy, occur, mud forms and flows, as does lava. Such mudflows are called *lahars*. Sometimes, explosions or open fissures breach the shores of crater or other lakes and release the water. The great flow of water, mixed with ash and other debris, may move rapidly, destroying everything in its path as it flows through valleys that extend downward from the volcano. This type of flow has been very common in Indonesia and is the type to which the term *lahar* was originally applied. Herculaneum, an important Roman town, was totally buried by a mudflow in the A.D. 79 eruption of Mt. Vesuvius. The lahar apparently did not trap many people—only about 30 skeletons have been found so far—but it completely buried the town. The burial was so complete, in fact, that Herculaneum was lost and forgotten. The town of Resina was built on top of the 65-foot thick lahar, which by then had formed a more or less coherent rock, called *tuff* (see Table 10–4). Not until almost 1600 years later did diggers discover Herculaneum, and even then, the discovery was accidental. In short, wet volcanic products, usually manifested as mudflows, constitute a definite geological hazard.

A notable aspect of many eruptions is the enormous quantity of material spewed out by the volcano. Pompeii, the famous Roman city, was,

10–35
Debris flow near Cartogo, Costa Rica. Eruption of Irazú, 1964. (Photograph by H. H. Waldron, U. S. Geological Survey)

in a few hours, completely and deeply buried by ash from the A.D. 79 eruption of Mt. Vesuvius. The ash fell so rapidly, making breathing so difficult, that approximately 16,000 people died.

Broad sheets consisting of ash and other materials have been deposited essentially as *flows* of the dense mixtures that moved out from volcanoes over the surrounding countryside. These *pyroclastic flows* have buried extensive areas, as shown in Table 10-5.

Ash from the 1883 eruption of Krakatoa was hurled 50 miles high and distributed around the entire world. The extra atmospheric dust from that eruption was noticeable, particularly as it caused brilliantly colored sunsets for a year or more. Tennyson noted this in *St. Telemachus*:

> *Had the fierce ashes of some fiery Peak been hurled so high*
> *they ranged round the world, for day by day through many*
> *a blood-red eve the wrathful sunset glared.*

The rain of ash in Java was so heavy that refugees could not see and became lost.

This situation was presumably similar to what had happened at Pompeii. Many reports of the great eruptions note the fact that total darkness occurred, even in midday, due to the ash in the air. Parícutin, in Mexico, first erupted on February 20, 1943. By the middle of September it had emitted nearly two billion cubic yards of solids. Tile roofs were cracked, and trucks had to continually haul away load after load of ash in Uruapan, 15 miles northeast of the volcano. Here is what one account tells:

> *Poor Uruapan! Not so long ago it was called the Vale of*
> *Flowers! Now its 20,000 residents endlessly swept dust*
> *from walks, patios, and roofs. They were fearful of the*
> *day when the rain of ash would be too deep for brooms —*

10–36
Destruction of vegetation by ashfall from eruption of Irazú Volcano, Costa Rica, 1964. (Photograph by H. H. Waldron, U. S. Geological Survey)

> *yes, even for the hoes some of them were using. Street lights were turned on in dusty daylight.*[*]

On a broader scale, the eruption of Irazú in Costa Rica spread ash widely over the countryside, including the city of San Jose. This rain of ash lasted from 1963 to 1965 and caused losses to Costa Rica of about $150,000,000.

Lava, too, is a hazard and a nuisance. Persistent breakouts of lava flows from Vesuvius, Etna, Mauna Loa, Kilauea, and many other volcanoes have sent torrents of lava down through fields and villages. In 1669 courageous men from Catania broke holes in the crust of lava flowing from Mt. Etna and diverted the flow. This was quite a feat, as they had only iron bars to poke holes with and were protected from the heat only by wet animal skins. Unfortunately, the diverted flow headed for the town of Paterno, whose citizens, understandably unhappy about this, attacked and drove away the Catanians before the stream was totally diverted. Guns, whisky, and women have been the cause of many a brawl,

---

*J.A. Green, "Paricutin, the Cornfield That Grew a Volcano," *National Geographic* **LXXXV**, 2, Feb. 1944, p. 136. Reprinted by permission.

but this seventeenth-century Sicilian brawl was probably the first one ever fought over a lava flow! The Catanians lost more than the fight, since the lava did destroy much of Catania.

In 1935 volcanologist T. A. Jaggar suggested bombing a lava flow in Hawaii. The Army Air Corps dropped 600-pound bombs on the flow, scoring some hits, and caused the lava to flow through the wound in the encrusted part of the flow. This saved the parts of the city of Hilo that lay in the initial path of the lava.

Dams and diversion walls have been built in various places to keep lava out, but their success has been only moderate. The best plan, it would seem, would be to place homes, buildings, and towns on ridges away from likely lava channels.

On a number of occasions, volcanic eruptions have dumped vast quantities of pumice into the sea. Pumice is a volcanic rock consisting essentially of glassy froth. Because it contains a large amount of air spaces, it floats and consequently may constitute a navigational hazard. Ships found it difficult to penetrate the waters near Krakatoa after the island blew up. Blocks of pumice, part of the approximately five cubic miles of rock hurled out in that eruption, covered miles of the sea's surface. Some of these blocks floated for months before they became sufficiently waterlogged to sink.

Dionisio Pulido, a Tarascarán Indian who owned a small farm in western Mexico, probably had the most intimate of all personal experiences with a volcano. Pulido and some others were working in his field when the Parícutin volcano was literally born beneath their feet. When ashes, gases, and then cinders began to spurt out of a crack in the ground, Pulido and his friends fled. By the end of the day, a cone had formed, and the ash had covered the field. Within weeks, first Pulido's farm and then those of his neighbors were buried. Within months the villages of Parícutin and San Juan de Parangaricutiro were buried under ash, cinders, and lava.

From the brief illustrations we have presented, you will probably conclude that volcanism can be a significant environmental hazard. Much property has been destroyed and about a quarter of a million people have died in historic eruptions. One should assess these figures in light of property losses and casualties suffered in the other natural and manmade disasters such as wars, famines, floods, and earthquakes. Although it is apparent that volcanoes have not caused the damage that these other disasters have, they are, nevertheless, significant hazards. What would happen, for example, if Mt. Rainier were to erupt today? We know that it erupted many times in the past 10,000 years — specifically, in 1820, 1843, 1846, 1854, 1858, 1870, and 1894. Mt. Lassen erupted in 1914, spreading ash as far away as Winnemucca, Nevada, 200 miles to the east. One writer admitted that there is still evidence of continuing volcanism in the Cas-

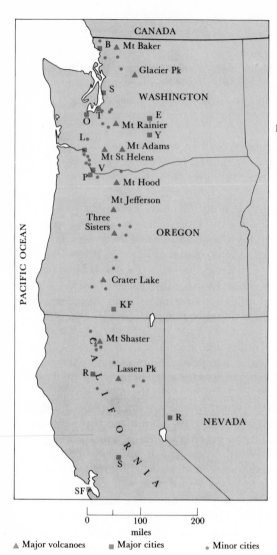

B: Bellingham
S: Seattle
T: Tacoma
O: Olympia
E: Enumclaw
Y: Yakima
L: Longview
V: Vancouver
P: Portland
KF: Klamath Falls
R: Redding, Calif.
R: Reno, Nev.
S: Sacramento
SF: San Francisco

10-37
Major volcanoes of the Cascade Range, showing locations of cities and towns in their vicinity. (D. R. Crandell and H. H. Waldron, "Volcanic Hazards in the Cascade Range," in Geologic Hazards and Public Problems, Conference Proceedings, ed. Robert A. Olson and Mildred M. Wallace, Santa Rosa, Cal.: Office of Emergency Preparedness, 1969.)

▲ Major volcanoes   ■ Major cities   ● Minor cities

cade Mountains, but said that "there is no concern that any of the Cascade volcanoes will reduce the communities about them to modern Pompeiis or Herculaneums."*

—————————
*Grant McConnell, The Cascades: Mountains of the Pacific Northwest, ed. Roderick Peattie, New York: Vanguard, 1949, p. 81.

10-38

Mt. Hood, Oregon, one of the large volcanic peaks of the Pacific Northwest. (Photograph courtesy Oregon State Highway Department)

No concern, indeed! Scientists regard future eruptions of Mt. Rainier or any of the other volcanoes of the Cascades (Fig. 10–37) as not only likely, but also certain. Understanding the possibilities for disaster posed by these volcanoes makes it impossible to express no concern. The main hazard appears to be that of mudflows, which we examined in Chapter 9. There is a direct relationship between what happened at Herculaneum and what *could* happen at Enumclaw, Washington, were Mt. Rainier to erupt and cause a mudflow. In addition to the hazards posed by mudflows are the hazards such as those of floods caused by the damming of rivers by mudflows or avalanches, or from ash falls, lava flows, and forest fires.

The next very logical question would be to ask what can be done about volcanic hazards. As seen in our study of earthquakes, courses of action relate to protection and prediction. No serious thought has been given to the prevention of volcanic eruptions, as stopping them would be about as difficult as preventing the continents from drifting. Protection from volcanic disasters should focus primarily on situating towns and other manmade structures in the safest places possible. Obviously, lava flows and mudflows normally travel down valleys. Topographic and geologic mapping should facilitate proper siting assessments. Better construction methods and building codes perhaps will be able to insure that structures will not collapse under the weight of ash falls. Certainly, buildings can be safeguarded against earthquakes that are often associated with eruptions.

A number of measurements and observations may be useful in predicting volcanic eruptions. Patterns of seismic activity and tiltmeter measurements are useful, as they may suggest the upward movement of magma in a reservoir. Accurate mapping of the types of rocks surrounding a volcano provides information about the past history of eruptions and perhaps clues as to the nature of future eruptions. Some volcanoes, such as those in Hawaii, emit mostly lava, and lava flows constitute the greatest danger from these volcanoes. At the other extreme, however, are volcanoes like Mt. Pelée (Martinique) and Soufrière (St. Vincent), which emit ash, scoriae, and other products in explosive eruptions.

We can also monitor the gases and lava in the craters of volcanoes. Changes in the kind or quantity of gases given off or in the levels of lava pools may indicate impending eruptions. Heat-flow measurements can also be made on a continuing basis; sudden increases in temperature may be warnings of eruptions.

In summary, we have seen that earthquakes and volcanoes are an integral part of man's environment. They are widespread and affect heavily populated areas. They have done much harm in the past, killing perhaps one million persons. As populations grow and spread, these natural phenomena will become potentially even more destructive. Nevertheless, these hazards can be reduced, mainly by educating, monitoring and studying the phenomena, establishing protective building codes and practices, and planning for emergencies.

REFERENCES

Adams, W.M. "Tsunami effects and risk at Kahuku Point, Oahu, Hawaii," *Engineering Geology Case Histories,* **8** (1970): 63–70.

Bernstein, J. "Tsunamis," *Scientific American,* **191,** 2 (1954): 60–64.

Bullard, F.M. *Volcanoes in History, in Theory, in Eruption,* Austin: University of Texas Press, 1962.

Crandell, D.R. "The geologic story of Mount Rainier," *U.S. Geological Survey Bulletin,* 1292, 1969.

Crandell, D. R. and D. R. Mullineaux. "Volcanic hazards at Mount Ranier, Washington," *U. S. Geological Survey Bulletin,* 1238, 1967.

Crandell, D.R. and H.H. Waldron. "Volcanic hazards in the Cascade Range," in *Geologic Hazards and Public Problems,* Conference Proceedings, 1969, pp. 5–18.

Eardley, A.J. *General College Geology,* New York: Harper & Row, 1965.

Evans, D. M. "The Denver area earthquakes and the Rocky Mountain Arsenal disposal well," *Engineering Geology Case Histories,* **8** (1970): 25–32.

————. "Man-made earthquakes—a progress report," *Geotimes,* **12,** 6 (July-August 1967): 19–20.

Gates, G. O. "Earthquake hazards," in *Geologic Hazards and Public Problems, op. cit.,* pp. 19–52.

Greensfolder, R. W. and Douglas Crice. "Geodimeter fault movement investigations in California," *California Geology* 24, 6 (June 1971): 105–109.

Gutenberg, B. and C. F. Richter. "Seismicity of the earth and associated phenomena," reprint, New York: Hafner, 1965.

Hagiwara, Takahiro. "Prediction of earthquakes," *The Earth's Crust and Upper Mantle,* Geophysics Monograph 13, American Geophysical Union, 1969, pp. 174–176.

Hansen, W.R. *The Alaska Earthquake, March 27, 1964: Field Investigations and Reconstruction Effort*, U. S. Geological Survey Professional Paper 541, 1966.

Heck, N. H. *Earthquakes*, New York: Hafner, 1965.

Hodgson, J. H. *Earthquakes and Earth Structure*, Englewood Cliffs, N. J.: Prentice-Hall, 1964.

Jaggar, T.A. *Volcanoes Declare War*, Honolulu: Paradise of the Pacific, 1945.

Lander, J. F. "National earthquake information center," in *Geologic Hazards and Public Problems, op. cit.*, pp. 197–198.

Leet, L.D. and Florence Leet. *Earthquake: Discoveries in Seismology*, New York: Dell, 1964.

Major, M. W. and R. B. Simon. "A seismic study of the Denver (Derby) earthquakes," *Quarterly of the Colorado School of Mines* 63, 1, pt. A (January 1968): 9–55.

Miyamura, Satumi. "Seismicity of the earth," *The Earth's Crust and Upper Mantle, op. cit.*, pp. 115–124.

Moore, Ruth. *The Earth We Live On*, New York: Knopf, 1958.

Niddrie, David. *When the Earth Shook*, London: Hollis and Carter, 1961.

Ollier, Cliff. *Volcanoes*, Cambridge, Mass.: M.I.T. Press, 1969.

Pakiser, L. C., Jr. "Earthquake prediction and modification research in progress," in *Geologic Hazards and Public Problems, op. cit.*, pp. 297–304.

Radbruch, D.H. *et al. Tectonic Creep in the Hayward Fault Zone, California*, U.S. Geological Survey Circular 525, 1966.

Report on earthquake hazard reduction, Washington, D.C.: U.S. Government Printing Office, September 1970.

Richter, C.F. *Elementary Seismology*, San Francisco: Freeman, 1958.

Rittmann, A. and E.A. Vincent. *Volcanoes and Their Activity*, New York: Wiley, 1962.

Steinbrugge, K. V. "Earthquake hazard abatement and land use planning," in *Geologic Hazards and Public Problems, op. cit.*, pp. 143–152.

United States Department of Commerce, E.S.S.A. *Earthquakes*, pamphlet 0-350-033, 1969.

Wiegel, R.L. "Seismic sea waves," in *Geologic Hazards and Public Problems, op. cit.*, pp. 53-76.

Wilcoxsen, K. H. *Chains of Fire: The Story of Volcanoes*, Philadelphia: Chilton, 1966.

Williams, H. "Volcanoes," *"Scientific American" Resource Library, Readings in the Earth Sciences*, San Francisco: Freeman 1, 1969, pp. 163–172.

There is very little land surface that man has not disturbed in some way. (Eureka Mill, Nevada, Edward Fairman, 19th century, courtesy Museum of Fine Arts, Boston)

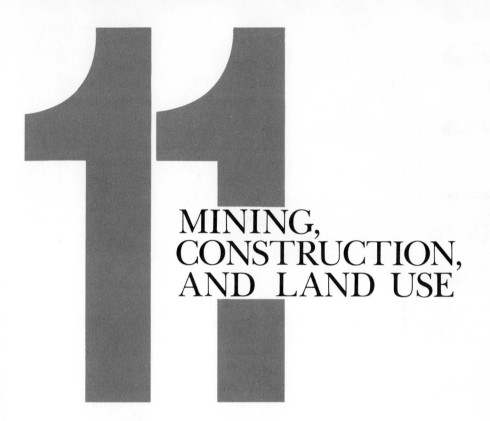

MINING, CONSTRUCTION, AND LAND USE

INTRODUCTION

Geology allies itself with the various branches of engineering when dealing with mining, removal of water or petroleum, and the building of tunnels, dams, large buildings, highways, and other structures. To some extent we have already referred to the relation between geology and man's works when we discussed such problems as river-channel control and landslides. There are no really clear-cut boundaries in the spectrum ranging from the science of geology at one end to the practical application of engineering knowledge at the other; rather, the spectrum consists of a gradation ranging from engineering geology to geological engineering.

There is very little land surface that man has not disturbed in some way. Man truly is the greatest of the digging creatures. We have plowed, tunneled, blasted, pumped, shoveled, drilled, scraped, hauled away and

11-1
Dredge tailings from gold mining operation in Yuba County, California. (Photograph by L. D. Clark, U.S. Geological Survey)

dumped, or at least made tracks across, nearly every square mile of ground on the face of our planet. Occasionally, we have done it for some kind of positive good: the growing of food, protection from floods, or the creation of a park. Many times, however, we have desecrated the land through such practices as coal mining and have thereby rendered it un-productive, ugly, or even useless. Whatever we have done to the face of the earth, we have always acted in the name of progress, often a somewhat fuzzy goal.

Again, we must think about population. When the population was small and man was just starting to learn how to disturb his earth, he had relatively little impact on the surface environment. Now, we are numer-ous and have spread over the entire globe, and our impact is becoming substantial. Additionally, we have machinery, explosives, and tools that we use to rearrange the ground. As indicated in Chapter 1, there is no reason to expect our population to diminish to the point where it can have little effect on the surface of the earth. Therefore, we must act wisely in carrying out such activities as farming, mining, building, and land use.

THE EXPLOITATION OF MINERAL RESOURCES

The extraction of such mineral resources as metals, nonmetals, and the fossil fuels has produced large-scale environmental effects. Surface mining has resulted in great scars where rock has been removed and in

11-2
Kennecott Copper Corporation's open pit mine at Bingham Canyon, Utah. (Photograph courtesy Kennecott Copper Corporation)

huge piles where the waste has been dumped. Underground mining has led to cave-ins and ground-settling. In either case, numerous problems concerning safety and the pollution of ground and surface waters have arisen. In addition to these effects of mining are the problems of reclamation and land use after the mines have been depleted. The withdrawal of oil and natural gas has caused the ground to subside in some areas. Petroleum exploration and development have also left scars on the land in the form of easily erodible temporary roads and tracks, well drilling sites and pipeline routes.

Surface Mines

Many of the great mines of the world, such as the iron mine at Hibbing, Minnesota, and the copper mines at Bingham Canyon, Utah, and Santa Rita, New Mexico, are surface mining operations (Fig. 11–2). Countless

11-3
Gold dredge near Fairplay, Colorado, about 1920. (Photograph by L. C. Huff, U.S. Geological Survey)

smaller mines, rock quarries, sand and gravel pits, coal strip mines, and placer mines are also involved in surface mining. The environmental impact of a surface mine, pit, or quarry may differ greatly from one operation to another, depending on its type and size.

One obvious impact is the radical alteration of the land surface, usually to the extent that it can never possibly look like it did prior to mining. The vegetation and soil cover are stripped off, a hole is created in the ground, and unsightly piles of waste rock, including the overburden (rock which overlies a minable ore body or coal bed) are left scattered about. The waste piles and the pit floors contain exposed, unweathered rock, so that the time required for vegetation to grow on them may be extremely long. If the ground water table is high enough, the bottom of the pits may be flooded, which creates a hazard. Waste piles are usually topographically steep, irregular, and easily eroded. The economics of mining are such that in most cases, companies cannot easily afford to fill in a pit or to level out waste piles. Abandoned surface mines thus become parcels of useless land.

If you recall our previous discussions concerning population growth and the increased demand for mineral resources and fossil fuels, you will understand that the problem of proliferating surface mines will certainly become increasingly significant. What should be done with an old mine? Part of the answer depends on the size and type of mine. Giant open-pit mines may be almost useless, especially if they are wet at the bottom. Other, smaller mines may be filled in. Depending on the type and condition of the rock comprising the walls and floor of the mine, the space may be utilized for sanitary landfill or outdoor industrial storage. Rock quarries and gravel pits can sometimes be filled with water and converted to recreational lakes. For example, after gravel was obtained from pits along Interstate 80 in Nebraska, the pits were filled with water and now form a chain of fishing and swimming ponds. In other places, suburban housing has been developed in clusters around attractive lakes, many of which are located on the sites of former quarries or clay pits that have furnished useful minerals in the past. Where careful attention is given to the development of these mines, the value of the property can actually be greater than it was prior to mining.

Coal Strip Mining

The stripping of coal from the earth is the greatest problem in surface mining, one reason being that there are vast areas underlain by coal that have yielded coal in the past or that contain the coal supplies of the future. Many thousands of acres in Kentucky, West Virginia, Illinois, and other states have been devastated, while thousands more remain, especially in Arizona, New Mexico, and in the Montana-Wyoming-North Dakota region. Another reason for the severity of the strip coal mining problem is the chemical damage created by this process. Coal invariably contains iron sulfide; water that percolates through coal waste dumps picks up the sulfur and carries it away in the oxidized form of sulfuric acid. This acidic mine water kills all life with which it comes into contact and renders the water supply useless. Other problems include the addition of other chemicals to the water's dissolved solid load and the excessive erosion resulting from the loss of vegetative cover.

In recent years ecology-minded citizens in many states have passed laws requiring reclamation of strip mines. Let us examine a typical strip mine and then discuss how it can be reclaimed. Typically, a bed of coal (coal seam) underlies an area close enough to the surface for a mining operator to dig it up and sell it for a profit. In order to do this, the operator must remove the overburden (Fig. 11-4). Shovels, some of them unbelievably huge, dig the rock away from the "highwall." The exposed coal is scooped up and loaded into trucks or trains, while the waste rock, called "spoil," consisting of overburden, the coal from thin seams uneconomical to mine, and rocks interbedded with the coal, is dumped to the rear in a spoil bank. A spoil bank usually consists of a series of steep, conical mounds not having any vegetation. The mining continues along a strip which extends as far as is practical; when the strip is depleted, a new strip, parallel to the first, is started. Spoil from the new strip is dumped in the preceding strip. In this manner the land is totally ravaged by the big shovels. The land they leave behind consists of mounds too steep to build on or to farm, denuded of vegetation and easily erodible, barren of fertile soil, and having acidic waters leaching out of them.

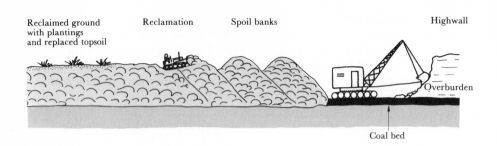

Reclaimed ground   Reclamation   Spoil banks   Highwall
with plantings
and replaced topsoil

Overburden

Coal bed

11-4
Coal strip mining and reclamation operation.

11-5
Coal strip mine operation, showing highwall (right), trench, and spoil bank (left).
(Photograph courtesy Environmental Protection Agency)

Reclamation involves smoothing out the topography, reestablishing a soil profile, and planting cover crops. Reducing the steep slopes and covering the ground with vegetation retards erosion and minimizes the leaching of sulfur. Once a workable surface has been reestablished, the land can once again be used. As stripping begins, the topsoil layers may be stripped first and stored. When reclamation is undertaken, the topsoil can be spread over the smoothed-over land. After a few seasons in which clover or similar crops have been planted and grown, sufficient restoration of soil minerals and organic matter is achieved to permit other crops to be planted. In some cases ground water may fill the remaining empty portions of the mines and thereby form recreational lakes.

It costs money for a coal company to grade down spoil banks, replace topsoil, and plant cover crops. This cost is not included in the mining cost itself—in other words, coal can be, and in many areas has been or still is, mined without incurring the additional cost of restoration. Ultimately, however, the additional cost will prove to be relatively cheaper than the cost of destroying the environment. Although laws in many states now require reclamation of strip mines, many thousands of acres where mining took place years ago now lie unused and useless. As the mining operations have already been completed, the cost of reclaiming these areas is prohibitive. In addition, such a cost cannot be conveniently added to that of current mining and reclamation. Nevertheless, some areas have been reentered and reclaimed. Yet generally, the economic and geological situations require that reclamation occur while the coal is being mined.

Underground Mines

Many valuable mineral products, including metals, coal, building stone, diamonds, sulfur, most salt and potash, and many other products are located at such depths that they cannot be mined from open pits. Gold and diamonds, for example, are being brought up from mines two miles deep.

Because it is more difficult to find the ore and develop the mining operations, underground mining is more challenging than is surface mining to the geologist and the engineer. Although there may be surface outcrops of ore minerals, sometimes ore bodies are entirely hidden. Gravity and magnetic surveys, core drilling for samples, structural mapping, and geochemical analysis of soils and cores are all necessary in the exploration and development of an underground mine.

Although underground mines do not disturb the surface as do surface mines, other problems, such as the disposal of waste rock and mine waters, do arise. The rock, called *tailings*, is removed with the ore, separated from it, and dumped, often forming large, unsightly mountains. Dust and chemical-laden waters from the tailings piles are a nuisance and are also occasionally hazardous (see Chapter 13).

Depending on both the nature of the rocks in an area and the care exercised in mining, underground mines sometimes cave in, creating hazards to anything on the ground over them. Subsidence has been a particular problem, especially in the coal-mining areas of the Appalachians. Mines lying beneath towns have on occasion collapsed, causing the houses and streets to be dumped into them.

Use of Underground Mine Space

The use of underground limestone quarries for storage and office space in the Kansas City area is one example of what can be done with depleted mines. At first, this space was used mostly for storage, and later it also served as office, salesroom, and manufacturing space (Figs. 11-6 and 11-7). About two dozen sites, which provide well over 100 million square feet of space, have been developed, most of the quarrying being done from the 20- to 25-foot thick Bethany Falls Limestone. In the early days of quarrying, the only goal was to remove as much rock as possible. Where this was done, weak pillars were left to support the roof, little attention was given to water drainage, and the limestone was removed up to the base of the overlying shales. In time the overlying shales, followed by the overlying limestone beds, collapsed. Since the beds are jointed (fractured), chemical weathering along the joints aided in the collapse, and even sinkholes may form. Failure of the rock up through the surface may make the surface area unsuited for land usage.

Currently, much attention during the development of a mine is given to its possible future use. Requirements such as future transportation, fire protection, heating and cooling, and vibration control are all considered. Larger pillars and walls are left in order to minimize the danger of collapse. A few feet of well-bedded, solid limestone are also left on the roof in order to minimize the likelihood of collapse. The structure of the rock is mapped so that the drainage of water along joint planes can be controlled. The space provided by an underground mine is inexpensive, saves valuable surface land, and has a relatively constant temperature and humidity. Facilities developed in these spaces are protected from the weather, noise, vibration, and other dangers present on the surface.

The headquarters of the North American Air Defense Command near Colorado Springs is probably one of the most notable underground facilities. This installation was constructed in the granite of the Pikes Peak area, and illustrates the role geology plays in the deliberate selection of a site for a specific purpose. Military defense is the major concern of this facility, and its requirements were therefore specialized and complex. Only a limited number of sites that had a large mass of structurally sound, unweathered rock, were safe from earthquakes, were safely workable, and were located in an area where the installation would be accessible to communi-

11-6
Entrance portals to underground limestone mine used for storage in the Kansas City area.
(Photograph courtesy Jerry D. Vineyard, Missouri Geological Survey)

11-7
View of interior of storage area in underground limestone mine in the Kansas City area.
(Photograph courtesy Jerry D. Vineyard, Missouri Geological Survey)

cation and transportation were found. The Pikes Peak granite was finally judged to have the optimum combination of features. Core drilling was carried out to determine the precise rock type, its structure, and water conditions. The actual tunneling began only after the core-drilling results were analyzed. An underground city composed of a series of giant rooms was built inside the mountain.

As underground mining continues and competition for surface space increases, underground space will be used increasingly. Thorough knowledge of geological conditions, proper mine engineering and development, and rational plans for land usage are required before this can be done safely and effectively.

Petroleum Development

The search for, and development of, oil and gas resources for an energy-hungry world have caused numerous environmental problems which will probably be even greater in the future. There is apparently no way to look for oil and later develop it without destroying the land. Bulldozing make-shift roads and trails to make way for seismograph crews and other ex-plorers, clearing sites for drilling rigs, and excavating routes for pipelines all contribute to the devastation of the land. Because the exploratory and drilling operations are short-term and temporary, few improvements are made. As the routes are likely to be direct and sometimes steep, the land is quickly eroded. Strict zoning laws and environmental controls which may exist in well-populated and developed areas, such as east Texas or even down-town Los Angeles, can often regulate operations conducted in these areas. In the less explored parts of the world, however, the potential damage to the environment is severe. The arid lands of the Southwest are badly scarred as a result of the search for oil. Exploration and development are now well underway in the tundra of Alaska and the Canadian sub-Arctic. The re-covery of the land surface is exceptionally slow in these fragile lands where tracks made by the passage of one vehicle may last for decades (Fig. 11–8).

Petroleum, Pack Ice, and Permafrost

Since 1968, when oil was discovered at Prudhoe Bay on the Arctic Coast of Alaska, exploration has intensified throughout the Arctic. Estimates ranging from 10 to 500 billion barrels of Arctic oil have been made, based on the discoveries to date and the presence of sedimentary rocks which could act as reservoirs. Although the latter figure is probably excessive, it is reasonable to assume that a large share of the world's petroleum lies in the Arctic. The demand for it will certainly require its production prior to that time when energy can be commercially obtained from solar or thermonuclear fusion sources. Even after other energy sources are developed,

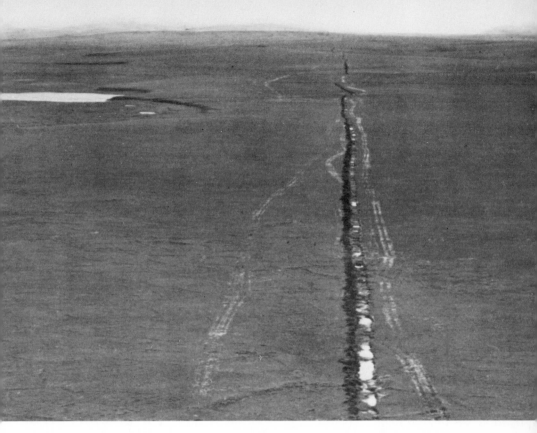

11-8

Tractor trail near Canning River, North Slope, Alaska. Thawing permafrost causes trench to form as roadway subsides. Small ponds form in roadway. (Photograph courtesy Bureau of Sport Fisheries and Wildlife)

petroleum will still be needed for petrochemicals. As its production is certain, the problem of transporting the petroleum from the Arctic must be dealt with.

Examination of a map indicates that the North Slope area of Alaska is very distant from the petroleum market. As it is a cold, remote, and inhospitable land, the only likely means of transporting the oil would be by ship or pipeline. Whereas moving the oil by ship around western Alaska to California is not totally impractical, transporting it by ship to the East Coast presents an enormous problem. The fabled Northwest Passage is more of a passage in name than in reality, for sea ice continually covers much of the Arctic Ocean and blocks the passage from west to east. Navigating the Northwest Passage has been a dream for centuries. John Cabot tried it twice, disappearing on his second try, in 1498. Henry Hudson vanished when he attempted the passage in 1610. There were no successful attempts until 1903, when Roald Amudsen finally crossed it in a 70-foot herring boat. There were no subsequent successes until the Royal Canadian

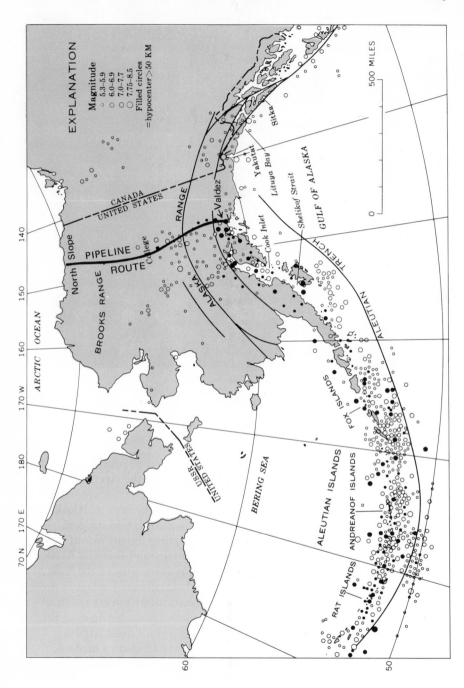

◀ *11-9*
Locations of major earthquakes in Alaska, 1898–1961, and approximate route proposed
for Alaska pipeline. (Source: W. R. Hansen, The Alaska Earthquake, March 27, 1964:
Field Investigations and Reconstruction Effort, *U. S. Geological Survey Professional Paper*
541, 1966, p. 6)

Mounted Police navigated the passage in 1942. Since then, icebreaker
vessels have negotiated their way through the passage a number of times.
In 1969 the Humble Oil Company, in a $43,000,000 experiment, sent
the tanker *Manhattan* from the east to Prudhoe Bay and back again to test
the feasibility of transporting oil via this route. The *Manhattan* failed to
make it through M'Clure Strait, but it did complete the trip on an alternate
route. Even though she was specially equipped with an icebreaker bow,
armor plate, and twice the horsepower of other tankers her size, the *Manhattan* was battered and succeeded only with the assistance of the Canadian icebreaker *John A. Macdonald.*

A pipeline would provide an alternative or additional means of
transporting North Slope oil. The construction of a four-foot diameter
pipeline from Prudhoe Bay to Valdez, Alaska, has been approved. This
pipeline will cover 800 miles of some of the coldest, wildest, and most
seismically active country in the world (Fig. 11-9). The route extends
from Prudhoe Bay south across the North Slope and the Brooks Range,
across the Yukon River, past Fairbanks, and over the Alaska Range to
Valdez, which is an all-year seaport on the Pacific.

Much of the route lies through permanently frozen ground called
permafrost, which is rock or soil constantly below freezing. Permafrost
covers about one-fifth of the world's land and about five-sixths of Alaska
(Fig. 11-10). Permafrost covers an extensive area of North America, but
isolated patches (discontinuous) of permafrost occur south of this area.
The rock or soil material may either contain water or be quite dry.

The surface ("active") layer of rock and soil thaws in the summer.
The depth at which the thawing occurs defines a *permafrost table* (analogous to a ground water table) below which the ground is constantly
frozen. The permafrost may be only a few feet thick, e.g., at the southern
limit, but may range to thicknesses of 1300 feet, e.g., at Barrow, Alaska.

Construction of oil pipelines, roads, railroads, and buildings on
permafrost is both difficult and expensive. If the vegetative cover is removed, thawing occurs in the summer and settling takes place. If slopes
are involved, downward mass movement also occurs. Construction on bedrock and well-drained or dry sediment like gravel is not difficult; however,
if it is attempted in poorly drained areas of fine sediment, construction is
difficult and beset with severe problems (Fig. 11-11). Much of the trans-
Alaska pipeline route will be in regions having the latter characteristics.

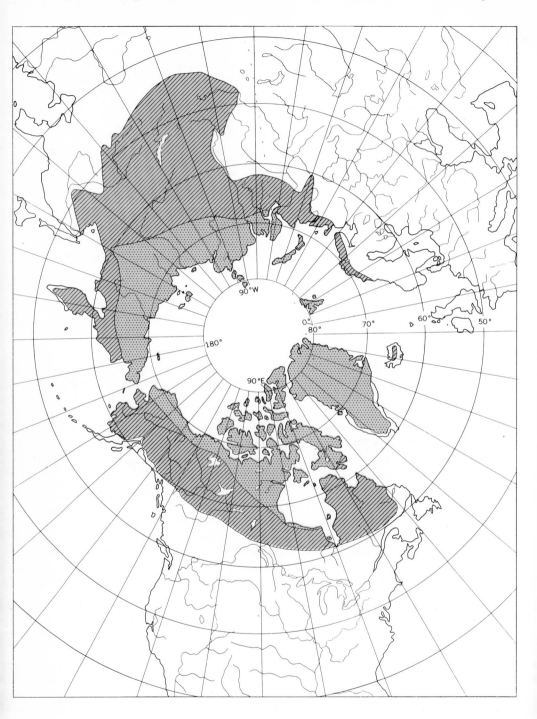

◀ *11–10*
Extent of permafrost zones in the Northern Hemisphere. (Source: O. J. Ferrians, et al., Permafrost and Related Engineering Problems in Alaska, *U.S. Geological Survey Professional Paper 648, 1969, p. 2)*

11–11
Differential subsidence of railroad near Strelna, 75 miles northeast of Valdez, Alaska. Permafrost began to thaw during construction and continued, forcing abandonment of the railroad. (Photograph by L. A. Yehle, U.S. Geological Survey)

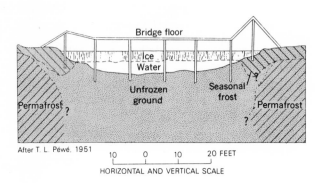

Bridge floor

Ice

Water

Unfrozen
ground

Seasonal
frost

Permafrost

Permafrost

?

?

?

After T. L. Péwé, 1951

10   0   10   20 FEET

HORIZONTAL AND VERTICAL SCALE

11–12

Photograph and sketch of bridge at Clearwater Lake, Alaska. Sketch shows relation-
ship of seasonally frozen gound to frost-heaved pilings. (Photograph by M. F. Meiser,
Geological Survey. Sketch from O. J. Ferrians, et al., Permafrost and Related Engineering Problems in Alaska, *U.S. Geological Survey Professional Paper 648, 1966,*
p. 31.)

11-13
Fracture due to frost heaving of concrete foundation of apartment house in Fairbanks, Alaska. (Photograph by T. L. Péwé, U.S. Geological Survey)

Although it depends on local conditions, the reestablishment of thawed permafrost almost never takes place, since the active layer between the ground surface and the permafrost table increases in thickness. In the summer, subsidence and flowage of sediments occurs; in the winter, freezing causes *frost heaving*. During frost heaving the material of the active layer is frozen to any structure, such as foundations, embedded in it, and upward heaving results from the expansion of the active layer. Foundations crack, bridge pilings are thrust upward, and utility poles are sometimes pushed up out of the ground (Fig. 11-12).

Oil produced from the North Slope would be hot when brought out of the ground and would be sent through the proposed pipeline at temperatures estimated to be about 158° to 176° F. The oil would not cool, as a result of the frictional heat created while it moves through the pipeline. If the proposed pipeline were to be buried six feet, the disturbance of the surface during its construction and upkeep would initiate thawing, which would be maintained by the heat from the pipeline. It is estimated that after 20 years, thawing might extend to depths of 35 to 50 feet in typical permafrost (Fig. 11-14). However, as there are many complications, subsidence will vary from place to place. Differential settling of the pipe could

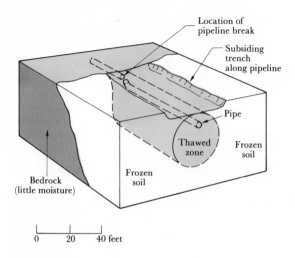

11-14
Thawing of permafrost around pipeline, with develop-
ment of subsiding trench and pipeline rupture.

place enough stress on it to break it. The resulting oil spill could flow over a sizable land area, causing serious damage to the environment.

To avoid thawing the permafrost, an overland pipeline could be built, even though this may also present problems. Because a maintenance road will still be required, the ground along the route will be disturbed in any case, both when the pipiline is first built and during its lifetime. Thus, thawing will still result. Stilts or any other structures supporting the line may be subject to frost heaving, which could possibly rupture the line.

Regardless of which parts of the pipeline are above ground or below, the route will pass through the seismically active area of southern Alaska. The potential damage could be severe enough to rupture the line. Access to the break could be prevented by other earthquake effects such as landslides or the destruction of roads or bridges.

Active oil exploration also continues in Canada's arctic territories. Partly because Canadian oil could possibly be picked up en route and partly because this line would avoid the active seismic zone, a pipeline that would extend from the Alaska North Slope through Canada has been proposed (Fig. 11-15). Much of this route would also lie in permafrost. Facilities at Inuvik and Sans Sault Rapids, Northwest Territories have tested the operation of a four-foot diameter experimental, above-surface pipeline. If it is designed properly, construction of a pipeline appears to be technically feasible.

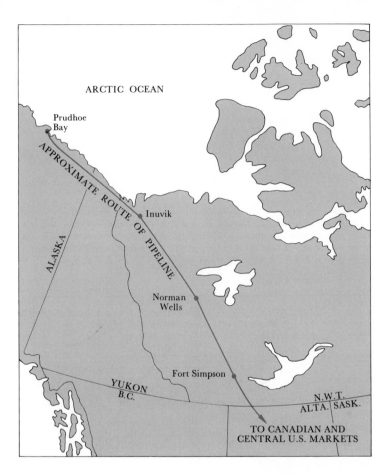

11–15
Location of possible pipeline from the Arctic through Canada.

Land Subsidence Due to Withdrawal of Fluids

The problem of land subsidence has become increasingly common as more and more petroleum and water are pumped from the ground. As fluids are withdrawn, the buoyant pressure falls, and the weight of the overlying sediments causes compaction and a reduction in volume (Fig. 11–16). A classic example is the land subsidence that occurred at Long Beach, California, where damage totaling approximately $100 million resulted from the withdrawal of oil from an oil field located directly under the city and its harbor. In a 40-year period beginning in 1928, subsidence occurred in an almost circular pattern and reached a maximum of about 29 feet at the

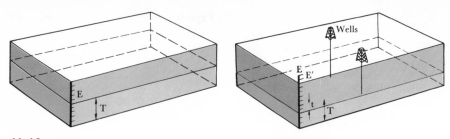

11-16
Subsidence caused by the withdrawal of fluid. A bed of thickness T *contains oil or water. When wells are drilled and the fluid is pumped out, the thickness of the bed is decreased by the amount* t. *The land subsides from its original elevation of* E *to* E′.

11-17
Subsidence at Long Beach, California due to withdrawal of oil, 1928–1962. (H. F. Poland and G. H. Davis, "Land Subsidence Due to Withdrawal of Fluids," in Reviews in Engineering Geology, II, *Boulder, Colorado: Geological Society of America, 1969. Modified by J. D. Martinez, "Environmental Geology at the Coastal Margin," in* Earth Sciences and the Quality of Life, *Symposium 1, Montreal: International Geological Congress, 1972, p. 55. Reprinted by permission.)*

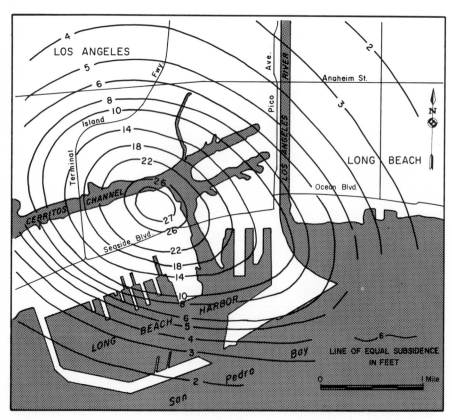

11-18
Land subsidence in California, with locations of Central Valley, Wasco area, and Long Beach. (Adapted from N. Prokopovich, "Land Subsidence and Population Growth," in Engineering Geology, *Section 13, Montreal: International Geological Congress, 1972, p. 48; and B. E. Lofgren and R. L. Klausing,* Land Subsidence due to Ground-Water Withdrawal Tulare-Wasco Area, California, *U. S. Geological Survey Professional Paper 437-B, 1969.)*

center of the area (Fig. 11-17). After 1958 a program of repressuring which consisted of pumping water into the reservoir to compensate for the withdrawal of the oil was begun, and subsidence was finally halted.

Another example of subsidence is found at Mexico City. Here, as was mentioned in Chapter 3, water withdrawn from highly porous, relatively uncompacted beds caused the city to subside and some larger buildings to sink more than ten feet below ground level. The subsidence in some areas was more than 20 feet.

Large areas of the Central Valley of California have subsided as the result of ground water being withdrawn from unconsolidated sediments for agricultural use (Fig. 11-18). In the Tulare-Wasco area, for example, subsidence has reached 12 feet (Fig. 11-19). In some cases, sinking has changed the grade on canals of the extensive aqueduct system to such an extent that the water overflows the top of the canal or the grade is re-

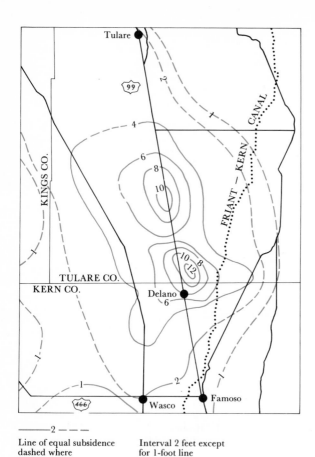

—————2 – – – –

Line of equal subsidence   Interval 2 feet except
dashed where   for 1-foot line
approximate

11–19
Subsidence in the Tulare-Wasco area, California, 1926–
1954. (Adapted from B. E. Lofgren and R. L. Klausing,
Land Subsidence due to Ground-Water Withdrawal
Tulare-Wasco Area, California, *U. S. Geological Survey*
Professional Paper 437-B, 1969.)

versed, forcing the water to overflow. Expensive repairs have had to be
made. Many other areas of the world, including Houston and surrounding
areas in Texas, Tokyo and Osaka, Japan, and northern Italy, suffer from
the problem of subsidence due to water withdrawal.

 The problem areas are those where fluids are withdrawn from loose,
uncompacted soil, clay, sand, silt, peat, or other material. Subsidence in

well-consolidated rock is negligible. Many of the world's cities are built on unconsolidated material, which has been, to a great extent, inherited from the Ice Age. Such material includes sediments from estuaries, ancient lakes, and flood plains. As more water is needed to supply expanding urban populations, more and more water is withdrawn from the sediments, causing subsidence to take place. Water for agriculture in arid regions must come mostly from ground water; therefore, subsidence will probably occur in areas such as the Central Valley of California, where ground water is obtained from geologically young, unconsolidated sedimentary deposits. The solution to the problem is not readily apparent. If the fluid being withdrawn is oil, the system can be repressurized through the addition of water, as has been done at Long Beach. However, nothing can be done to replace withdrawn ground water.

CONSTRUCTION

In order to illustrate his teaching, Jesus told the parable of the wise man who built his house on the rock. When rain, floods, and wind came, the house did not fall, but the house of the foolish man collapsed because it had been built on sand. The problem of having proper foundations for his works has apparently confronted man for a long, long while. In fact, four millennia ago, Hammurabi of Babylon issued a provision for rascally builders who cheated the public: "If a builder builds a house for man and does not make its construction firm, and the house which he has built collapses and causes the death of the owner of the house, that builder shall be put to death." Modern laws do not mandate so drastic a punishment, but in recent times the engineers who built the Vaiont Dam were put into prison for their responsibility in the disaster (Chapter 9).

We have already discussed such problems as flooding, shoreline stabilization, earthquake risks, and construction in permafrost. Let us now examine in more detail the problems of construction. Because of advances in technology, we build dams, tunnels, highways, large buildings, canals, and other works on a scale never dreamed possible a hundred years ago. One could start by noting that the Pyramids of Egypt were built about the same time as Hammurabi lived. However, these edifices were exceptional, for after all, not just anyone could order one built. Now, however, structures of grand proportions are common, and because so many people in the world are building so many structures, it is more essential than ever to build them properly. We want our structures to be safe, to last, and to function well. The Pyramids met these requirements. Why? They were well planned, placed in the right spot, and well constructed with good materials. No engineer or geologist could ask for more.

11-20
Gullying in a newly developed urban area. Improper disposal of surface water has resulted in a threat to this home. (Photograph courtesy USDA - Soil Conservation Service)

The California Aqueducts

The aqueduct system in California illustrates how geological conditions of an area affect engineering plans for a major piece of construction. About two-thirds of the state's water supply comes from the northern third of California; yet, the need for water is greatest in the southern two-thirds, where the large majority of people are concentrated and where most of the agriculture and industry occur. Water must therefore be transported over long distances. A variety of geological conditions, including large differences in elevation, types of rock in various degrees of consolidation or weathering, and the presence of many faults, some of them active, occur over these distances.

11-21
The Los Angeles aqueduct system.

The system is comprised of three major aqueducts (Fig. 11-21): the Los Angeles (Owens River) Aqueduct, the Colorado River Aqueduct, and the California Aqueduct. The California Aqueduct, alone, represents an investment of about two billion dollars, which is indicative of the importance attached to supplying water to southern California. These costly facilities had to be constructed so that the supply of water would continue, even if an earthquake which might disrupt a canal, tunnel, or other portion of the aqueducts were to occur.

There is no way that such an extensive system in California could avoid crossing major and possibly active faults. Aqueducts extending into Los Angeles must cross the San Andreas and other faults. Crossing a fault through a tunnel many feet underground risks a shearing or blocking of the tunnel in the event of an earthquake. Repairs might be difficult and dangerous, perhaps involving long delay, and a critical water shortage could ensue. Because of these considerations, the more recent construction has been made to cross many major fault zones at the surface. For ex-

11-22
The St. Francis Dam before collapse, Los Angeles County, California. (Photograph courtesy H. T. Stearns, U.S. Geological Survey)

11-23
The wreck of the St. Francis Dam five days after failure. The contact between the schist and sedimentary rock can be seen in the photo. (Photograph by H. T. Stearns, U.S. Geological Survey)

ample, the west branch of the California Aqueduct crosses the San Andreas fault through a sag pond, a lake located in a depression along the fault. The main aqueduct north of there crosses the San Andreas via a canal and traverses the Tehachapi Mountains, a problem zone, by way of a series of high-level, short tunnels. Although pumping the water up to these tunnels is expensive, it would have been more risky to build a long, low-level tunnel.

The Mono Craters tunnel, located east of the Sierra Nevada, is part of the First Los Angeles Aqueduct. It passes through the volcanic necks underlying the Mono Craters. When the tunnel was built, much difficulty was caused by water entering the tunnel under high pressure, by carbon dioxide gas, and by the flowing and squeezing of ground in the tunnel. Floods have often wrecked siphons where the aqueduct crosses canyons, and one of the most recent problems was the damage done to the Van Norman dam and reservoir during the 1971 San Fernando earthquake.

How Not to Build a Dam: The St. Francis Dam

The failure of the St. Francis Dam near Saugus, California, is a classic example in geology and engineering of how ignorance and disregard of scientific facts can lead to disaster. It is difficult to imagine how this structure could have been located in a worse place; geologically, there probably was no worse place, even had it been looked for deliberately. The dam, built from 1924 to 1926, was placed across a canyon which, on one side consisted of schist and, on the other, of clastic sedimentary rocks. The

11-24
Wreckage in Santa Paula, Ventura County, California, after the St. Francis flood. (Photograph by H. T. Stearns, U.S. Geological Survey)

layers of the schist (planes of schistosity) lay parallel to the canyon wall, and numerous landslide scars present in the area indicated that the rock slid readily along these planes. The sandstones, siltstones, and other rocks across the canyon appeared to be resistant, although it was noted too late that they disintegrated in water! In short, the reservoir in the canyon lay over the clastic rocks which fell apart when wet and over schists which could slide, especially when wet, along planes of schistosity.

In addition, a fault ran through the west side of the canyon and formed the contact between the schist and the clastic rocks. The 1922 Fault Map of California showed the fault, but the dam was built across the fault in spite of this. California Institute of Technology student Thomas Clements was mapping the area; when he found the fault he "began to doubt his own competence, for . . . surely no one would build a dam across so large and obvious a fault." He tried to check the area below the dam for more information, but since it was raining and rather than camp there in his favorite place, he went back to Los Angeles after getting his Model T unstuck from the mud. This was just as well for Mr. Clements, because the dam broke that night. Thirty-eight thousand acre-feet of water cascaded down the valley and killed over 500 people. A review suggested that soft material along the fault had washed out, followed by washout of the weak clastic rock which caused the west section of the dam to drop. Having no support, the east section collapsed as the rushing water undercut the schist. Only the center remained standing.

Dams and Reservoirs

Various problems may arise in the building of dams and reservoirs. It is evident that many dams have failed when they have been geologically stressed beyond their capacities. We have previously noted the problems with the Van Norman, St. Francis, Hebgen, and Vaiont dams. Earthquakes, landslides, and unusually heavy rains have contributed to their failures, as has poor siting. Obviously, very careful geological, meteorological, and seismological studies should be a prerequisite of dam construction.

We have also seen (at Aswan) that the erosive and sediment-bearing capacities of rivers are changed below dams. Great changes occur in a stream, its channel, and its flood plain. Another problem which some reservoirs have is leakage of water into porous, permeable strata. Unless what happens to the water in a reservoir is irrelevant—a rare situation— careful geological investigations must be made to determine the extent to which leakage may occur. A problem related to leakage is the entry of water into clays and other rocks. Core drilling and surface mapping can reveal unique structural or lithological conditions that might be troublesome, e.g., Vaiont and St. Francis.

Sedimentation in reservoirs is yet another problem. As streams decrease their velocity upon entering a large body of water (ocean, lake, or reservoir), their ability to carry sediment diminishes greatly, causing the particles to settle out. As sediment accumulates, a delta forms at the upper end of the reservoir and expands until the reservoir is filled, at which point the reservoir is no longer useful for water supply, flood control, or hydroelectric power generation. In just 14 years after the Hoover Dam was built, sediments filled the upper end of Lake Mead on the Colorado River for 43 miles into the reservoir from the river's original point of entry. Fine sediment spread over the entire reservoir bottom. The total sediment accumulation in 24 years was about four billion tons. If this rate continues, the reservoir will lose half of its storage capacity in 120 years, and the usefulness of the dam and reservoir will be greatly decreased. Similarly, the reservoirs along the upper Missouri are filling with sediment; when they, too, lose their effectiveness, the multiple uses of the Missouri basin will have been greatly altered (see Chapter 9).

Geologists and engineers base their estimates of the life of a dam and reservoir on the sedimentation rate and then balance the life against the cost. Sometimes, however, the estimates are proved to be erroneous. Lake Nasser, behind the Aswan High Dam, was supposedly designed to have a 500-year life (the history of civilization along the Nile is 5000 years). It was originally judged that the currents in the river and reservoir would distribute sediment to certain areas of the reservoir which would serve as settling basins, while the rest of the reservoir would have other uses. Whether the Nile is behaving accordingly is now questionable. Estimates suggest both that the reservoir is being filled with sediment faster than originally planned and that the sediment is being deposited in areas not previously foreseen.

Proponents assert that large dams and reservoirs provide flood control, recreation, water supply, hydroelectric power, and attendant employment. Opponents argue that floods can be controlled by reforestation and by small dams upstream, that hydroelectric power is an insignificant fraction of the energy supply, that water can be obtained by other means, that too much water is lost by leakage and evaporation, that valuable land is inundated, and that final silting will leave us with useless, costly structures. It is possible that some major (and many minor) dams should never have been built. It is also likely that long-range geological studies have been inadequate. Studies of river systems in which the behavior of a river basin and its stream flows is charted through time (hundreds of years) should be carried out before it is decided to build a major dam. It is essential to bear in mind that stream systems are dynamic and that the construction of a dam and reservoir is at best an attempt to stabilize only part of the system. Even if this attempt is successful, control over the rest of the system is either not achieved or is lost because of the changes in sediment supplies and gradients. Man cannot truthfully speak of "harnessing" streams.

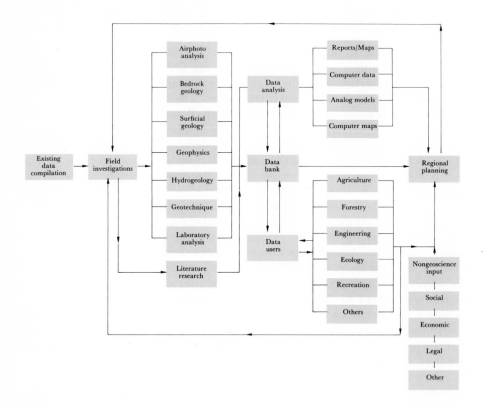

11-25
*Relationships between geoscience activities and other studies in regional planning. (Denis
A. St-Onge and John S. Scott, "Geoscience and Ste-Scholastique," Canadian Geographical
Journal LXXV, 1, July 1972, p. 233. Reprinted by permission.)*

Foundations

Our observations regarding the construction of dams suggest the broader topic of geology and its application to the foundations of structures. Dams, large buildings, homes, and highways are some of the structures that must be anchored firmly in order to prevent their shift or collapse. We attempt to achieve structural stability on a geologically dynamic earth. We have already alluded to some of the problems, such as construction in earthquake zones (Chapter 10) and in permafrost. How do we decide where and how to place our structures?

We need to know the soil or rock makeup of the proposed location, the susceptibility of the materials to slippage, the character of the drainage of the area, the seismicity of the area, the location of areas where erosion or deposition is taking place, and the pattern of land development and use in the area. Maps of the topography, of the rock types and ages, of seismic zones, and of other pertinent data help us to plan. Generally, field examination of the surface rocks and soils is carried out, and depending on the kind of structure to be built, test holes may be drilled to obtain data that provide a three-dimensional picture of the ground.

The geological investigations completed for the new Montreal airport at Ste-Scholastique, Quebec (Fig. 11–25) illustrate the steps taken to study an area. This comprehensive investigation, carried out in 1971, included soil and rock analyses, the study of fluid movement through them, and an examination of the processes at work both on the surface and in the subsurface. Seismic surveys yielded information about the depth to bedrock (Fig. 11–26), and studies were made of the cuttings obtained from seismic drilling and of samples obtained from subsequent core drilling. Devices that provide electrical logs were run down the drill holes to obtain information about the porosity, density, salinity, and conductivity of the materials. This entire body of information was fed into a computer that produced maps on which interpretations and decisions about development and construction could be based.

Geological structures, such as joints and faults, sometimes present problems in foundation surveys. Joints or faults may so seriously weaken the rock that it is likely to slide or cave in. Variations in bearing strength across zones of crushed rock or between areas having different structural orientations may occur. Open joints and faults may allow water to pass through, which may create a problem in the construction or support of the structure.

Clay, especially illite and montmorillonite clay, is a particularly difficult material with which to work. Because these clays swell when they soak up water, they sometimes cause the structures placed on them to fail.

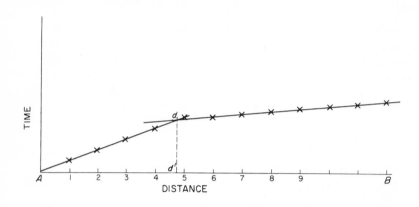

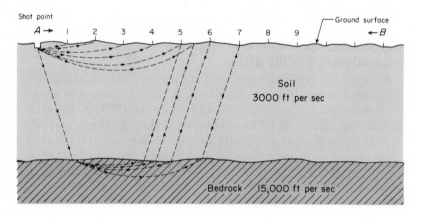

11-26
Determination from seismic information of depth to bedrock. (Source:
L. W. Currier, The Seismic Method of Subsurface Exploration of Highway
and Foundation Sites in Massachusetts, *U.S. Geological Survey Circular*
426, 1960, p. 4)

When subjected to stress, clays may also flow plastically. They form unstable slopes, and stresses may cause landslides along clay beds, even very
thin clay seams.

Highways

One of the distinguishing marks of modern civilization is its transportation
system, particularly automobile transportation. The network of paved
roads constructed in the past 50 years probably surpasses in cost and scope

11-27
Crack and heave in facing beneath the Interstate 25 Freeway bridge, over 2nd Street in Pueblo, Colorado, caused by swelling clay in the Niobrara Formation. (Photograph by G. R. Scott, U.S. Geological Survey)

11-28
Loose, open-textured crust produced upon exposure of swelling clay, El Paso County, Colorado. (Photograph by D. J. Varnes, U.S. Geological Survey)

11-29
Construction of highway interchange, Prince Georges County, Maryland. The knowledge of geological conditions help insure proper construction. (Photograph courtesy USDA–Soil Conservation Service)

any other class of man's works. The problems encountered in the construction and maintenance of roads are daily indicating the obvious need for highway geology. The construction of roadbeds, bridges, interchanges, and tunnels involves such typical problems as the makeup and behavior of earth materials. In many cases, special problems of permafrost, limestone solution sinks, landslides, and earthquakes also arise. Let us look at a typical geological study of a highway route.

A 70-mile section of the Illinois Toll Road was planned to extend from Aurora to Rock Falls. Engineering reports prepared in 1970 described the geological conditions along the route and the design of the highway. Nearly all of the route crosses Pleistocene glacial deposits, which are mostly till, composed of silt, clay, and sand; very small portions cross Cambrian and Ordovician deposits. Loess and *outwash deposits* (deposits laid down by meltwater streams) of gravel, sand, and silt are also present, as are recently formed (Holocene) soil and stream deposits which blanket

11-30
Remedial treatment of slide area along highway. Note exit pipes for subdrains in center of background and left foreground. Rock material loads the toe of the slide and prevents washout by stream. (Photograph by C. R. Tuttle, U.S. Geological Survey)

much of the area. Bedrock lies 20 to more than 100 feet under the glacial deposits.

It was discovered that one part of the highway would cross a bowl-shaped area with poor drainage. To protect farms in the area, planners designed a drainage ditch, parallel to the highway, which would drain water away from the bowl and into the natural stream drainage. Unstable soil conditions, including the presence of wet soils highly susceptible to frost heaving, created special problems. Some of the soils were removed and replaced by gravel, and drainage systems which would allow the removal of water from the soil were installed at appropriate locations.

Tunnels

Some geological problems encountered in the construction of tunnels are of little significance in most surface projects. In some areas, tunnels penetrate rocks having tremendous *stored pressures*, which means that at

11-31
Damage to house caused by soil slippage on a hillside. (Photograph courtesy USDA—Soil Conservation Service)

some time in the earth's past, the rocks may have been buried deeply and were therefore subjected to great pressure. Energy was thus stored in the rock. When a tunnel is driven into the rock, exposing it to atmospheric pressure, the rock may expand by heaving or bulging (common in clay-stone and shale) or even by bursting, with slabs of rock being violently spalled off the rock surface.

Tunnels may penetrate water-bearing rocks that yield so much water that drainage becomes a problem. Sometimes the water originates in fault zones and sometimes from permeable formations. Construction is often delayed. If it is possible to map the fault zones on the surface or through drill holes, the builders of the tunnel may be able to predict where, and possibly how much, water will be encountered. With this information preparations for drainage can be made.

The type of rock through which the tunnel is driven greatly affects the construction. Any kind of material may be tunneled—igneous, sedimentary, or metamorphic, all of which have varying degrees of strength, induration, and structural character, or loose sediments such as glacial till, river gravel, silt, mud, or soil.

Canals and Waterways

Canals are built for one of two reasons: for either ship and barge navigation or water transportation as part of an aqueduct or irrigation system. Although no record of the first canal exists, it was certainly built many centuries ago. Depending on its primary use, different geological problems can arise in the construction of a canal. Navigation canals are designed to connect two bodies of navigable water by a level route. Leakage may not be of concern, but level gradient, the construction of necessary locks, and the provision of a water supply for the locks may be problems. In the case of water-supply canals, gradient is critical if water flow is to be assured. Leakage from unlined canals into porous earth materials is a problem, as is excessive evaporation. Examples of the problems of maintaining a gradient and of crossing seismic zones were mentioned in our descriptions of the California Aqueduct system.

The Suez Canal, constructed across the isthmus between the Mediterranean and the Red seas, shortened the distance from Europe to India and the East by thousands of miles (Fig. 11–32). This canal was used for nearly 100 years, from its opening in 1869 to its closing (by war) in 1967. As it is a sea-level canal, its construction did not present as many problems as did the construction of the Panama Canal. After completing the Suez, its French builder, Ferdinand De Lesseps, next attempted to construct the Panama Canal; however, fever, the jungle, and the mountains defeated him. It was not until 1914 that the Panama Canal opened.

There were two major geological problems in the construction of the Panama Canal. One was landslides. The famous Culebra (now Gaillard) Cut, an eight-mile stretch of the canal, had to be dug deeply through rugged mountains. Slopes that were oversteep and temporarily denuded of their jungle cover absorbed water from the tropical rains and began to give way. Persistent digging back of the slopes finally stabilized them.

Another problem was that of supplying water to the locks. Each ship that passes through the canal requires enough water to supply a city of one-third million people. To provide this enormous quantity, a dam was built across the Chagres River, creating Gatun Lake, a reservoir that fills during the rainy season and stores enough water to last through the year. As the lake level is the highest part of the route, fresh water flows down from it through the locks to both oceans. No salt water enters the route, and no mixing between the Atlantic and Pacific occurs.

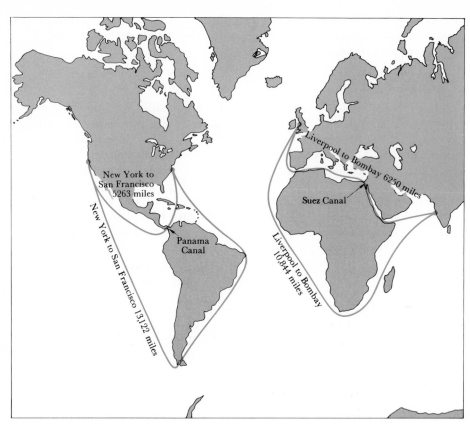

11-32
Suez and Panama Canals, with examples of mileage differences between canal and Southern Hemisphere routes.

Because of the increased traffic through the Panama Canal and because some ships are too large for the canal, a new sea-level route has been proposed. Some environmental objections to this route can be mentioned: excavation would probably have to be completed through nuclear blasting, and mixing of Atlantic and Pacific waters might disturb the biological regimen in either ocean.

Two very significant waterways in North America are the St. Lawrence Seaway and the Intracoastal Waterway (Figs. 11-33 and 34). The St. Lawrence Seaway includes the St. Lawrence River, the Welland Canal and locks between Lakes Erie and Ontario, the Great Lakes, and the locks at the Sault Ste-Marie Canal. The Sault Ste-Marie Canal and locks carry more tonnage than does any other waterway system in the world.

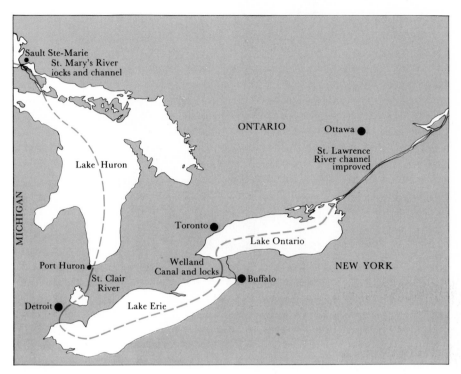

11-33
Location of the St. Lawrence Seaway.

The Intracoastal Waterway extends along the Gulf and Atlantic coasts of the United States. These coasts are characterized by the presence of offshore *barrier islands* which are separated from the shore by lagoons, estuaries, and other bodies of water (Fig. 11-35). Barrier islands form on gently sloping sea floors in rather shallow water, probably as a result of sand piled up by waves and currents, although the reasons for their formation are not fully understood. The islands protect the inner lagoon from large waves, and small craft can travel safely along this body of water. Dredging, navigation buoys and lights, and other improvements along the Intracoastal Waterway created a channel deep enough to accommodate larger vessels. Continual maintenance, including dredging, is necessary along the St. Lawrence River and the Intracoastal Waterway. Tropical storms

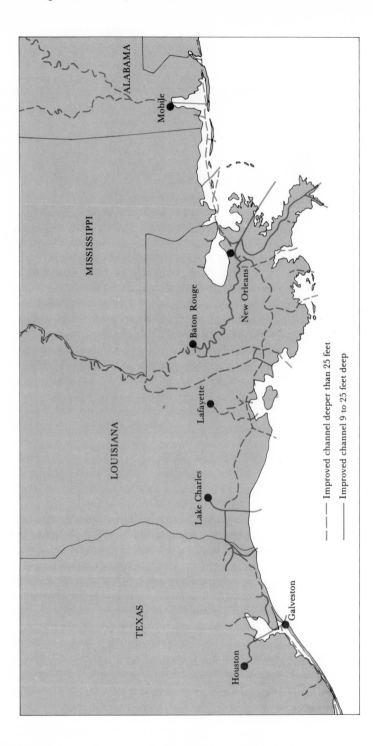

11-34
The Intracoastal Waterway, from Houston to Mobile.

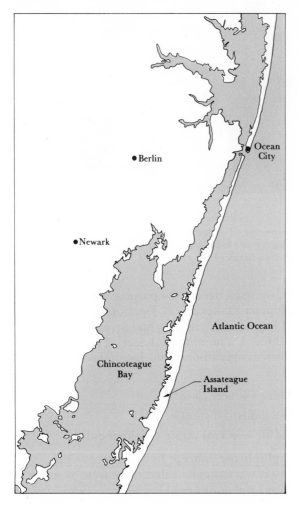

11-35
The Atlantic Coast of Maryland. Assateague Island
is a barrier island; Chincoteague Bay is a lagoon.

present a problem along the coast, as large storm waves can move enough
bottom sediment to damage the channel severely.

GEOLOGY AND LAND USE

The intensification of land use due to increases in population, industrial
expansion, and increasing agriculture implicitly necessitates wise planning
and development of our land resources. If we are to have orderly growth

11-36
Making a soil survey for urban development. The soil was found to be unsuitable for buildings; houses in background, however, are located on a different soil type. (Photograph courtesy USDA—Soil Conservation Service)

and optimum usage of the land, we must decide the purpose for which the land is best suited, before development takes place. Development should be an evolutionary process that begins with planning and proceeds accordingly—perhaps from one usage through a number of usages. The concept of multiple usage is becoming increasingly important. Examples might include:

1. farmland ⟶ strip mining ⟶ recreation
2. underground quarry ⟶ industrial storage
3. surface mine ⟶ landfill ⟶ real estate development.

We can demonstrate the multiple-use concept by using an example from Denver, Colorado. In 1960 it was estimated that there were about 240 million tons of aggregate (sand and gravel used for construction) available for mining within a 15-mile radius of downtown Denver; urban growth, however, had covered up more than twice that amount. Thus, valuable mineral deposits were found to be situated where they would be the most useful but where they were unobtainable. It is expensive to transport high-bulk, low price-per-volume aggregate from farther distances. Another facet of the story, however, concerns the multiple use of one area in Denver where sand and gravel had been removed years ago. The abandoned pits were used for sanitary landfill, and after they were filled in this manner, the Denver Coliseum and parking lots were built on the site (Fig. 11-37). Thus, for a single plot of ground, we have seen three uses, arranged in an orderly evolution.

It is particularly vital to plan wisely when the development of an area involves the alteration of land to the extent that it cannot be used for any

(a)

11-37
(a) Aerial view of abandoned gravel pits used as sanitary landfill sites by the City of Denver. (Photograph, taken in 1948, courtesy U.S. Bureau of Mines); (b) Aerial view of site of former gravel pits and landfill, now occupied by the Denver Coliseum, parking lots, and streets. (Photograph, taken in 1967, courtesy U.S. Bureau of Mines)

(b)

other purpose. Examples of such alteration are: coal strip mining without reclamation, inundation of land by reservoir waters, and construction of urban areas over valuable mineral deposits. Nuclear blasts produce radioactivity in amounts sufficient to prevent use of the land for some years.

Another important field in which land-use planning is critical is urban development. F. V. Kotlov of the Soviet Union has pointed out that in 1800 only 2 percent of the world's people lived in urban areas, but that by the year 2000, 60 percent will live in urban areas. There is no way that we can maintain quality in urban living or insure stability and effectiveness in the day-to-day operations of the cities if careful planning is not carried out. We have already seen how the problems of space usage, transportation, water supply, and waste disposal have increased in the haphazard growth of many large cities. In recent years we have heard of the development of "urban geology," which is geology applied to high-use areas with concentrated populations. Although urban geology embodies the same scientific principles as do other aspects of geology, it differs from these aspects in its application to a unique situation which requires special treatment.

Mapping is the key to any land-use planning. When most people think of a map, they think of a two-dimensional piece of paper on which are printed the locations of roads and towns. However, almost anything can be shown on a map; for example, maps prepared for land-use planning show many geological features of the mapped areas. These maps may show:

1. earth materials—rock types, bedrock distribution, soils, and mineral deposits;

2. hydrology—amount and location of water, quality of water, and precipitation;

3. hazards—flood or hurricane potential, earthquake risks, and landslides and unstable slopes;

4. engineering geology—construction materials, soils susceptible to frost, folded rocks, faults, joints, and strength of materials;

5. processes—areas of erosion, areas of deposition, and stable areas;

6. geological formations—ages, distribution, and outcrops;

7. topography—landforms and elevations;

8. evaluation maps—slopes, suitability for agricultural, industrial, or housing development, transportation, waste disposal, and other usage.

In addition to geological maps, maps of population, present land use, future projections, and economic factors will be developed. The preparation of environmental atlases may involve seismic surveys, geophysical logging, drilling of core holes, surface sampling, laboratory analyses, aerial photography, and monitoring of natural phenomena on a continuing basis.

11-38
Measuring the dip (angle of tilt) of sedimentary strata by means of a clinometer. (Photograph by J. R. Stacy, U.S. Geological Survey)

11-39
Preparing a countour map. This instrument, a Kelsh Plotter, by stereoscopic projection of aerial photographs helps the operator to draw topographic countour lines. Topographic maps are of great importance in geology, engineering, and land-use planning. (Photograph courtesy USDA— Soil Conservation Service)

An example of the many varied approaches to land-use planning is the work of the Geological Survey of Alabama. The Survey analyzed the population and development trends of the state and determined the location of expansion corridors where, because of geology, topography, transportation, and other factors, future growth was expected to occur (Fig. 11-40). These corridors contain 40 percent of the land, but also 70 percent of the population. Maps showing environmental geological characteristics of certain areas, especially those in the centers of growth, are being prepared. Basic data maps which illustrate technical information are the first to be prepared and are subsequently used in the preparation of *use maps,* which interpret the feasibility of various developmental activities.

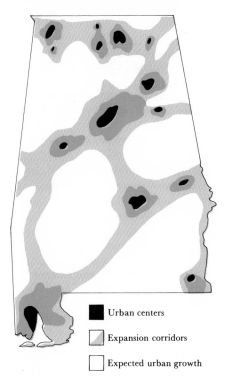

11-40
Expansion corridors in Alabama. (Paul H. Moser, "Environmental Geology Studies in Alabama," in Engineering Geology, International Geological Studies, *Section 13, Montreal: International Geological Congress, 1972, p. 38. Reprinted by permission.)*

Similar work has been done in Texas, where the Bureau of Economic Geology has prepared an environmental atlas of the Texas Gulf Coast. The coast is an area of greatly varied and intensive use, and, as is typical, is an area of rapidly dynamic geological processes (see Chapter 9). Planning, therefore, must be careful, thorough, and appropriate. Using the map as the basic geologic document, the Bureau has prepared maps that show the great varieties of geological features, including their environmental aspects, along with existing man-constructed works. The maps are available to agriculturists, industrialists, land developers, recreation and wildlife agencies—in short, to a great number of parties who can use them in their work.

REFERENCES

Beam, E. F. "Engineering geology of highway location, construction, and materials," in *Application of Geology to Engineering Practice*, Berkey Volume, Geological Society of America, 1950.

Burwell, E. B., Jr. and B. C. Moneymaker. "Geology in dam construction," *op. cit.*

Cochran, William. *Mine Subsidence – Extent and Cost of Control in a Selected Area*, U. S. Bureau of Mines Information Circular 8507, 1971.

Currier, L. W. *The Seismic Method of Subsurface Exploration of Highway and Foundation Sites in Massachusetts*, U. S. Geological Survey Circular 426, 1960.

Dapples, E. C. *Basic Geology for Science and Engineering*, New York: Wiley, 1959.

Dean, T. J., Gomer Jenkins, and J. H. Williams. "Underground mining in the Kansas City area," *Missouri Mineral Industry News* 9, 4 (1969): 37–56.

Earth Sciences and the Quality of Life, Symposium 1, International Geological Congress, Montreal, 1972.

Ellis, W. S. "North Slope: will Alaska's oil and tundra mix?" *National Geographic* 140, 4 (Oct. 1971): 485–517.

Ferrians, O. J., Jr., Reuben Kachadoorian, and G. W. Greene. *Permafrost and Related Engineering Problems in Alaska*, U. S. Geological Survey Professional Paper 678, 1969.

Grimes, W. W. "Congruence geomorphology, geology and highway engineering," Abstract and paper presented at the 15th Annual Meeting of the Association of Engineering Geologists, Kansas City, Mo., 1972.

Hayes, W. C. and J. D. Vineyard. *Environmental Geology in Towne and Country*, Missouri Geological Survey and Water Resources, Education Series No. 2, 1969.

Lachenbruch, A. H. *Some Estimates of the Thermal Effects of a Heated Pipeline in Permafrost*, U. S. Geological Survey Circular 632, 1970.

Lofgren, B. E. and R. L. Klausing. *Land Subsidence due to Ground-Water Withdrawal Tulare-Wasco Area, California*, U. S. Geological Survey Professional Paper 437-B, 1969.

Louderback, G. D. "Faults and engineering geology," in *Application of Geology to Engineering Practice, op. cit.*

Lung, Richard and Richard Proctor, eds. *Engineering Geology in Southern California*, Los Angeles: Association of Engineering Geologists, 1969.

McKenzie, G. D. and R. O. Utgard, eds. *Man and His Physical Environment*, Minneapolis: Burgess, 1972.

Nichols, T. C., Jr. *Engineering Geology of the Paducah East Quadrangle in Kentucky*, U. S. Geological Survey Bulletin 1258-A, 1968.

Price, D. G. and J. L. Knill. "Reservoirs and environmental management," in *Earth Sciences and the Quality of Life, op. cit.*, pp. 69–73.

Sanborn, J. F. "Engineering geology in the design and construction of tunnels," in *Application of Geology to Engineering Practice, op. cit.*

Schwab, G. O., R. K. Frevert, K. K. Barnes, and T. W. Edminster. *Elementary Soil and Water Engineering*, 2nd ed., New York: Wiley, 1971.

Scott, G. R. *General and Engineering Geology of the Northern Part of Pueblo, Colorado*, U. S. Geological Survey Bulletin 1262, 1969.

"Special issue on environmental problems faced by the mineral industry," *Mineral Industry News* 7, 4 (1967).

Stall, J. B. "Man's role in affecting sedimentation of streams and reservoirs," ed. K. L. Bowden, *Proceedings of the 2nd Annual American Waters Resources Congress,* 1966, pp. 79–105.

Task Force on Northern Development. *Pipeline North—The Challenge of Arctic Oil and Gas,* Ottawa, Information Canada, Northern Pipelines Report 72-1, Cat. No. R-72-7572.

Thomas, H. E. *First Fourteen Years of Lake Mead,* U. S. Geological Survey Circular 346, 1954.

Thornbury, W. D. *Principles of Geomorphology,* 2nd ed., New York: Wiley, 1969, pp. 557–566.

Trefethen, J. M. *Geology for Engineers,* 2nd ed., New York: Van Nostrand, 1959.

Turner, D. S. *Applied Earth Science,* Dubuque: Wm. C. Brown, 1969, pp. 50–79.

The health and well-being of man and animal are linked to the intake in proper amounts of certain essential trace elements. (Photograph courtesy United States Department of Agriculture)

MEDICAL
GEOLOGY

INTRODUCTION

The inclusion of a chapter on health and disease might initially seem inappropriate. However, only a little reflection should be sufficient to make
the association between geology and medicine. Our health and well-being
depend on our intake of chemical elements derived from the crustal rocks
of the earth. These elements become available to us through a variety of
geologic processes, the most common of which is the cycle whereby elements are released by weathering, enter the soil and water, and are absorbed by endemic plants which are later eaten by animals in the area. Our
use of the water for drinking and the plants and animals for food provides
ample routes for our ultimate contact with the elements which were originally bound up as rock material.

439

Table 12-1 Element distribution in total body of the standard man (average chemical composition of the adult human body)

Element		Per cent by weight	Approximate amount in 70 kg man (g)
Oxygen	(O)	65.0	45,500
Carbon	(C)	18.0	12,600
Hydrogen	(H)	10.0	7000
Nitrogen	(N)	3.0	2100
Calcium	(Ca)	1.5	1050
Phosphorus	(P)	1.0	700
Sulfur	(S)	0.25	175
Potassium	(K)	0.2	140
Sodium	(Na)	0.15	105
Chlorine	(Cl)	0.15	105
Magnesium	(Mg)	0.05	35
Iron	(Fe)	0.0057	4
Zinc	(Zn)	0.0033	2.3
Rubidium	(Rb)	0.0017	1.2
Strontium	(Sr)	2×10^{-4}	0.14
Copper	(Cu)	1.4×10^{-4}	0.1
Aluminum	(Al)	1.4×10^{-4}	0.1
Lead	(Pb)	1.1×10^{-4}	0.08
Tin	(Sn)	4.3×10^{-5}	0.03
Iodine	(I)	4.3×10^{-5}	0.03
Cadmium	(Cd)	4.3×10^{-5}	0.03
Manganese	(Mn)	3×10^{-5}	0.02
Barium	(Ba)	2.3×10^{-5}	0.016
Arsenic	(As)	$<1.4 \times 10^{-4}$	<0.1
Antimony	(Sb)	$<1.3 \times 10^{-4}$	<0.09
Lanthanum	(La)	$<7 \times 10^{-5}$	<0.05
Niobium	(Nb)	$<7 \times 10^{-5}$	<0.05
Titanium	(Ti)	$<2.1 \times 10^{-5}$	<0.015
Nickel	(Ni)	$<1.4 \times 10^{-5}$	<0.01
Boron	(B)	$<1.4 \times 10^{-5}$	<0.01
Chromium	(Cr)	$<8.6 \times 10^{-6}$	<0.006
Ruthenium	(Ru)	$<8.6 \times 10^{-6}$	<0.006
Thallium	(Tl)	$<8.6 \times 10^{-6}$	<0.006
Zirconium	(Zr)	$<8.6 \times 10^{-6}$	<0.006
Molybdenum	(Mo)	$<7 \times 10^{-6}$	<0.005
Cobalt	(Co)	$<4.3 \times 10^{-6}$	<0.003
Beryllium	(Be)	$<3 \times 10^{-6}$	<0.002
Gold	(Au)	$<1.4 \times 10^{-6}$	<0.001
Silver	(Ag)	$<1.4 \times 10^{-6}$	<0.001
Lithium	(Li)	$<1.3 \times 10^{-6}$	$<9 \times 10^{-4}$
Bismuth	(Bi)	$<4.3 \times 10^{-7}$	$<3 \times 10^{-4}$
Vanadium	(V)	$<1.4 \times 10^{-7}$	$<10^{-4}$
Uranium	(U)	3×10^{-8}	2×10^{-5}
Cesium	(Cs)	$<1.4 \times 10^{-8}$	$<10^{-5}$
Gallium	(Ga)	$<3 \times 10^{-9}$	$<2 \times 10^{-6}$
Radium	(Ra)	1.4×10^{-13}	10^{-10}

(Source: K. Z. Morgan, *et al.*, *Report of Committee II on Permissible Dose for Internal Radiation*, ICRP Publication 2, New York: Pergamon, 1959, p. 146.)

The elements found in living material, both plant and animal, can be divided into three groups (see Table 12-1). These are:

1. Bulk elements—the familiar elements that make up the bulk of living material, e.g., hydrogen, oxygen, carbon, nitrogen, sodium, magnesium, phosphorus, sulfur, chlorine, potassium, calcium, and, for those species that have hemoglobin, iron.

2. Trace elements—those that occur in concentrations of one or two parts per million (ppm) and occasionally a few parts per billion (ppb) and comprise less than 0.01 percent of the organism. These elements seem to relate mostly to enzyme activity, in which they usually act as regulators. Among these are such elements as cobalt, molybdenum, chromium, iodine, copper, and zinc.

3. Age elements—elements that are accumulated as tissues age. In a few cases the physiological consequence of this accumulation is known, but for most of the elements in this category, the physiological action is either not fully understood or completely unknown. These elements include beryllium, titanium, nickel, arsenic, aluminum, silver, barium, and gold.

Neither the second nor the third list includes all of the elements that fall under the respective headings. Instead, the lists are meant to be guides which indicate the type of elements occurring in each division. It should also be noted that since some elements fit equally well in two categories, there is some overlap between the divisions.

Although we are still gathering new information about the bulk elements, in many areas interest is shifting to the study of the trace elements and their role in controlling our health. It is becoming increasingly clear that the nature and quality of our lives is often controlled by the presence or absence of trace elements in our dietary intake and, in some cases, even by the form in which certain trace metals are available to us. In spite of the growing interest in trace elements and the increasing frequency with which interdisciplinary symposia and conferences on them are being held, the role of many of these elements in the life processes is either poorly understood or completely unknown. Currently, we do not have a clear comprehension of the interrelations among the approximately 40 mineral elements which have been found in living tissue.

Many aspects of trace elements are being studied. For example, projects are now under way to better define the geologic and geographic areas which show either abnormally high or abnormally low incidence of such diseases as hypertension, arteriosclerosis, and cancer. It is important to identify and delineate these geographic distributions so that research in a variety of disciplines can then be focused on these "problem" areas. This research has led to the marriage of geology, geography, agriculture, and botany on the one hand, and geology, pharmacology, medicine, and nutrition on the other.

So far, we have emphasized the activity concerning the subject of trace elements in health. But what is known of the geologic aspects of trace elements? The study of their geologic aspects focuses on three areas: (1) the availability of the trace elements to the biosphere from the crustal material; (2) the concentrations of trace elements in the environment; and (3) our ability to detect trace elements in earth materials.

AVAILABILITY OF TRACE ELEMENTS TO THE BIOSPHERE

All elements, including the trace elements, are released into the environment through the process of chemical weathering. (Refer to Chapter 4 for a detailed discussion of chemical weathering.) However, we should point out that since the concentration of trace elements is not the same in all rocks, no weathering, no matter how intensive, can release an element which is not present.

The availability of elements to plants is further affected by the mode in which the elements occur in the rocks and minerals. Most plants are severly limited in the amount of any element — whether trace or not — available to them because of their inability to extract inorganic nutrients from the tightly bonded silicate structures usually associated with igneous rock. You may remember having visited areas in which fresh, igneous rocks crop out and may further remember that the surfaces of these rocks were covered only by the most primitive forms, mostly lichens. These plants have demonstrated their ability to extract needed inorganic nutrients from these inhospitable sonrces. However, once these complex silicates have been broken down by chemical weathering and the debris has been deposited as sedimentary rocks, the availability of the elements increases, and many more plants are able to extract the needed nutrients.

The release of essential elements may be very slow. If the elements are bound to a silicate structure, this release may be so slow that the plants and animals suffer from a deficiency of an element, even though it is abundant in the rocks of the area. This is often the case in areas underlain by iron-rich rocks in which the iron occurs as insoluble Fe_2O_3. If, however, the element is associated with a less stable structure, as is the magnesium found in a carbonate compound, the availability may present no problem at all.

A word of caution about the methods of assaying rock compositions is perhaps appropriate at this point. When using the results of analysis to determine an element's availability to plants, one should have some idea of the type of analysis involved. It is very possible to get a distorted picture of what nutrients are actually available to the plants. Some methods of analysis are comprehensive and report even the most minor constituents of the most stable and insoluble minerals. A person using such a method might break down tourmaline, for instance, into boron and other constituent elements; however, a plant nutritionist interested in the availability of boron to local crops would know that this boron would be available only after a few million years of weathering.

CONCENTRATIONS OF TRACE ELEMENTS IN THE ENVIRONMENT

When considering trace element concentrations, it is impossible to speak of the "environment" in general terms. There is often a marked difference in these concentrations among bedrock, soil, and ground water in a given area. In addition, there is a wide variation of element distribution and concentration within a particular rock group. For example, differences in elements occur between granites and gabbros, which both belong to the igneous group. Concentrations can be reasonably considered only in terms of averages for the lithosphere, hydrosphere, and atmosphere.

Lithosphere

As is true of most aspects of trace element studies, the available data about the abundances in the lithosphere are far from complete. Much of the data collected has been based on studies of limited geographic areas, and many geographic areas have not yet been studied. The nature of the older analyses also hinder our ability to obtain a clear picture of lithosphere concentrations. In most cases, only the eight or ten most common elements were determined, and any trace metals were either ignored or merely listed as being present, while their quantities were omitted.

Table 12-2 represents the results of an extensive task which, based on available published sources, attempted to indicate the abundances of elements in a variety of rock types. One major disadvantage of such a table is that it does not adequately indicate the variations in abundance within major rock groups. For instance, chromium has an average value, shown in Table 12-2, of 100 ppm for igneous rocks: however, the variation within igneous rocks ranges from 2-10 ppm in granite to 1000-4000 ppm in peridotite. Barium, with an average value of 425 ppm for igneous rocks, varies from about 3 ppm in pyroxenite to nearly 2000 ppm in syenite. This is also true of other rock types: copper, having an average value of 45 ppm for shales, actually varies from 1-10 ppm in nonmarine muds to nearly 130 ppm in marine muds. Thus, although tabulations such as those shown in Table 12-2 are invaluable in giving us general estimates of composition,

Table 12-2 Elementary composition of igneous and sedimentary rocks. All values in parts per million (ppm).

Element	Igneous rocks[1]	Shales[2]	Sand-stones[2]	Lime-stones[2]	Coal[3]
Ac	5.5×10^{-10}				
Ag	0·07	0·07	0·05	0·05	0·1
Al	82300	80000	25000	4200	
Ar	3·5				
As	1·8	13	1	1	25
Au	0·004	0·005	0·005	0·005	≤0·125
B	10	100	35	20	100
Ba	425	580	50	120	1–3000
Be	2·8	3	<1	<1	0·1–1000
Bi	0·17	1	0·3		1
Br	2·5	4	1	6·2	
C	200	15300	13800	113500	800000
Ca	41500	22100	39100	302000	
Cd	0·2	0·3	0·05	0·035	0·25
Ce	60	59	92	12	
Cl	130	180	10	150	3000
Co	25	19	0·3	0·1	15
Cr	100	90	35	11	60
Cs	1	5	0·5	0·5	1·3
Cu	55	45	5	4	300
Dy	3	4·6	7·2	0·9	
Er	2·8	1·9	1	0·36	
Eu	1·2	1·1	0·55	0·2	
F	625	740	270	330	80
Fe	56300	47200	9800	3800	
Ga	15	19	12	4	5·5
Gd	5·4	4·3	2·6	0·7	
Ge	5·4	1·6	0·8	0·2	25–3000
H	1400	5600	1800	860	50000
He	0·008				
Hf	3	2·8	3·9	0·3	
Hg	0·08	0·4	0·03	0·04	
Ho	1·2	0·61	0·51	0·17	
I	0·5	2·2	1·7	1·2	6
In	0·1	0·1	0·05	0·05	<0·1
Ir	0·001				
K	20900	26600	10700	2700	
Kr	0·0001				
La	30	20	7·5	6·2	≤1000
Li	20	66	15	5	≤25
Lu	0·5	0·33	0·096	0·067	
Mg	23300	15000	7000	47000	
Mn	950	850	50	1100	
Mo	1·5	2·6	0·2	0·4	10
N	20				15000
Na	23600	9600	3300	400	480

Table 12–2 (cont'd)

Element	Igneous rocks[1]	Shales[2]	Sand-stones[2]	Lime-stones[2]	Coal[3]
Nb	20	11	0·05	0·3	
Nd	28	16	11	4·3	
Ne	0·005				
Ni	75	68	2	20	35
O	464000	482600	491700	496800	50000
Os	0·0015				
P	1050	700	170	400	
Pa	$1·4 \times 10^{-6}$				
Pb	12·5	20	7	9	5
Pd	0·01				0·005
Po	2×10^{-10}				
Pr	8·2	6	2·8	1·4	
Pt	0·005				$\leqslant 0·035$
Ra	9×10^{-7}	11×10^{-7}	7×10^{-7}	4×10^{-7}	2×10^{-7}
Rb	90	140	60	3	15
Re	0·005				
Rh	0·001				0·01
Rn	4×10^{-13}				
Ru	0·001				
S	260	2400	240	1200	10000
Sb	0·2	1·5	0·05	0·2	
Sc	22	13	1	1	3
Se	0·05	0·6	0·05	0·08	$\leqslant 7$
Si	281500	73000	368000	24000	
Sm	6	5·6	2·7	0·8	
Sn	2	6	0·5	0·5	10
Sr	375	300	20	610	1000
Ta	2	0·8	0·05	0·05	
Tb	0·9	0·58	0·41	0·071	
Te	0·001				
Th	9·6	12	1·7	1·7	
Ti	5700	4600	1500	400	$\leqslant 20000$
Tl	0·45	1·4	0·82	0·05	0·05–10
Tm	0·48	0·28	0·3	0·065	
U	2·7	3·7	0·45	2·2	0·005–200
V	135	130	20	20	40
W	1·5	1·8	1·6	0·6	
Xe	0·00003				
Y	33	18	9·1	4·3	5
Yb	3	1·8	1·3	0·43	
Zn	70	95	16	20	40
Zr	165	160	220	19	$\leqslant 250$

[1] Taylor, S. R. (1964).
[2] Turekian and Wedepohl (1961); Haskin and Gehl (1962) for lanthanides.
[3] Rankama and Sahama (1950), Stutzer (1940) and Bethell (1962); Balashov *et al.* (1964) for additional data on lanthanides.

(H. J. M. Bowen, *Trace Elements in Biochemistry*, New York: Academic Press, 1966, pp. 16–17. Reprinted by permission.)

they are not as useful in providing detailed information. In a situation such as that represented by boron, the table is very useful because it can indicate that this element is approximately ten times more abundant in shale than in igneous rocks. In some instances, apparently, the weathering process tends to enrich rather than to remove certain elements.

Table 12-3 Abundances of selected trace elements in sea water having 35 parts per thousand salinity

Element	ppb = micrograms/liter	Element	ppb = micrograms/liter
Boron	4,450.00	Molybdenum	10.00
Cadmium	0.11	Nickel	6.60
Chromium	0.20	Selenium	0.09
Copper	0.90	Strontium	8,100.00
Gallium	0.03	Tin	0.33
Lead	0.02	Vanadium	1.90
Manganese	0.04	Zinc	5.00

(Data after K. K. Turekian, *Handbook of Geochemistry*, New York: Springer-Verlag, 1969, pp. 309–311.)

Table 12-4 Trace element composition of streams[*]

Element	ppb = micrograms/liter	Approx. est. ppb	Region
Boron	13.00	10.00	U.S.S.R.
Cadmium	—	—	—
Chromium	1.40	1.00	U.S. streams
Copper	12.00	7.00	Maine Lakes and streams
Gallium	0.089	0.09	Germany
Lead	2.30	3.00	Maine lakes and streams
Manganese	4.00	7.00	Maine lakes and streams
Molybdenum	1.80	1.00	U.S. streams
Nickel	0.30	0.30	Maine lakes and streams
Selenium	0.20	0.20	U.S. streams
Strontium	46.00	50.00	Eastern U.S.
Tin	—	—	—
Vanadium	0.90	0.90	Japan
Zinc	16.00	20.00	Columbia River

[*]In several cases multiple studies were listed for one element. Only the reading for the United States was taken (if available). The approximate estimate is the average of all the values and so may differ from the value given in the first column.

(Data after K. K. Turekian, *Handbook of Geochemistry*, New York: Springer-Verlag, 1969, pp. 314–316.)

Table 12–5 Average chemical composition of precipitation in Japan
as reported by Sugawara, 1967

Element	Precipitation (ppm)	Element	Precipitation (ppm)
Na	1.1	Si	0.83
K	0.26	Fe	0.23
Mg	0.36	Al	0.11
Ca	0.94	P	0.014
Sr	0.011	Mo	0.00006
Cl	0.0018	V	0.0014
I	0.0018	Cu	0.0008
F	0.08	Zn	0.0042
S	1.5	As	0.0016

(From K. K. Turekian, *Handbook of Geochemistry*, New York; Springer-Verlag,
1969, p. 320. Reprinted by permission.)

Hydrosphere

Karl Turekian has tabulated the available data up to 1969 for concentrations
of trace elements in both sea water (at a standard 35 parts per thousand
salinity) and fresh-water streams. Both Tables 12–3 and 12–4 are modifi-
cations of this tabulation. The lists are much more extensive than what is
given here, and you should refer to them for data on other elements.

Atmosphere

In conjunction with the recent surge of interest in air pollution, much
work is being done to determine the concentrations of trace elements in
the atmosphere. Although a variety of data may soon be available, there
is presently little to which one can refer for information on trace element
abundances in the air. Turekian provides data obtained from a study done
in Japan on the trace element content of precipitation. These data are
given in Table 12–5.

DETECTION OF TRACE ELEMENTS

Although it is beyond the scope of this book to detail the problems involved in trace element detection, we will briefly discuss some of the difficulties encountered in this branch of science. A major problem, which
is being solved to some extent, has been the low degree of sensitivity of
the instruments used in the analysis. Not all elements can be detected at
equally low levels of concentration. Since there is no single instrument
that can determine low concentrations of all elements with equal pre-

12-1
Collecting soil for geochemical survey, Shoshone
County, Idaho. (Photograph by L. C. Huff, U.S. Geo-
logical Survey)

cision, it is necessary to choose the one technique most suited to the
particular element in question. Some techniques are more sensitive to
low concentrations than are others; however, if an element's concentration
is low enough, it is possible to fail completely to detect its presence. An
additional problem, also being solved to some extent, is the processing of
samples so that multiple elements can be detected simultaneously. Some
techniques, e.g., atomic absorption, are capable of analyzing for only a
single element at a time.

Perhaps the biggest current problem in the field of trace-element
measurement is the sophistication and specialization required of those
individuals who use these analytical instruments. Because it has become
necessary that those in the laboratory develop expertise, often in only a
single phase of the analytical technique, the number of people who can
work on many of the current trace element problems has been greatly
restricted. Some of the methods currently used in trace analysis are:
chemistry, polarography, emission spectroscopy, flame emission spectrometry, atomic absorption spectrometry, and neutron activation analysis.

EFFECTS OF TRACE ELEMENTS

Because our knowledge of the roles of trace metals in various biological systems has expanded rapidly, we have started to focus our attention on their role in human metabolism. We are finding that trace elements are necessary in growth, the healing process, and many metabolic processes — in short, they are essential if man is to function at an optimum. Indeed, they are probably involved in longevity and in the aging process and, hence, with life itself.

Evidence indicates that the essential trace elements (like vitamins, their organic counterparts) participate in a variety of enzymatic actions. However, since they cannot be synthesized or metabolized, the trace elements are in many respects more important than vitamins. If they are not present in the environment in usable forms and in the quantities needed, life suffers and dies; either excesses or deficiencies of trace elements can be lethal.

Since there is an inverse relationship between atomic number and cosmic abundance — a relation which is largely applicable to the earth's crust — and since life depends on availability in the environment of the essential elements, it is not accidental that over 99 percent of living structure is composed of 12 bulk elements which occur within the first 20 elements of the Periodic Table (Fig. 12-2). Furthermore, it is not too surprising that the trace elements that participate in biological activity are found in the first half of the table, i.e., prior to atomic number 48, with the exception of iodine, whose atomic number is 53. Looking at the first 20 elements, one finds three inert ("noble") gases (helium, neon, and argon), two elements that have a limited function in living things (boron and fluorine), three apparently functionless (at least so far) elements (lithium, beryllium, and aluminum), and twelve that are bases of organic molecules, structures, or electrolytes (hydrogen, carbon, nitrogen, oxygen, sodium, magnesium, silicon, phosphorus, sulphur, chlorine, potassium, and calcium). The next 14 elements contain 11 that are essential for some form of life, not necessarily mammalian. Only two elements that are known to be essential (molybdenum and iodine) have atomic numbers greater than 34.

It is difficult to classify trace elements according to biological group, since they share few common characteristics. Currently, trace elements are most commonly grouped according to the following scheme: (1) those that are essential in diet, (2) those that seem to participate in some metabolic activity, and (3) those that, although present in the organism, seem to be inert. Table 12-6 lists the trace metals according to this classification. However, we must emphasize that those elements in the middle and right-hand columns may be, and many undoubtedly are, misplaced; they may

^{1}H									
^{3}Li	^{4}Be								
^{11}Na	^{12}Mg								
^{19}K	^{20}Ca	^{21}Sc	^{22}Ti	^{23}V	^{24}Cr	^{25}Mn	^{26}Fe	^{27}Co	^{28}Ni
^{37}Rb	^{38}Sr	^{39}Y	^{40}Zr	^{41}Nb	^{42}Mo	^{43}Tc	^{44}Ru	^{45}Rh	^{46}Pd

12-2
Partial periodic table, showing element position and associated atomic number.

Table 12-6 Classification of trace elements

Essential for nutrition	Active, but not essential	Inert
Iron	Vanadium	Gold
Zinc	Barium	Silver
Selenium	Arsenic	Aluminum
Manganese	Bromine	Tin
Copper	Strontium	Bismuth
Iodine	Cadmium	Gallium
Molybdenum	Nickel	Lead
Cobalt	Boron	Antimony
Chromium		Lithium
Fluorine		Mercury
		Niobium
		Rubidium
		Silicon
		Titanium

be there simply because research to date has not been able to determine their proper place.

A quite different classification has been suggested by Dr. Walter Mertz, a researcher of trace elements for the United States Department of Agriculture. He suggests that recent trace-element discoveries indicate that the "beneficial" and "toxic" labels applied to individual elements are inappropriate. Instead, he believes that scientists should refer to the toxic and beneficial *levels* of various elements in soil, the atmosphere, food, and tissue. The margin between toxic and beneficial levels varies sharply from element to element; in many cases, the two levels are only narrowly separated. Our knowledge of these margins, especially those occurring in man, is presently too limited to implement this kind of classification, but it is

							^{2}He
		^{5}B	^{6}C	^{7}N	^{8}O	^{9}F	^{10}Ne
		^{13}Al	^{14}Si	^{15}P	^{16}S	^{17}Cl	^{18}Ar
^{29}Cu	^{30}Zn	^{31}Ga	^{32}Ge	^{33}As	^{34}Se	^{35}Br	^{36}Kr
^{47}Ag	^{48}Cd	^{49}In	^{50}Sn	^{51}Sb	^{52}Te	^{53}I	^{54}Xe

12–2
(cont'd)

Table 12–7 Content of chromium in igneous rocks

Rock type	Average chromium content (ppm)
Peridotite	1000–4000
Gabbro	100–400
Diorite	25–80
Granite	2–10

(Data from V.M. Goldschmidt, *Geochemistry*, London: Oxford University Press, 1958.)

useful and may become more practical in the future. This classification does seem to be more natural, since anything an organism takes in can be toxic if absorbed in sufficient quantities: water, essential for man, is toxic if drunk in excessive quantities.

To enable you to understand better the function and availability of trace elements in the natural environment, several of the "essential" trace elements are covered in some detail in the following pages.

Chromium

Environmental availability Chromium has an average value of 100 ppm in igneous rocks, but its abundance varies widely within this group, as is shown in Table 12-7. This table indicates that chromium is more abundant in crystalline rocks that crystalize early in the process. Of the sedimentary rocks, shales contain chromium in about 90 ppm, indicating that little chromium is lost in the weathering process that forms shale. Limestones and sandstones, having averages of 35 ppm and 11 ppm, respectively, contain considerably less.

Soils, like igneous rocks, vary considerably. Although an average of 100 ppm is commonly given, this actually represents a range from 5 ppm to more than 3000 ppm. Soils having the highest concentrations of chromium are those derived from the more basic rocks, such as basalts and serpentinites. It is not unusual to find floras that have adapted to the high chromium content in such soil. The presence of chromium in concentrations of nearly 800 ppm in coal ash seems to indicate that it has been concentrated by plant material.

Function of chromium   The primary work suggesting chromium's function in the body has been done in the laboratory with rats. However, there is reason to believe that the function of this element in rats is similar to its function in man. Chromium apparently enhances the stimulatory effects of insulin. Although the role insulin plays in metabolism is not completely understood, it apparently stimulates the incorporation of glucose into lipids (fats) which store energy.

In addition to its role in glucose intolerance in man, studies indicate that chromium deficiency could be a factor in abnormal lipid metabolism and in atherosclerosis (the most common form of arteriosclerosis, or "hardening of the arteries") in man. Increasing evidence indicates that severe atherosclerosis is usually associated with abnormal glucose metabolism in man. For many years diabetes mellitus has been known to be associated with atherosclerosis, and the greatest single cause of death in diabetic subjects today is atherosclerosis.

Consequence of excess or deficiency   Rats raised on a chromium deficient diet showed definite impairment of glucose utilization, accompanied by depression of growth rates and the approach of a "diabetes mellitus-like" condition. In one experiment, five generations of rats were raised, and each generation showed an increasing incidence of diabetes. This condition was reversed when chromium was added to the rats' daily diet. In other circumstances this strain of rats would have been considered to be suffering from an hereditary disorder; however, as this was a controlled experiment, the diabetes could only have been a dietary-induced condition.

Atherosclerosis is believed to be less severe in areas where tissue chromium is high, presumably due to the presence of high amounts of chromium in that area's soil. This relationship has not been "proved," but there is evidence to suggest its existence. American adults are low in tissue chromium compared to similar samples of people from the Orient, Africa, and the Middle East. It has been suggested that the superheating inherent in much of our commercial food processing causes chromium to assume a chemical form which cannot be utilized by the human body, in effect, making it unavailable to us. There are additional suggestions that food processing may be the cause of our bodies' low chromium content.

Chromium is among the least toxic of the trace metals. Mammals can tolerate as much as 200 times the total body content of chromium taken orally without undergoing adverse reactions. Even large daily doses of this element do not tend to accumulate in tissue. Because of this, the existence of an intestinal and hepatic transport system which rejects excessive doses of chromium has been inferred. Such systems exist for other elements (iron, copper, and manganese), and one must exist also for chromium.

Cobalt

Environmental availability   The highest concentrations of cobalt (Fig. 12-3), approximately 20 to 25 ppm, occur in igneous rocks and shales. Sedimentary rocks other than shale show an average of only a few tenths of one part per million. Soils are reported to have an average of about 8 ppm, but those soils derived from basalt or serpentine-rich rocks usually show somewhat higher percentages. Some areas in the world, e.g., New Zealand, Western Australia, Scotland, and some areas in the eastern portion of the United States (Florida and the coastal portions of Georgia and South Carolina), have soils essentially devoid of cobalt. Its presence in forest soils may range up to 12 ppm. Earlier analyses of water, both fresh and marine, usually showed no cobalt to be present. This was probably the direct result of using instruments unable to detect the low levels present. More recent studies show that cobalt does occur, ranging from 0.0009 ppm in fresh water to 0.0003 ppm in sea water.

Function of cobalt   Cobalt is one of the central molecules in the vitamin B_{12} structure (making this a unique vitamin, as it is the only one using a metal ion) and, as such, functions as an essential element. As far as is currently known, man's daily requirement is about one microgram of vitamin B_{12}, containing only 0.04 micrograms of cobalt.

Ruminants require a few micrograms of inorganic cobalt daily in order for the microorganisms in the rumen to have a supply for the manufacture of B_{12}. Nonruminants also need a daily supply of B_{12}, although they cannot utilize inorganic cobalt directly.

Consequences of excess or deficiency   Diseases caused by cobalt deficiency have been apparent primarily in animals. In the 1930s it was found that cobalt deficiency was the cause of a condition in cattle and sheep known variously as coast disease, pining, vinquish, enzootic marasmus, Morton Mains disease, and salt sick disease. The usual symptoms of this condition are a gradual loss of appetite, emaciation, roughening of the

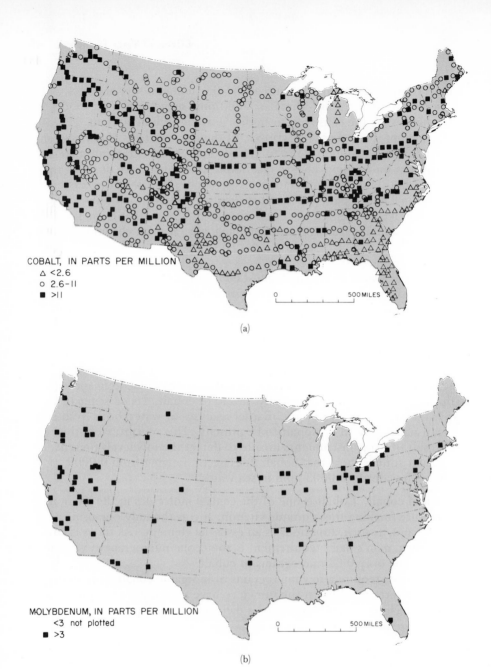

COBALT, IN PARTS PER MILLION
△ <2.6
○ 2.6–11
■ >11

0   500 MILES

(a)

MOLYBDENUM, IN PARTS PER MILLION
  <3 not plotted
■ >3

0   500 MILES

(b)

12–3
Abundance of molybdenum and cobalt in soils and other surficial materials of the conter-
minous United States. (H. T. Shacklette, "A U.S. Geological Survey Study of Elements
in Soils and other Surficial Materials in the U.S.," in Trace Substances in Environmental
Health–IV, *ed. Delbert D. Hemphill, Columbia: University of Missouri, 1971, pp. 41, 44.*
Reprinted by permission.)

coat, scaly skin, reproductive failure, and anemia. In the past, animals in some areas of New Zealand, the Australian coast, and Florida have been particularly susceptible to this problem. Although cobalt is very toxic to plants, it is only moderately toxic to mammals when injected intravenously; oral doses up to 30 times the daily requirement were needed to bring about visible signs of toxicity in cattle. Although no cobalt deficiency condition has been recognized in man, some work has been done with B_{12} and inorganic cobalt in the treatment of various human anemias. The results of these studies have not, however, been promising. The small amount of cobalt required daily may account for the lack of deficiency conditions.

Fluorine

Environmental availability   Table 12-2 shows that both igneous and sedimentary rocks contain large amounts of fluorine, ranging from 330 ppm to 740 ppm. The figures given for shale indicate that fluorine tends to accumulate during the weathering process. Fluorine occurs in all deposits of phosphate rock in the United States, as well as in such minerals as fluorapatite, fluorite, and cryolite. It also occurs in certain silicate minerals such as topaz, tourmaline, and some micas. Fluorine is widely distributed in fresh water in many parts of the United States and the world. Soil compositions indicate that fluorine is retained during the weathering process, and some soils in India and in Africa contain quantities sufficient to be toxic to grazing animals.

Function of fluorine   Although the element functions, so far as is presently known, in two areas which interest us — bone formation and tooth formation — its exact function is not clearly understood. The presence of fluorine increases the "crystallinity" of the apatite crystals present in the teeth and thus works to prevent tooth decay. This increased crystallinity appears as larger and more nearly perfect crystals. These changes would reduce the surface area relative to the weight, which consequently would tend to reduce the reactivity of the crystals. Whether this is the only function that fluorine serves in reducing tooth decay is still open to question. Its role in both teeth and bones will be explored later in the chapter.

Consequence of excess or deficiency   There are no known conditions of fluorine deficiency which are truly detrimental to man or animals. This may be due to the fact that fluorine is nearly ubiquitous, coupled with the small need that most animals, including man, have for this element. Although it is true that adequate fluorine drastically reduces the occurrence of dental caries and may help to alleviate a bone condition common in old age (osteoporosis), it is not completely accurate to label these conditions "deficiency diseases."

Usually, excess fluorine is detrimental to man only in industrial situations and to animals only as a result of ingesting apatitic food supplements. Under natural conditions the result of living for long periods of time in areas of high fluorine (8-20 ppm) is excessive bone formation in the periosteum (the tough, fibrous material which covers all bones) and calcification of ligaments which normally do not tend to calcify. However, in these high-concentration areas no detrimental effects to soft tissue have been identified. There appears to be little to fear from natural fluorine toxicity. For example, let us consider the possibility of fluoride poisoning caused by drinking water to which 1.0 ppm of fluoride had been added (the recommended level to give optimum protection against tooth decay). In order to receive the approximately one-quarter gram of fluoride necessary to make him ill, an average-sized man would have to drink nearly 45 gallons of water in a 24-hour period. He would get very ill if he ingested one gram, but for this to occur he would have to drink 275 gallons of water. For fluoride to be lethal, he would have to ingest 4 to 8 grams; this would necessitate his drinking between 1000 and 2200 gallons of water in one day. Water intoxication would kill him long before he could become ill from the fluoride in the water.

Molybdenum

Environmental abundance   Of the essential trace elements, molybdenum is among the least abundant (Fig. 12-3). Igneous and sedimentary rocks normally contain 1-3 ppm; soils normally contain 3-5 ppm, indicating a possible accumulation during the weathering process. Several studies have indicated that there is very little molybdenum in water, either fresh or marine, or in the atmosphere.

Function of molybdenum   All plants, except perhaps some blue-green algae, require molybdenum. It forms an indispensable group (that is, its position cannot be occupied by any other element) in an enzyme, *nitrate reductase*, that reduces NO_3 nitrogen to the ammonia form which can be utilized in plant metabolism. Among molybdenum's other functions in plants are: making iron physiologically available within the plant; serving in the synthesis of ascorbic acid; and helping to abate plant injury caused by the presence of excess amounts of copper, boron, nickel, cobalt, manganese, and zinc.

Animals need molybdenum because, in addition to nitrate reductase, a second necessary flavoprotein enzyme, *xanthine oxidase*, contains this metal. The activity of both of these enzymes depends on the presence of molybdenum.

Consequences of excess or deficiency   Deficiency symptoms in plants are associated with reduced yields, abnormally low content of protein and, in some cases, a significant reduction in the plants' total supply of ascorbic acid, vitamin C.

There is substantial evidence to support the suggestion that areas of low molybdenum (0.03 ppm or less) are related to the occurrence of xanthine "kidney stones" in some animals. Xanthine ($C_5H_4N_4O_2$) is a white, crystalline substance resembling uric acid.

Excess molybdenum is responsible for an animal disease once called "teart," but now called *molybdenosis*. This condition affects ruminants, especially cows, which graze on forage grown in high-molybdenum areas. Whereas normal herbage contains 3–5 ppm (on the dry basis), the levels may range from 20–100 ppm in areas showing "teart." The disease is characterized by extreme diarrhea, loss of weight, reduced milk yields, rapid deterioration of general condition, and the development of harsh, discolored coats.

INFLUENCE OF GEOLOGY AND GEOGRAPHY ON DISEASE

The existence of significant geographical variation in the incidence of many noninfectious diseases has been known for decades and, in the case of some diseases, for centuries. More recently many of these conditions have been related to the geology of these areas. More sophisticated epidemiological studies are revealing subtle geographic relationships in other diseases which have heretofore been unsuspected. Many of these studies are being approached from a geological point of view (Table 12-8).

A diverse group of people must cooperate in order to accomplish these studies. Currently, the United States Geological Survey is involved in three major studies in conjunction with various health agencies and has just recently finished two studies of a similar nature. The studies being made in direct cooperation with medical groups are:

1. *A geochemical survey of Missouri,* being done in conjunction with the Environmental Health Surveillance Center of the University of Missouri, will estimate the distribution and abundance of elements in the rocks, soil, vegetation, ground water, and surface waters of Missouri. These data are to be used by the Center in support of epidemiological studies currently being carried out in the state.

2. *Geochemical investigations in areas having extreme cardiovascular death rates* have only recently been initiated and are being done in collaboration with the United States Public Health Service. These studies will attempt to relate geochemical information to high death rates caused by heart diseases in some areas and to low death rates in others. As in the Missouri study, rocks, soil, vegetation, and water will be analyzed.

Table 12-8 Excesses and deficiencies to be expected in diverse geologic environments

Unique geologic environments	Possible deficiencies in geologic unit	Maximum measured concentrations in geochemical environment reported in text			
		Possible excess in geologic unit	In soils (ppm, dry wt.)	In plants (ppm, dry wt.)	In water (ppm)
Limestones	Molybdenum, strontium, calcium, magnesium,	Phosphorus Potassium			
Glaciated areas: Drift	Zinc, strontium, molybdenum, cobalt, boron, magnesium, phosphorus, iodine	Titanium Iron Chromium Nickel Zinc Lead Strontium			
Peat bogs	Copper (unavailable)	Zinc Copper Lead Cadmium	161,800 250 190	10,000 58 0.96	8.3
Coastal plain sands	Iron, manganese, copper, selenium, cobalt, boron, chromium				
Serpentine	Nitrogen, phosphorus, calcium, molybdenum, manganese.	Nickel Chromium Magnesium Iron	6,000	630	
Shales of N Central Plains	Phosphorus	Selenium Molybdenum	130	8,000	

Region	Element			
Evaporative basins:				
Nevada and California	Strontium			
	Chlorine		83,000	
	Sodium	1,500		
	Boron		400	
	Potassium			
	Lithium			12
	Fluorine			.025
Arizona, Gila River basin Chromium, barium	Molybdenum		2.25	
	Magnesium		>3,000	66
	Lithium			.380
	Chlorine			1,080
	Boron			.430
	Strontium			3.8
Phosphate rock:				
Idaho and Wyoming	Vanadium		55	
	Uranium		.34	
	Nickel			
	Chromium			
	Zinc			
	Manganese			
	Phosphorus			
Complex ores:	Copper			
	Lead		1,370	
	Zinc			
	Gold			
	Arsenic		140	
	Cadmium		25	
Contamination	Cadmium	>100	2.7	
	Lead		342	

(Helen L. Cannon, "Trade Element Deficiencies in Some Geochemical Provinces of the U. S.," in *Trace Substances in Environmental Health–III*, ed. Delbert D. Hemphill, Columbia: University of Missouri, 1970, pp. 38–40. Reprinted by permission.)

3. *Geochemical investigations of Washington County, Maryland, and San Juan County, New Mexico* are being done with the Cancer Institute and an environmental Public Health unit. Southeast Washington County is a high-cancer area and San Juan County is a low-cancer area; the rock-soil-plant-water relationship is being investigated in both areas, and the scientists anticipate that it will provide information of value in the study of cancer.

Recently, two studies, both done in conjunction with health authorities, were completed. These were:

1. *Geochemical environments and cardiovascular mortality rates in Georgia.* Done in conjunction with the United States Public Health Service, this study described the geochemical environments in two areas in Georgia having extremely different cardiovascular mortality rates. This study is presented in more detail later in this section.

2. *Smelter contamination in the Helena Valley, Montana.* This work was done at the request of the National Air Pollution Control Administration of the United States Public Health Service. The report was released recently.

In addition to these studies, the Geological Survey has a number of independent studies under way. The data will undoubtedly be used by medical groups in the future.

Having completed this prelude, we shall now review in detail some specific cases of geologic and geographic control of disease. These cases are not exhaustive, but rather have been chosen to give you an impression of the diversity of this area of study.

Water Composition and Cardiovascular Health

In 1957 a Japanese agricultural chemist, Jun Kobayashi, published a study that suggested a possible relationship between the composition of river waters in Japan and a high death rate from "apoplexy." This study had been going on since 1942 and had described the composition of about 600 rivers in Japan. These rivers were characterized by "scantiness of calcium carbonate," and many contained more sulfate than carbonate. Kobayashi concluded that "higher contents of sulfuric acid or sulfate in river water as compared with alkaline substance (calcium carbonate) should cause apoplexy."

In 1960 an American physician-researcher, H. A. Schroeder, published a study comparing mortality from cardiovascular disease with water composition in 163 of the largest municipalities in the United States during the period from 1949 to 1951. Schroeder found definite negative correlations between cardiovascular death rates and magnesium, calcium, sulfate, fluoride, and total dissolved solids, as well as with other parameters which he used. His data suggested that a definite inverse relation exists between hardness of water and occurrence of cardiovascular disease, i.e., the harder the water, the lower the death rate from some types of cardiovascular disease.

To date, studies concerning this relationship have been completed in seven countries (Japan, the United States, England, the Republic of Ireland, Sweden, the Netherlands, and Canada), and the results have been essentially the same. However, the results of a study done in Oklahoma and of another done in Ireland have introduced some discordant evidence. The Oklahoma study was designed to study county units, and its results showed a *positive*, though statistically insignificant, correlation between water hardness and cardiovascular death rates. These data, however, were later reworked, using only well-defined water supplies, i.e., the majority of the population served by public water supply, and the inverse relationship, though still not statistically significant, was reestablished. In the Irish study no correlation was found, but when the same worker later reworked the study and included the urban areas of Northern Ireland (which had not been included in the earlier study) the inverse correlation, though again not statistically significant, emerged.

At present, there is no way to assess whether the apparent inverse relation between hard water and heart attacks is real or spurious (Fig. 12–4). Part of the problem is inherent in the nature of the analyses involved in the various studies. The results of these studies may not be strictly comparable. Even if one accepts, for the sake of argument, that the relationship is real, there is still no explanation for this relationship. However, several possibilities do exist. First, the excessive calcium in the "hard" water may preclude the "harmful" substance (whatever it is) from going into solution. Second, it has been suggested that the absence of calcium in the water may be responsible for allowing the water to attack the galvanized pipes through which it flows and to leach out some harmful trace metal, such as cadmium. If this is true, a decrease in the relationship should occur as more copper pipe replaces the older, galvanized pipes. A third possibility is that water hardness and cardiovascular mortality may not be related to each other at all; rather, both may be varying in response to a third, unknown, factor. Whether there is a protective agent in hard water, a deleterious factor in soft water, or a completely unknown, controlling factor is still a matter of conjecture; therefore, based on the information known today, it is still impossible to make recommendations for action. The suggestion, made by some, that a family with a past history of heart disease should move to an area having hard water is unrealistic, given the present evidence. Future

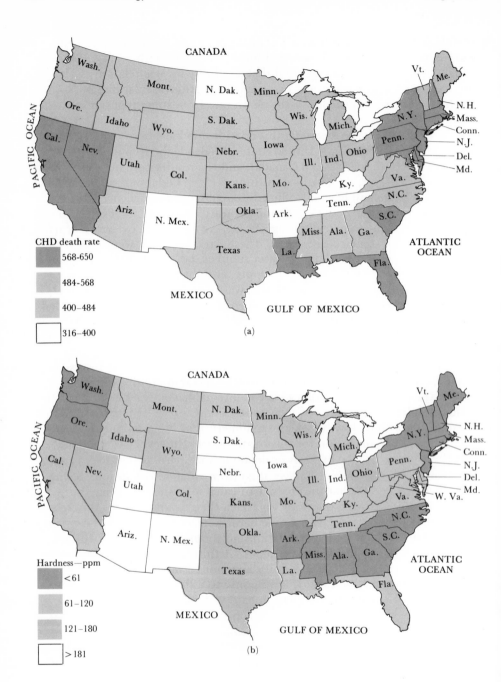

(a)

(b)

◀ *12-4*
Comparison of death rates in the United States from coronary heart disease with average
water hardness by state. Note that the states with the hardest water are usually the same
as those with the lowest death rates from coronary heart disease. (a) Average annual age
adjusted death rates from coronary heart disease for white males ages 45–64, 1949–1951;
(b) weighted average water hardness of finished water from public supplies for 1315 of the
larger cities in 1952. (E. F. Winton and L. J. McCable, "Studies Relating to Water Miner-
alization and Health." Reprinted from Water Quality and Treatment, *3rd ed., 1971,*
1950, 1940 and the Journal American Water Works Association, *1970 by permission of*
the Association. Copyrighted 1971, 1950, and 1970 by the American Water Works Asso-
ciation, Inc., 2 Park Avenue, New York, N. Y. 10016.)

studies may enable us to recommend an optimum range of water hardness
for humans; however, our work to date has only provided good indications
of where additional work is needed.

Soil and Cancer

Since cancer incidence is not uniform in some parts of the world (Fig. 12–5),
some have speculated that trace elements may play a part in certain types
of cancer. Several studies have established that the distribution of trace
elements in man is upset by this disease. However, each element is affected
differently: some increase markedly above their average ranges in healthy
individuals, and some undergo marked decreases.

At the same time, some studies have produced data that strongly
suggest that some trace elements, such as manganese, copper, and vanadium,
induce inhibiting or therapeutic actions in certain kinds of cancer. Little is
known about the reactions involved, but the suggestions of a definite relationship are present.

The maps in Fig. 12–5 show the results of work done in the rural
districts of North Wales. One map illustrates the incidence of stomach cancer,
while the other shows the organic content of the soil. The close association between the two factors shown by these maps strongly supports the
suggestion that the high mortality rate from stomach cancer in North Wales
may be related to the high organic content of the soil.

In 1955 further work was started in North Wales under the British
Empire Cancer Campaign. In this study researchers established correlations
between certain trace elements and the frequency of stomach cancer. It
was found that organic matter, zinc, and cobalt had a significant, positive
relation to the incidence of stomach cancer, but not to that of intestinal
cancer. Chromium was positively correlated to both types. Although the
geographic distribution of zinc- and cobalt-rich soils seemed to have no significant correlation to stomach cancer rates on the whole, the occurrence

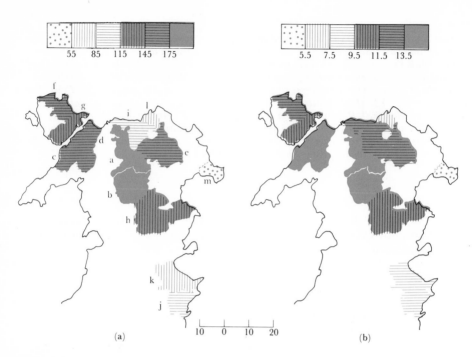

12-5
Incidence of stomach cancer in North Wales compared with the organic content of the soil: (a) gastric cancer standardized mortality ratios; (b) ignition loss of organic material of cultivated soil. (C. D. Legon, "The Aetiological Significance of Geographical Variations in Cancer Mortality," British Medical Journal, 27 September 1952, p. 701. Reprinted by permission.)

of high zinc- and cobalt-rich soils in the immediate vicinity of the occurrence of stomach cancer was "excessive." This may suggest that a lower-order relationship between the two exists. Although nickel, titanium, and lead showed no connection with any form of cancer, vanadium and iron were found to be somewhat related to stomach cancer in one area that was studied.

The work completed in the general area of soils and trace elements and their relationship to certain types of cancer strongly supports the concept of geological control of certain types of cancer. For example, the geographic distribution of all cancers in Sweden demonstrates that incidence very definitely increases from north to south across the country. A form of cancer common in children is so widespread in an east-west belt across equatorial Africa, that this area has been designated the "Lymphoma Belt." Although no real link between this condition and the soil has been definitely established, there is a strong apparent relationship. For these problems to be

12-6
Areas of the United States having low iodine content
(shaded portion), known as the "goiter belt."

solved will ultimately require interdisciplinary cooperation between earth
scientists and medical scientists.

Endemic Goiter

Goiter is an enlargement of the thyroid gland, which is located in the neck,
and is caused by the thyroid gland's failure to manufacture thyroid hormones
at a normal rate. In the thirteenth century Marco Polo noted this condition
while traveling through Chinese Turkestan on his journey between Venice
and the court of the Great Khan. Goiter occurs in almost every country of
the world, and no race or nationality is exempt from this disease. There
are, however, areas in the world in which this disease is more prevalent than
in others. These areas are located in high mountain regions such as the
Alps, the Pyrenees, the Himalayas, and the Andes; however, it does occur
in low-lying areas such as The Netherlands and the Great Lakes region of
the United States.
    Among the reasons why the thyroid fails to manufacture thyroid
hormones at a normal rate are iodine deficiency and the presence of
goitrogens. Goitrogens are chemical agents that interfere with the synthesis
of the thyroid hormones. Since iodine is a constituent of both of these
hormones (*tetraiodothyronine* and *triiodothyronine*), a deficiency brought
about by low concentrations of iodine in soil and water is the main cause
of endemic goiter. Apparently, glaciation and flooding during the Pleistocene
removed much iodine from the soils. The Great Lakes area and the mountainous regions of the western part of the United States (Fig. 12-6) are

generally the areas in which the soil, water and, hence, the diet have low iodine content. For obvious reasons this area has been called the "Goiter Belt," but even this is only one of many such areas in the world in which goiter is endemic.

However, goiter does not always respond to treatment consisting of iodine-supplemented diets. In the Cauca Valley of Colombia, South America, endemic goiter has been correlated with the source of drinking water. The agents supposedly responsible for causing goiter in this area are sulfurated hydrocarbons, whose exact structure are unknown at present. Their presence results in typical thyroid enlargement, reduced iodine intake by the thyroid, decreased thyroidal iodine content, and impaired formation of the thyroid hormones. All of these symptoms persist in spite of high supplementary iodine intake; here, this prophylactic diet supplement was totally ineffective. It should also be noted that this area reports a greater incidence of thyroid cancer than do adjacent areas.

Osteoporosis

This disease, which primarily affects people of late middle-age or older, causes a gradual decrease in both the amount and the strength of bone. The most common site of bone collapse due to this disease is the spine; the next highest number of fractures occurs in the thigh bone (femur). The disease accounts for many of the fractures in the elderly, since the affliction tends to weaken the bones so that fractures occur even under the most minor stresses. This disease has been ranked as one of the major causes of disability in the elderly.

A study of over 1000 individuals living in two areas of North Dakota having significantly different amounts of fluoride in the ground water was conducted in the mid-1960s. In the towns studied in the southwestern part of the state (Mott and Hettinger), the fluoride content ranged from 4 ppm to 5.8 ppm. In the towns surveyed in the northeastern part of the state (Grafton, New Rockford, and Carrington), the range was from 0.15 ppm to 0.3 ppm. The authors of this study concluded that fluoride concentrations between 4.0 and 5.8 ppm had significantly decreased the occurrence of osteoporosis. Apparently, the greater fluoride consumption significantly reduced the prevalence of reduced bone density in women, while a similar, although not statistically significant, trend was indicated for men.

At least a half-dozen studies have provided evidence indicating that the "fluoride bone effect" developed by persons living in high-fluoride areas is, in fact, beneficial and desirable in adult bone. Fluoride apparently counteracts the effects of osteoporosis in the aged and the other diseases induced by calcium loss, without bringing on any harmful skeletal effects.

Dental Caries

The controversy over the addition of artificial fluoridation to municipal
drinking waters as a measure to counter dental caries has continued for 25
years. The consensus developed after 25 years of extensive experience is
that the adjustment of the fluoride content of community water supplies
is safe, effective, and practical. Fluoridation has been shown to reduce dental decay among children by 64 percent, and no known deleterious effects
have resulted from the consumption of fluoridated water.

The relationship between fluoride and teeth was first established in
1931 and was a result of 30 years' work by many researchers to determine
the cause of "mottled" enamel which had been described in various parts
of the United States as well as in many other regions of the world. During
the 1920s many investigators observed that there was less decay among
children whose teeth were "mottled." Once the relationship between fluoride and "mottling" had been established, the next logical step was to
test the suggested relationship between fluoride and tooth decay. This was
done in a study of 7257 children in 21 cities whose water supplies had
varying levels of fluoride. Three facts emerged from this study:

1. Fluoride levels greater than 1.5 ppm do not significantly reduce the
   decay of teeth, but do increase the occurrence and severity of
   mottling.

2. The optimum balance between maximum reduction of caries with
   no esthetically significant mottling occurs at a level of about 1.0 ppm.

3. Although some benefit occurs at levels below 1.0 ppm, the reduction of caries is not great and gradually decreases to zero as the
   fluoride level is reduced to zero.

Next, controlled studies in which fluorides were added to a water
supply were started and their results were carefully monitored. The out-
break of World War II delayed these studies, but in 1945 three cities (Grand
Rapids, Michigan; Newburgh, New York; and Brantford, Ontario) began
programs of artificial fluoridation of their water supplies. At that time,
these programs affected slightly over 230,000 persons; at the end of 1969,
just under 90 million persons were using controlled-fluoridation municipal
water in the United States.

It is interesting to compare the similarity between a graph plotted from
the results of the original study and one plotted from the results of about
57 subsequent studies. The graph in Fig. 12-7(a), which was developed in
the first study, shows that as the fluoride content increased up to an optimum level of about 1.5 ppm, dental caries decreased. Figure 12-7(b), based

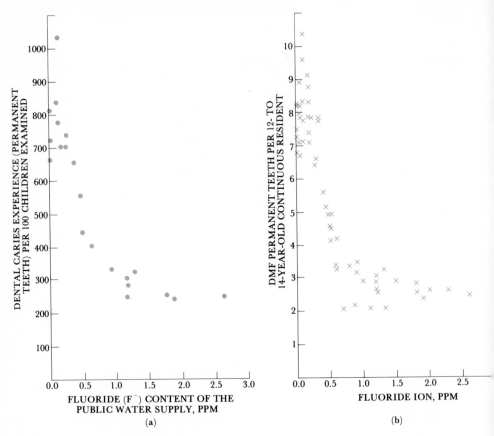

12-7
Relationship between flouride concentration and incidence of dental caries. (a) Relation-
ship between dental caries in 7257 children and the flouride content of the public water
supply; (b) relationship between tooth decay and flouride levels, based on data from 57
studies (DMF = decayed, missing, and filled). These data show that an increase in flouride
from 0.5 ppm to 1.0 ppm produces a decrease of approximately two DMF teath/person;
an increase in flouride from 0.7 ppm to 1.0 ppm produces approximately 1 DMF/person.
(F. J. Maier, "Flourides in Water." Reprinted from Water Quality and Treatment, *3rd ed.,*
1971, 1950, 1940 by permission of the Association. Copyrighted 1971, 1950, and 1940
by the American Water Works Association, Inc., 2 Park Avenue, New York, N.Y. 10016.)

on the results of another 57 studies, shows a very similar inverse relationship.
Thus, the additional research has only confirmed and refined the results
found in the original study.

The relationship between fluoride and dental caries and the overall
problem of fluoridating water supplies has been discussed so much that
other relationships have been mostly overlooked. Researchers have noted

the relation of soil, trace elements, and dental caries for about 30 years but only recently have displayed a renewed interest in this relationship. Two accidental discoveries provided strong evidence that molybdenum was instrumental in reducing caries. In two separate instances, one in Hungary and one in New Zealand, a preliminary study of a pair of towns was undertaken prior to a fluoridation study; in both, a marked difference in the prevalence of caries was discovered and was traced to a higher molybdenum content in the domestic water supply of the town in which there was a lower incidence of caries. Subsequent studies on rats suggest that molybdenum and fluoride, combined, have a strong preventive effect on the occurrence of caries.

Some studies indicate that both boron and lithium sharply reduce caries in rats. A possible synergistic effect for boron and fluoride, similar to that for molybdenum and fluoride, exists.

The discovery that significant numbers of new navy recruits from contiguous counties in northwestern Ohio had never experienced a "clinically detectable caries attack" led to comparative studies of the water of that area and the finished water (water leaving the city water plant) of the seven largest cities in Ohio. Although no definite conclusions have been reached about the cause of this condition, it was found that the water of northwest Ohio contained amounts of boron, lithium, molybdenum, strontium, titanium and vanadium which were, statistically, significantly greater than the amounts of these elements in the finished waters.

However, another series of studies showed a negative correlation between tooth decay and dietary consumption of small amounts of selenium during early tooth formation. Studies of farm people who had lived most of their lives in highly seleniferous regions, such as South Dakota, Wyoming, and Nebraska, showed that a high incidence of dental caries was the most frequent symptom of disease.

Cardiovascular Mortality

Geographic variations in cardiovascular disease have been much studied in recent years in an attempt to delineate a cause-and-effect relationship. Much work still remains to be done, but progress has been made. In general, the highest cardiovascular mortality rates occur along the eastern seaboard, in the Great Lakes region, and in the Mississippi River delta area. The lowest rates occur in the midcontinent. Figure 12-8 shows the areas having the lowest and highest mortality rates based on deaths of whites, aged 45-74, for the years 1949 through 1951.

A recent study carried out in Georgia considered this problem in some detail, but on a smaller scale. The state is divided geologically (Fig. 12-9) and is comprised of a northern part, which is about one-third of the state,

12-8
*Areas of lowest and highest death rates from cardiovascular diseases for whites, ages 45–
74, for the years 1949-1951. (H. I. Sauer, "Epidemiology of Cardiovascular Mortality—
Geographic and Ethnic,"* American Journal of Public Health **52,** *1, January 1962, p. 96.
Reprinted by permission.)*

and a southern part, consisting of the remaining two-thirds. The physical
environments of the two parts are drastically different. The northern
portion lies in the Appalachian Highlands and is characterized by valleys,
ridges, and highlands underlain by a variety of Precambrian and Paleozoic
metamorphic rocks and igneous intrusive rocks formed during various
geologic ages. Much of the northern area has been strongly weathered,
and the soils are characterized by sandy loams with thin subsurface clay
layers and appreciably weathered material.

The Coastal Plain (southern) portion is comprised of a variety of soils,
the most typical of which are sandy soils that overlie relatively young
marine rocks. The landscape in this area has little relief, and swamps and
sluggish streams are common.

A relationship between the environments represented in the state and
the human health of these areas was recognized even in pre-Civil War days.
The United States Geological Survey and the Heart Disease and Stroke
Control Program of the United States Public Health Service conducted a

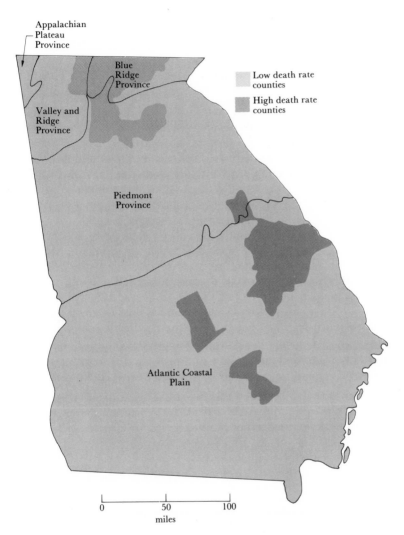

Appalachian
Plateau
Province

Blue
Ridge
Province

Valley and
Ridge
Province

Piedmont
Province

Atlantic Coastal
Plain

Low death rate
counties

High death rate
counties

0   50   100
miles

12-9
Physiographic regions and provinces in Georgia, with locations of coun-
ties with high and low death rates per 100,000 population for cardiovas-
cular diseases among males, ages 35–74 during the period 1950–1959.
(Source: H. T. Shcklette, et al., Geochemical Environments and Cardiovascular Mortality Rates in Georgia, *U.S. Geological Survey Professional*
Paper 547–C, 1970.)

Table 12–9 Elements analyzed for in all samples and the results
compared for the two areas

Aluminum	Iron	Potassium
Barium	Lanthanum	Scandium
Boron	Lead	Sodium
Calcium	Magnesium	Strontium
Cerium	Manganese	Titanium
Chromium	Neodymium	Vanadium
Cobalt	Nickel	Ytterbium
Copper	Niobium	Yttrium
Gallium	Phosphorus	Zinc
		Zirconium

joint study in which native trees, uncultivated soils, endemic vegetables,
and cultivated garden soils were analyzed for those elements listed in
Table 12–9.

When the results for the two areas were compared, several interesting
points were revealed:

1. The mortality rates for the cardiovascular diseases are roughly twice
   as high in the southern portion of the state as in the northern portion.

2. High cardiovascular death rates occur in an area located on soils derived
   from unconsolidated or slightly consolidated sands, sandy clays, and
   clays. The area of low cardiovascular death rate is located on soils
   derived from the weathering of igneous and metamorphic rocks.

3. The concentrations of the elements studied in garden soils and in
   uncultivated soils from both areas were significantly different in
   the two areas and tended to be higher in the area of low death rate.

4. If geochemical differences between the two areas do, indeed, present
   a causal relationship for cardiovascular deaths, the cause apparently
   would be due to deficiencies, rather than excesses, of elements.

REFERENCES

Anderson, R. J. "Dental caries prevalence in relation to trace elements," *British Dental Journal,* **120** (1966): 271-275.

Aykroyd, W. R. *Conquest of Deficiency Diseases,* United Nations, World Health Organization, 1970.

Beeson, K. C. "Soil management and crop quality," in USDA 1957 Yearbook, *Soil,* Washington: U. S. Department of Agriculture, 1956, pp. 258-267.

Bernstein, D. S., N. Sadowsky, D. M. Hegsted, C. D. Guri, and F. J. Stare. "Prevalence of osteoporosis in high- and low-fluoride areas in North Dakota, *Journal of the American Medical Association,* **198**, 5 (1966): 85-90.

Bowen, H. J. M. *Trace Elements in Biochemistry*, New York: Academic Press, 1966.

Dubos, R. and M. Pines, eds. *Health and Disease*. New York: Life Science Library, Time-Life, 1965.

Fleischer, M. "Fluoride content of ground water in the conterminous United States," in *Relation of Geology and Trace Elements to Nutrition*, eds. H. L. Cannon and D. F. Davidson, Geological Society of America Special Paper 90, 1967.

Gaitan, E., D. P. Island, and G. W. Liddle. "Identification of water-borne goitrogens in the Cauca Valley of Colombia," University of Missouri-Columbia: 5th Annual Conference on Trace Substances in Environmental Health, June 29–July 1, 1971.

Goldschmidt, V. M. *Geochemistry*, London: Oxford University Press, 1958.

Hadjimarkos, D. M. "The role of selenium in dental caries," in *Proceedings of University of Missouri's 4th Annual Conference of Trace Substances in Environmental Health*, 1970, pp. 301–306.

Harris, H. C. "Effect of micronutrient deficiencies on mineral composition of certain plants and on animal and human nutrition," in *Relation of Geology and Trace Elements to Nutrition, op cit.*

Henschen, Folke. *The History and Geography of Diseases*, New York: Delacorte, 1966.

Hill, A. C., S. J. Toth, and F. E. Bear. "Cobalt status of New Jersey soils and forage plants and factors affecting the cobalt content of plants," *Soil Science*, 76 (1953): 273–284.

Kidson, E. B. "Cobalt status of New Zealand soils," *The New Zealand Journal of Science and Technology*, 18 (1937): 694–707.

Kobayashi, J. "On geographical relationship between the chemical nature of river water and death-rate from apoplexy," *Berichte Ohara Institute Landwirtschaft Biologie*, 11 (1956): 12–21.

Lamb, C. A., O. G. Bentley, and J. M. Beattie, eds. *Trace Elements—Proceedings of the Conference Held at the Ohio Agricultural Experiment Station*, Wooster, Ohio: Academic Press, 1958.

Legon, C. D. "The aetiological significance of geographical variations in cancer mortality," *British Medical Journal* (September, 1952): 700–702.

Leone, N. C. "Areas of the USA with a high natural content of water fluoride," in *Fluorides and Human Health*, World Health Organization Monograph 59, 1970.

Losee, F. L. and B. L. Adkins. "Anti-cariogenic effect of minerals in food and water," *Nature*, 219 (1968): 630–631.

Maier, Franz J. "25 years of fluoridation," *Journal of the American Water Works Association* 62 (January 1970): 3–8.

————— . "Fluorides in water," in *Water Quality and Treatment*, American Water Works Association, 1971, pp. 397–412.

Malthus, R. S., T. G. Ludwig, and W. B. Healy. "Effect of trace elements on dental caries in rats," *New Zealand Dental Journal*, 60 (1964): 291–297.

Marienfeld, C. J., S. L. Silberg, R. W. Menges, W. T. Crawford, and H. T. Wright. "Multispecies study of congenital malformations in Missouri," *Missouri Medicine* (March, 1967): 230–233.

Sauchelli, V. *Trace Elements in Agriculture*, New York: Van Nostrand Reinhold, 1969.

Sauer, H. I. and P. E. Enterline. "Are geographic variations in death rates for the cardiovascular diseases real?" *Journal of Chronic Diseases* 10, 6 (1959): 513–524.

Sauer, H. I. "Epidemiology of cardiovascular mortality—geographic and ethnic," *American Journal of Public Health*, 52, 1 (1962): 94–105.

Schroeder, H. A. "Relation between mortality from cardiovascular disease and treated water supplies," *Journal of the American Medical Association* 172, 17 (1960): 1902–1908.

————— . "The biological trace elements or peripatetics through the periodic table," *Journal of Chronic Disease* 18 (1965): 217–228.

————— . "Cadmium, chromium, and cardiovascular disease," *Circulation* 35 (1967): 570–582.

Schroeder, H. A., J. J. Balassa, and I. H. Tipton. "Abnormal trace metals in man: chromium," *Journal of Chronic Disease*, 15 (1962): 941–964.

Schroeder, H. A., A. P. Nason, B. S. Tipton, and I. H. Tipton, "Essential trace metals in man: cobalt," *Journal of Chronic Disease* 20 (1967): 869–890.

Schroeder, H. A., J. J. Balassa, and I. H. Tipton. "Essential trace metals in man: molybdenum," *Journal of Chronic Disease* 23 (1970): 481–499.

Schroeder, H. A., A. P. Nason, B. S. Tipton, and I. H. Tipton. "Chromium deficiency as a factor in atherosclerosis," *Journal of Chronic Disease* 23, (1970): 123–142.

Schroeder, H. A. and A. P. Nason. (1971) "Trace-element analysis in clinical chemistry," *Clinical Chemistry* 17 (1971): 461–474.

Schutte, K. *The Biology of the Trace Elements—Their Role in Nutrition*, Philadelphia: Lippincott, 1964.

Shacklette, H. T. "U. S. Geological Survey study of elements in soils and other furficial materials in the United States," in *Proceedings of the University of Missouri's 4th Annual Conference on Trace Substances in Environmental Health*, June 23–25, 1970, pp. 35–45.

Shacklette, H. T., H. I. Sauer, and A. T. Miesch. *Geochemical Environments and Cardiovascular Mortality Rates in Georgia*, U. S. Geological Survey Professional Paper 574–C, 1970.

Silberg, S. L., C. J. Marienfeld, H. Wright, and R. C. Arnold. "Surveillance of congenital anomalies in Missouri, 1953–1964," *Archives of Environmental Health* 13 (1966): 641–644.

Soane, B. D. and D. H. Saunder. "Nickel and chromium toxicity of serpentine soils in southern Rhodesia," *Soil Science* 88 (1959): 322–330.

Sognnaies, R. F. "Fluoride protection of bones and teeth," *Science* 150 (1965): 989–993.

Spencer, J. M. "Geological influence on regional health problems," *Texas Journal of Science* 21, 4 (1970): 459–469.

Stiles, W. *Trace Elements in Plants*, 3d ed., Cambridge, England: Cambridge University Press, 1961.

Stocks, P. and R. I. Davies. "Epidemiological evidence from chemical and spectrographic analyses that soil is concerned in the causation of cancer," *British Journal of Cancer* 14 (1960): 8–22.

Stookey, G. K., R. A. Roberts, and J. C. Muhler. "Synergistic effect of molybdenum and fluoride on dental caries in rats," *Proceedings of the Society of Experimental Biological Medicine* **109** (1962): 702–705.

Underwood, E. J. *Trace Elements in Human and Animal Nutrition,* 2nd ed., New York: Academic Press, 1966.

Winton, E. F. and L. J. McCabe. "Studies relating to water mineralization and health," *Journal of the American Water Works Association* **62** (January 1970): 26–30.

Material useless to or discarded by man is waste, the disposal of which becomes critical in modern industrialized society. (Photograph by Bruce Anderson)

WASTE
DISPOSAL

INTRODUCTION

Waste is useless or discarded material; it can be in the form of heat, spent
chemicals, excreted matter from people and other animals, useless rock
from mining, or products that man has made and no longer wants or needs.
Waste disposal in an industrialized society (Table 13-1) which has millions
of affluent people is an overwhelming problem. The United States annually
generates more than 250 million tons of waste from homes and businesses.
It costs nearly five billion dollars a year to have this waste taken to the city
dump for disposal. Nearly half of this waste consists of paper or paper products, about 12 percent is glass, nearly 10 percent is metal or metal products,
10 percent is garbage, about 5 percent is plastic and rubber products, and
the remainder is wood, grass, leaves, and dirt.

Table 13–1 Comparison of wastes generated by major
contributor

Waste source	Waste quantity (million tons/year)
Residential, commercial and institutional wastes (1969)	250
Industrial wastes	110
Mineral wastes	1,700
Agriculture wastes	2,280
	4,340

(Source: National Industrial Pollution Control
Council, 1971)

Most methods of waste disposal involve some aspect of geology, and
it is these in which we are interested. In order to give you a more comprehensive picture of the problem, the chapter will begin with a discussion of
the more common kinds of waste products encountered today; hopefully,
this will give you a sense of the enormity and complexity of the problem.
In the latter part of the chapter we will discuss some of the more important methods of waste disposal that are in some way connected with geology. Burning as a disposal method has been omitted because it is not related to geology and because it is being used less frequently. Discussion of
the use of rivers for waste disposal is also not included. Although rivers
are definitely linked to geology, the issue of dirty rivers has been so widespread that our discussion of it would be superfluous. The methods we do
discuss have received less public attention and are, therefore, less likely to
be generally known.

CHEMICAL WASTE

Industry and agriculture are largely responsible for generating chemical
waste. Chemical waste from industry is usually the by-product of some
manufacturing process. Agricultural chemical waste may be the result of
drainage from animal feed lots (nitrates) or drainage from cultivated lands
(fertilizers and pesticides). Although both industry and agriculture have
disposed of their waste mainly into streams, which may ultimately affect
subsurface water supplies, in the past few years the number of deep wells
used for the disposal of industrial wastes has risen.

Industry

The nature of industrial wastes depends on how they have been processed
and from what they have been produced; therefore, they have few common

Table 13-2 Concentrations of soluble organics in the waste of various industries

Source	Average BOD within 5 days (mg/1)
Domestic sewage	200
Paper mill	130,000
Slaughter house	1,785
Commercial bakery	3,745
Tannery wastes	7,500
Textile mills	37,500
Brewery	1,090
Brewer's spent grain processor	23,762
Cannery	75,000

characteristics. However, it may be worthwhile to consider some of the more common types of wastes generated by industry.

Soluble organics The soluble organic compounds are those carbon compounds that are soluble in water; adding soluble organics to a river is equivalent to putting raw sewage into the water. Table 13-2 lists the effects of the soluble organic wastes which are discharged by a variety of industries. The BOD (biological oxygen demand), which is the amount of oxygen required to oxidize the organic matter into its inorganic components, gives a measure of the amount of soluble organics. These inorganic components are carbon dioxide, sulfur oxide, nitrous oxides, and water. The BOD, then, is a measure of the organic waste being dumped. The wastes from these sources are carbohydrates from sugar and paper mills, hydrocarbons from oil refineries, and fatty acids and proteins from such industries as dairies and packing houses.

Toxic materials Acids, nondegradable chemicals, and heavy metals are among the toxic substances found in industrial waste. Cyanide is used in the extraction of gold and is also a by-product of producing coke from coal. Concentrated brines high in potassium chlorides result from the preparation of table salt. Acid baths are used by many industries in their processes, and this spent acid plus the materials it contains can be toxic. For instance, it is not uncommon to find heavy concentrations of copper, zinc, and chromium in the acid bath used in making brass.

Hydrocarbons It has been estimated that approximately 150,000 metric tons of hydrocarbons are carried to the sea annually by American rivers. Although the sources are difficult to pinpoint, these hydrocarbons come from used lubrication oil, solvents used to clean tanks, and the waste and spills associated with the manufacture and retailing of petroleum products from the refinery to the local service station.

Table 13-3 Production of animal waste in 1965 and the impact this represents in human population equivalents

Animal	1965 population[1] (millions)	Annual production of waste – million tons[1] Solid	Liquid	Equiv. human BOD in millions of humans[2]
Cattle	107	1,004.0	390.0	1,754.8
Horses (work stock only)	3	17.5	4.4	33.9
Hogs	53	57.3	33.9	100.7
Sheep	26	11.8	7.1	63.7
Poultry	490	48.0	—	52.5
Total		1,138.6	435.4	2,005.6

[1]U. S. Department of Agriculture, Miscellaneous Publication 1065, 1968.
[2]National Industrial Pollution Control Council, 1971b, p. 8.

Agriculture

Fertilizers, manure, and pesticides are the main wastes associated with agriculture. The fertilizers and pesticides are unique because they are not wastes *per se,* but rather become wastes by not being fully used at the location at which they were intended to be used. Thus, the problem inherent in these wastes is not how to dispose of them; the objective is how to prevent them from becoming a waste.

Manure   The primary difference between animal waste and human waste (Table 13-3) is volume rather than composition. Farm animals in the United States generate an amount of waste about ten times greater than human waste. A cow generates waste equal to that generated by about 16.5 people each day; a hog generates about twice as much waste as does a human, while seven chickens are required to produce waste equal to that of one person.

The rate at which the number of beef cattle feedlots is increasing annually is about 10 percent nationwide and about 30 percent in the plains area of Texas, Oklahoma, Colorado, Nebraska, and Kansas. At these rates, the number of feedlots doubles about every seven years nationwide and in slightly over two years in the plains area. This growth has occurred mainly in the 5000- to 100,000-head capacity range. A feedlot having 100,000 cattle has a waste-producing capacity equal to that of a city with a population of over one and one-half million, and this waste is untreated before being put into the streams or infiltrating the ground. There are only about 26 cities in the United States that are large enough to have a problem of these dimensions.

Fertilizers   As much as 90 percent of the fertilizer distributed on fields in some areas is not utilized by the plants and is therefore carried away by surface and ground water. In some of these areas fertilizer is applied in concentrations 50 times greater than the national average. Runoff from such heavily fertilized areas contributes significantly to the problem of chemical waste.

Approximately 24 million tons of commercial fertilizers were used in the United States in 1960; by 1969 approximately 39 million tons were being used, an increase of over 60 percent. However, as crop output increased only slightly more than 10 percent during the same period, it is likely that a significant portion of the additional fertilizer was not utilized by the plants.

The United Nations Food and Agriculture Organization (FAO) indicates that fertilizer use in North America increased at a rate of 5.5 percent during the period 1954–1964, while during the last five years of this period it increased at a rate of over 7.0 percent. If this latter rate were to continue, the use of chemical fertilizer would double in a ten-year period. If past experience is indicative, we would not expect a doubling in crop production during the same period.

Pesticides   Since DDT was introduced on the market in 1943, thousands of pesticide products and formulas have been developed. The United States Department of Agriculture estimated that more than one billion pounds of chemical pesticides were produced in 1969. Many of these are nondegradable or require such a long period of time to degrade that they are essentially nondegradable.

The amount of pesticide that remains unused after application, thereby becoming an agricultural waste product, can be reduced by several means, as follows:

1. Don't overuse the pesticide. Such overuse can be moderated by designing equipment which will not apply the pesticide in excessive amounts.

2. Modify the formulas. Pesticides should be recomposed to make them both more effective and more degradable to reduce their residence time in the soil or water.

3. Eradicate pests. If the pest is eliminated, so is the need for the pesticide. Pest eradication could cause a drastic reduction in the need for pesticides in the United States. This can be, and in many cases has been, accomplished by such nonchemical methods as the use of sterile males, the introduction of predators, and the quarantine and physical destruction of the habitat.

The total effect of pesticides on man and his animals has not yet been fully assessed. Pesticides frequently accumulate in particular species of wildlife and have been found in concentrations as much as 100,000 times the normal environmental background levels. However, according to the American Chemical Society's report in 1969: "There is no evidence at present that long-term, low-level exposure to pesticides at concentrations approximating those found in the diet or the environment in the U.S. has any deleterious effect on man. At this time, therefore, the net effect of pesticides on human health in the broad sense is positive." As pesticides are, nevertheless, an agricultural waste product, they present a variety of potential problems and should be treated accordingly.

SEWAGE

Domestic sewage, averaging from 100 to 150 gallons per person per day, amounts to between 20 and 30 billion gallons per day in the United States. This effluent can be divided into two major components—water and solids. Water from the community water supply into which we add urine, feces, paper, soap, detergents, grease, garbage and food waste from the kitchen, starch, and dirt is the vehicle for our sewage. Some used petroleum products from the local service station are also usually dumped into the sewers; if the storm sewers join the sanitary sewer system, as is often the case, a good deal of suspended soil is added, too. Table 13-4 lists the products emitted by two major sources along the west coast of the United States.

Because many industrial wastes are discharged through the municipal sewer system, the sewage problem becomes more complicated. It has been estimated that the production of industrial waste in this country is growing at a rate of about 4.5 percent annually. This is four times the rate at which the population is increasing and twice the rate, estimated at a little over 2 percent, at which disposal and treatment facilities are growing. Urbanization is adding to this growth in waste. The increase in domestic waste due to rapid urbanization is estimated to be approximately 6 percent, which is three times the rate at which facilities to handle this waste are currently being built.

HEAT

Waste heat is generated and emitted in a variety of processes, by far the largest of which is the generation of power by power plants. This heat generation is a result of the inefficient conversion of fuel to energy when work is done. If cooling towers are available, this heat is released to the

Table 13-4 The combined discharge for 1968 of the Los Angeles County Sanitation Districts (360 mgd) and City of Los Angeles (340 mgd) to the Pacific Ocean

Constituent	Mass emission rate (tons/day)
Dissolved solids	3,600
Chloride (Cl)[a]	1,150
Sodium (Na)[a]	880
Sulfate (SO$_4$)	610
Suspended Solids (includes digested sludge)	565
BOD	560
Total Nitrogen (N)	165
Phosphate (PO$_4$)	100
Grease	90
Potassium (K)	82
Thiosulfate (S)	50
Detergents	21
Phenols	9
Iron (Fe)	7
Fluoride (F)	5
Boron (B)	4.4
Zinc (Zn)	2.4
Chromium (total Cr)	1.3
Copper (Cu)	1.0

Selected trace constituents	(lb/day)
Nickel (Ni)	1,125
Cyanide (CN)	930
Lead (Pb)	730
Manganese (Mn)	400
Cadmium (Cd)	270
Arsenic (As)	100
Selenium (Se)[b]	55
Chromium (hexavalent, Cr^{+6})	30
Barium (Ba)[b]	~ 0
Silver (Ag)[b]	~ 0

[a]Does not include direct discharge of oilfield brines to ocean.

[b]Data for City of Los Angeles only.

(Data summarized by permission from *Wastes Management Concepts for the Coastal Zone*, Publication ISBN 0-309-018552, Committee on Oceanography and Committee on Ocean Engineering, National Academy of Sciences-National Academy of Engineering-National Research Council, Washington, D. C., 1970, p.7.)

atmosphere; but water, which is more commonly used to carry the heat away, is either dumped into the surface streams or pumped into the sub-surface to mingle with the ground water.

Today, the average efficiency of converting the energy in fuel to electricity is about 33 percent. The common steam-power plant fueled by fossil fuels has an efficiency of about 40 percent, whereas present nuclear power plants have an efficiency of only 30 percent. This means that about 60 percent of the energy in the fossil fuel appears as waste heat, compared to about 70 percent for nuclear plants. This heat is commonly released into streams. Because a nuclear plant requires the burning of about 33 percent more fuel per kilowatt hour to produce as much electricity as does a conventional fossil-fuel plant in the same amount of time and since about 70 percent of the fuel then goes to waste heat, the nuclear plants actually produce about 50 percent more waste heat than do fossil-fuel plants. Thus, the difference between the two is significant.

As water is heated, it loses its ability to absorb oxygen; therefore, water that carries waste heat into a stream is oxygen-poor and potentially damaging to the organisms living in that stream. Because the organics from sewage and industrial discharge may be additionally depriving that water of the little remaining oxygen, it is obvious that a very undesirable situation can develop. Currently, industry uses more than 3 percent of the United States' annual precipitation; however, since the demand for power is currently doubling every ten years while industrialization is increasing, that figure will be even larger in the future. Based on present trends in power usage and on existing technology, it has been estimated that approximately 50 percent of all the water flowing on the surface of the United States in the year 2000 will have been used to carry the waste heat generated by power plants.

RADIOACTIVE WASTES

Radioactive wastes are produced whenever radioactive materials are used. In recent years the major source of this type of waste has been the processing of nuclear fuels for nuclear power plants, which includes the mining, milling, and the other processes involved in fuel fabrication, as well as the reclamation of the usable portions of spent fuel. An additional source of radioactive wastes is the irradiation of nonfuel materials which are near the fuel being used in nuclear reactors. These materials may include structural materials, impurities in the coolant, as well as the coolant itself. A number of radioisotopes which, because they are useful, are purposely produced by irradiation of nonfuel materials are another source of radioactive waste.

The wastes may be in liquid, gas, or solid form and are handled according to the concentration of radioactive material they contain—hence, their potential hazard. They are commonly grouped into low-, intermediate-, and

Table 13-5 The principal fission products which occur in radioactive
wastes from a nuclear installation

Radioisotope	Atomic number	Half-life	Radiation emitted*
Krypton–85	36	4.4 hr (IT)† → 9.4 yr	β, γ, e⁻ → β, γ
Strontium–89	38	54 days	β
Strontium–90	38	25 yr	β
Zirconium–95	40	65 days	β, γ
Niobium–95	41	90 hr (IT) → 35 days	e⁻ → β, γ
Technetium–99	43	5.9 hr (IT) → 5×10^5 yr	e⁻, γ → β
Ruthenium–103	44	39.8 days	β, γ
Rhodium–103	45	57 min	e⁻
Ruthenium–106	44	1 yr	β
Rhodium–106	45	30 sec	β, γ
Tellurium–129	52	34 days (IT) → 72 min	e⁻, β → β, γ
Iodine–129	53	1.7×10^7 yr	β, γ
Iodine–131	53	8 days	β, γ
Xenon–133	54	2.3 days (IT) → 5.3 days	e⁻, β → β, γ
Cesium–137	55	33 yr	β, γ
Barium–140	56	12.8 days	β, γ
Lanthanum–140	57	40 hr	β, γ
Cerium–141	58	32.5 days	β, γ
Cerium–144	58	590 days	β, γ
Praseodymium–143	59	13.8 days	β, γ
Praseodymium–144	59	17 min	β
Promethium–147	61	2.26 yr	β

*β: beta particle, an electron. γ: gamma ray, similar to
X rays. e⁻: internal electron conversion.
†IT: isomeric transition, internal.

(Source: U. S. Atomic Energy Commission)

high-level wastes. Low-level wastes are those whose radioactivity content is
low enough so that they may be released into the environment after some
dilution or simple processing. The range of this category extends up to
those wastes which have about 1000 times the radioactivity considered

Table 13-6 Principal activation-produced radioisotopes produced by neutron irradiation of nonfuel materials

Radioisotope	Atomic number	Source and reaction*	Half-life	Radiation emitted†
Air and Water				
Tritium (H-3)	1	^{2}H (n,γ)	12.3 yr	β, γ
Carbon-14	6	^{14}N (n,p)	5700 yr	β
Nitrogen-16	7	^{16}O (n,p)	7.3 sec	β, γ
Nitrogen-17	7	^{17}O (n,p)	4.1 sec	β
Oxygen-19	8	^{18}O (n,γ)	30 sec	β, γ
Argon-41	18	^{40}A (n,γ)	1.8 hr	β, γ
Sodium				
Sodium-24	11	^{23}Na (n,γ)	15 hr	β, γ
Sodium-22	11	^{23}Na (n,2n)	2.6 yr	β^+, γ
Rubidium-86	37	^{85}Rb (n,γ) (impurity in sodium coolant)	19.5 hr (IT)‡ → 1 min	β, γ → K, γ
Alloys				
Aluminum-28	13	^{27}Al (n,γ)	2.3 min	β, γ
Chromium-51	24	^{50}Cr (n,γ)	27 days	β^+, K, γ
Manganese-56	25	^{56}Fe (n,p)	2.6 hr	β, γ
Iron-55	26	^{54}Fe (n,γ)	2.9 yr	K
Iron-59	26	^{59}Co (n,p)	45 days	β, γ
Copper-64	29	^{63}Cu (n,γ)	12.8 hr	β, γ, β^+, K
Zinc-65	30	^{64}Zn (n,γ)	250 days	β^+, e^-, γ
Tantalum-182	73	^{181}Ta (n,γ)	115 days	β, γ
Tungsten-187	74	^{186}W (n,γ)	24 hr	β, γ
Cobalt-58	27	^{58}Ni (n,p)	71 days	β^+, γ
Cobalt-60	27	^{59}Co (n,γ) (also purposeful irradiation)	5.3 yr	β, γ
Phosphorus-32	15	^{31}P (n,γ) (purposeful irradiation)	14.3 days	β

*(n,γ): absorbs neutron, emits gamma. (n,p): absorbs neutron, emits proton.

†β: beta particle, an electron. γ: gamma ray, similar to X rays. β^+: positively charged electron. K: orbital electron capture into nucleus. e^-: internal electron conversion.

‡IT: isomeric transition, internal.

(Source: U. S. Atomic Energy Commission)

safe for direct release into the environment without any dilution. Intermediate-level wastes range up to those having about 1000 times the concentration of low-level wastes. Wastes above this range are considered to be high-level wastes and pose a severe health-hazard potential. Tables 13–5 and 13–6 list the products which commonly occur as radioactive waste.

MINING WASTES

Mining wastes (Table 13–7) present two major problems: the first is the physical volume of waste rock to be disposed of, and the second is the undesirable elements often associated with this waste rock. Rock volume increases with crushing and may increase as much as 40 percent. Crushed rock is normally disposed of by piling it up on the surface. This is often dangerous, as well as unsightly. In 1966 a 400-foot high pile of rock which had accumulated on a mountainside during a period of nearly a century of coal mining slid down the mountainside and engulfed part of the town of Aberfan, Wales. It was estimated that the avalanche, which killed 144 people, consisted of nearly two million tons of coal-mining waste.

The chemistry of mine wastes is perhaps even more of a problem. Two distinct problems related to the chemistry of mine wastes have arisen in recent years: one is acid mine drainage, mostly from coal mines and associated waste piles, and the other is the accumulation of radon associated with dumpings from uranium mining and processing. The first is the most common of the two. It has been estimated that from 5000 to 10,000 miles of streams in this country are being polluted by acid drainage related to mining operations; abandoned mine workings account for as much as two-thirds to three-fourths of this pollution. The acid is formed when the mineral pyrite, an iron sulfide commonly associated with coal deposits, is oxidized as it is exposed to moist air and forms sulfuric acid as well as a series of sulfates and iron oxides. This oxidation occurs both in underground mines when tunneling exposes new surfaces to the air and in surface mining when the rock material associated with the coal is stripped off and dumped, exposing it to the atmosphere. As water passes through the mines or the rock dumps, it dissolves these oxidation products and often disperses them into either the groundwater or the surface drainage where they can cause much damage.

Other chemical problems, though not so widespread as the acid problem, are associated with mining. Some of these potentially harmful chemicals are arsenic compounds associated with silver deposits, cyanide used to extract gold, and high concentrations of potassium chlorides associated with the purification of common table salt. Many heavy metal ions are highly soluble and are frequently found in rivers near mining operations.

Table 13-7 Land disturbed in each state by strip and surface mining, Jan. 1, 1965

State	Clay	Coal (bituminous, lignite and anthracite)	Stone	Sand and gravel	Gold	Phosphate rock	Iron ore	All other	Total
Alabama [1]	4,000	50,600	3,900	21,200	100	....	52,600	1,500	133,900
Alaska [2]	....	500	....	2,000	8,600	....	....	....	11,100
Arizona [1]	2,700	....	1,000	7,200	1,200	....	....	20,300	32,400
Arkansas [2]	600	10,100	900	2,600	....	....	100	8,100	22,400
California [2]	2,700	20	8,000	19,900	134,000	....	900	8,500	174,020
Colorado [1]	2,000	2,800	6,200	15,500	17,100	....	25	11,400	55,025
Connecticut [1]	....	....	100	16,100	....	....	....	100	16,300
Delaware [2]	200	....	200	5,200	....	....	100	10	5,710
Florida [1]	13,200	....	25,300	3,900	....	143,600	....	2,800	188,800
Georgia	*1,300	*300	*6,800	*1,200	....	....	*100	*12,000	[1]21,700
Hawaii [2]	....	....	....	....	....	....	....	10	10
Idaho [2]	500	....	700	11,200	21,200	3,100	35	4,200	40,935
Illinois [2]	1,400	127,000	5,700	9,000	....	....	....	....	143,100
Indiana [2]	1,500	95,200	10,200	18,000	....	....	....	400	125,300
Iowa [1]	1,300	11,000	12,200	17,600	....	....	6	2,300	44,406
Kansas	*1,100	[2]45,600	*7,500	*5,100	....	....	....	*200	59,500
Kentucky	[1][2]2,400	[1][2]119,200	*3,900	*1,700	....	....	....	*500	127,700
Louisiana [1]	900	....	100	29,700	....	....	50	....	30,750
Maine [1]	400	....	4,400	28,200	12	....	100	1,700	34,812
Maryland	[1][2]1,200	[2]2,200	*2,200	*18,800	....	....	*20	*800	25,220
Massachusetts [1]	700	....	1,200	36,400	....	....	1,100	900	40,300
Michigan [2]	600	....	7,700	25,200	....	....	2,200	1,200	36,900
Minnesota [1]	600	....	3,900	41,600	3	....	67,700	1,600	115,403
Mississippi [2]	2,700	....	400	26,500	....	....	30	....	29,630
Missouri [2]	6,600	31,800	8,400	3,800	....	....	200	8,300	59,100
Montana [2]	....	1,500	10	*13,500	5,600	100	10	6,200	26,920
Nebraska [2]	900	....	4,300	23,700	....	....	....	....	28,900
Nevada [1]	100	....	1,600	5,500	5,600	....	600	19,500	32,900
New Hampshire [2]	....	....	100	8,000	....	....	....	200	8,300
New Jersey [2]	1,400	....	2,000	27,600	....	....	1,000	1,800	33,800
New Mexico [2]	13	1,200	100	400	40	....	100	4,600	6,453
New York [1]	1,700	....	12,500	42,200	5	....	700	600	57,705
North Carolina [1]	5,800	10	6,000	18,400	2,200	300	100	4,000	36,810
North Dakota	*800	*7,700	*300	*26,100	....	....	....	*2,000	36,900
Ohio	[1]10,200	[2]212,800	*21,000	*28,100	....	....	*4,000	*600	276,700
Oklahoma [2]	....	23,500	....	*2,500	....	....	....	1,400	27,400
Oregon [2]	100	....	300	1,300	6,300	....	10	1,400	9,410
Pennsylvania	*10,400	*302,400	*24,400	*23,800	*2	....	*8,800	*400	370,202
Rhode Island [1]	....	....	20	3,600	....	....	....	....	3,620
South Carolina [1]	10,900	....	1,400	10,400	200	8,100	100	1,600	32,700
South Dakota	*2,000	*900	....	*28,000	....	....	....	*3,300	34,200
Tennessee [2]	2,700	29,300	4,400	18,400	....	27,000	5,300	13,800	100,900
Texas [1]	6,800	2,900	21,900	122,300	....	....	9,600	2,800	166,300
Utah [2]	600	....	200	2,200	....	10	500	2,000	5,510
Vermont	....	....	*2,300	*4,000	....	....	....	*400	6,700
Virginia	[1][2]1,100	*29,800	*4,300	*13,100	*600	*100	[1][2]7,700	[1][2]4,100	60,800
Washington [2]	500	100	1,300	5,700	400	....	20	800	8,820
West Virginia [2]	300	192,000	2,800	300	....	....	100	....	195,500
Wisconsin [2]	100	....	9,000	26,400	5	....	49	....	35,554
Wyoming	[1][2]3,500	[2]1,000	[1][2]300	[1][2]200	....	[2]800	[1][2]300	[2]4,300	10,400
Total	108,513	1,301,430	241,430	823,300	203,167	183,110	164,255	162,620	3,187,825

* Estimate.
[1] Data obtained from Soil Conservation Service, U. S. Department of Agriculture.
[2] Data compiled from reports submitted by the States on U. S. Department of the Interior form 6-1385X.

(Source: "Surface Mining and Our Environment," U. S. Department of the Interior, 1967)

Radium-226 is a waste product of uranium mining which forms Radon-222, an inert gas. In open areas having adequate air circulation, there is little danger of significant accumulation of this gas. However, the gas can be extremely dangerous if it is allowed to collect in closed spaces such as basements and rooms where people are. Such a situation occurred in 1971 in Grand Junction, Colorado. The local populace collected uranium tailings

13-1
Mine waste piles surrounding Richer, Oklahoma. (Photograph by H. I. Smith, U. S. Geological Survey)

which were remnants of the uranium extraction process for a variety of uses. Between 1952 and 1966, local residents carried away about 200,000 tons of the gray, powdery tailings "sand" which was used as fill material around foundations, soil conditioner, and even as a replacement for sand in mortar used in masonary walls. In 1966 the Colorado Health Department ordered the people to stop this practice. A survey of structures in the area has revealed evidence of tailings either under or against many of the structures.

This condition best illustrates the potential for possible dangerous accumulations within the structure. The final word is not in yet on this situation, and the extent of the damage to the health of the population, if any, has not been assessed. At other processing sites this type of waste is now being buried under several inches of soil to avoid removal by either man or wind. The effects of having the wind scatter such low-level radioactive materials over wide areas and into streams is not known.

The mining and processing of oil shale, although not a serious problem today because of the size of oil shale operations, can become a serious problem in the near future. Oil shale is totally mined and removed to a location where as much as possible of the oil contained in the shale is removed. The bulk of what was originally mined becomes waste and must be disposed of. The estimates of the extent and thickness of oil shales vary, but commercial production from these shales will undoubtedly generate many million tons of retorted oil shale residue. It has been suggested that even the smallest plant will probably generate more than 50,000 tons of retorted shale per day.

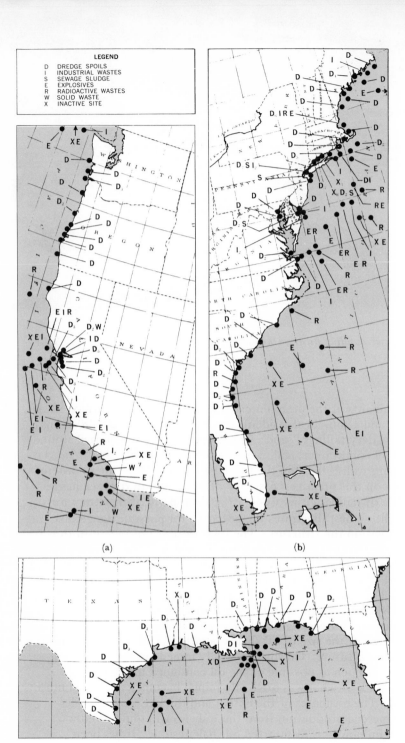

LEGEND
D   DREDGE SPOILS
I   INDUSTRIAL WASTES
S   SEWAGE SLUDGE
E   EXPLOSIVES
R   RADIOACTIVE WASTES
W   SOLID WASTE
X   INACTIVE SITE

(a)

(b)

(c)

◀ 13-2
Known disposal sites of the coasts of the United States. (Source: Council on Environ-mental Quality, Ocean Dumping—A National Policy, *Washington, D.C.: U.S. Government Printing Office, 1970)*

DISPOSAL METHODS

Now that we have discussed the generation of various kinds of waste, we will use the last part of the chapter to discuss some of the current methods of waste disposal and their relation to geology. We will emphasize ocean dumping, underground cavern disposal, and fluid-injection wells because you are probably less familiar with these methods of disposal. Landfills (city dumps) and sewage treatment facilities will be covered in order to give you a comprehensive view and review of the major points of these methods. Perhaps the ultimate waste disposal technique is recycling, but we are not yet close to achieving this.

Ocean Dumping

Ocean dumping as a means of waste disposal is becoming increasingly important. In October 1970 there were 246 known disposal sites off the coast of the United States (Fig. 13-2). Not included in this total are more than 100 artificial reefs built of old car bodies and tires, which act as fish shelters. Of the 246 sites, 122 (50 percent) were along the Atlantic coast, 56 (22 percent) were along the Gulf coast, and 68 (28 percent) were along the Pacific coast; however, these data are incomplete, for little good information exists for sites off Alaska and Hawaii or outside the 12-mile limit. These disposal sites were depositories for dredge spoils, industrial wastes, sewage sludge, explosives, radioactive wastes, and solid wastes. Table 13-8 gives the percentage of each of these.

Table 13-8 Waste dumped in the ocean in 1968

Waste type	% of total (1968)	Amount in tons
Dredge spoils	80	38,428,000
Industrial wastes	10	4,690,500
Sewage sludge	9	4,477,000
Construction and demolition debris	1	574,000
Solid waste	1	26,000
Explosives	1	15,200
TOTAL	100	48,210,700

(Source: Council on Environmental Quality, 1970)

The continuous dredging operations to keep the harbors and rivers navigable are the largest single source of waste. It has been estimated that about 34 percent, or 13 million tons, of these dredge spoils are polluted by contamination from industrial, municipal, and agricultural sources. Analyses of these pollutants indicate that toxic heavy metals such as cadmium, chromium, lead, and nickel occur in concentrations up to three million times the natural concentrations found in sea water and several thousand times the concentrations toxic to marine life.

Currently, almost all the waste dumped into the ocean is taken to a selected spot in either a barge or ship. Because of transportation cost, these spots are usually not very far from the shore (normally between 5 and 25 miles), and often some of the waste migrates back to shore in a relatively short period of time. Some of the more highly toxic wastes are taken out at least 300 miles before they are put overboard. Apparently because of the large capital outlay and maintenance costs necessary for large-capacity barges, it is very unlikely that barging will be the solution to the waste disposal problem. This method will still be used, undoubtedly, to dispose of small quantities of highly toxic wastes, but that this will expand into large operations seems improbable.

It has been proposed recently that old cargo vessels be equipped with modern incinerating facilities so that the ship could carry a load of municipal waste out to the open sea, incinerate it, and release the ashes. Supposedly, the ship would be out far enough that the smoke would not create an air pollution hazard and that the released ashes would be adequately diluted by the sea. It has been suggested that three, 10,000-ton vessels could adequately handle the 23,000 tons of garbage that the city of New York generates daily.

An alternative method of disposal which seems promising is to pipe the waste out to the open ocean. This method entails laying large pipes in the ocean to the edge of the continental shelf. Apparently, two such pipes, one in Washington state and one in San Diego, California, are presently in operation. The city of Los Angeles is building two such pipes to help solve its waste problem. Government cost analysis of this method indicates that it is economically feasible if the pipe is less than 30 miles long or if the pipe serves more than one community. Cost-benefit calculations estimate that the cost of handling 1000 gallons of sewage would be less than half the cost of current methods. In addition, the cost analyses indicate that if an entire area within a radius of 1000 miles is serviced by a large-diameter pipe, the cost would be reduced even further.

Deep-Well Injection

This method of disposal consists of injecting, usually under some pressure, liquid wastes into permeable subsurface beds. (Fig. 13–3). At one time

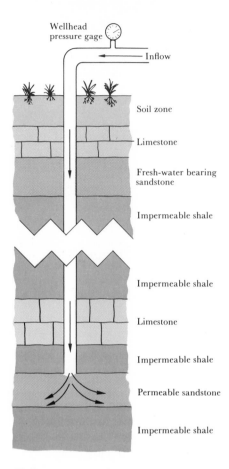

Wellhead
pressure gage

Inflow

Soil zone

Limestone

Fresh-water bearing
sandstone

Impermeable shale

Impermeable shale

Limestone

Impermeable shale

Permeable sandstone

Impermeable shale

13–3
Waste-injection well with a sandstone aquifer.

these beds may have served as reservoirs for oil and gas. They are usually several thousand feet deep and are separated from possible fresh-water aquifers above by relatively thick beds of impermeable material such as shale or salt.

Deep-well disposal for liquid wastes is not a new method of disposal. The petroleum industry has been using this technique to dispose of brines associated with oil production since the 1930s. There are about 40,000 of these brine-injection wells in the country today. Other industries have been following the lead of the petroleum industry and are turning increasingly to this method of liquid-waste disposal. In the early 1950s fewer than six

13-4
Cumulative number of injection wells in operation from January 1950 to January 1967. (D. L. Warner, "Subsurface Disposal of Liquid Industrial Wastes by Deep-Well Injection," in Subsurface Disposal in Geologic Basins—A Study of Reservoir Strata, *ed. John E. Galley, Tulsa, Oklahoma: American Association of Petroleum Geologists, Memoir 10, 1968, p. 18. Reprinted by permission.)*

deep wells (Fig. 13-4) were used for industrial waste. While conservative estimates place the number of fluid-injection wells today at more than 1100, only about 246 of these are used for industrial waste (Table 13-9). Some of these are used for ground water recharge and as shields against salt water encroachment, but increasing numbers are used for waste disposal.

There are at least four important factors to consider before initiating deep-well disposal: suitability of the site, suitability of the waste, economics, and legal aspects.

Site suitability   The suitability of a site to be used for injection and storage of liquids is dictated by the geology of the disposal area (Fig. 13-5). Two prime geological conditions are crucial: first, the site must have beds which will accommodate the liquid waste; second, these beds must be below, and separated from, the fresh-water bearing beds in the area.

Either beds having a natural porosity and permeability such as sandstone, limestone, and dolomite, or beds which have developed porosity through solution or through natural or man-made fracturing can fulfill the first condition. If the waste is acidic, the use of limestone or dolomite as

Table 13–9 Tabulation of industrial waste-injection wells as of 1972

State	Number of wells	State	Number of wells
Alabama	5	Nevada	1
California	4	New Mexico	1
Colorado	2	New York	4
Florida	5	North Carolina	1
Illinois	5	Ohio	8
Indiana	12	Oklahoma	9
Iowa	1	Pennsylvania	8
Kansas	27	Texas	71
Kentucky	3	Tennessee	4
Louisiana	40	West Virginia	7
Michigan	27	Wyoming	1
			246

(Source: D. L. Warner, *Survey of Industrial Waste-Injection Wells: Final Report to the U. S. Geological Survey*, 1972)

reservoir rock has worked especially well, since the permeability of these carbonate rocks often improves with acidization.

Although beds that permit the flow of fluids are essential because they act as reservoirs for the liquid waste, impervious beds are equally important, since they can act as a seal to prevent the upward migration of the waste which would contaminate the groundwater. Beds of shale, clay, marl, gypsum, and anhydrite satisfy this specification and, if unfractured, form effective reservoir seals. In addition to this type of containment, known as stratigraphic traps, certain geologic features might prove to be useful as waste disposal units. Wastes, especially those with high specific gravity, could be collected in synclines (downwarped beds of rock) which form closed subsurface basins.

Other conditions that might affect the selection of a site are the hydrodynamic gradient within the injection bed and the hydrodynamic dispersion. The former could cause uneven distribution within the injection zone, while the latter, which is the mixing of the fluids being injected with those already in the bed, could result in a much wider distribution of injected fluids than anticipated.

Waste suitability Only the most concentrated and most objectionable liquid wastes which are, for some reason, untreatable are suitable for well injection. Since the receiving area (that is, the permeable bed) has a limited storage capacity and the waste can be injected at only a limited rate, commonly 100 to 400 gallons per minute, this mode of disposal is not appropriate for large quantities of waste.

13–5
Geologic features significant in deep waste-injection well-site evaluation and locations of industrial waste-injection systems. (D. L. Warner, "Subsurface Disposal of Liquid Industrial Wastes by Deep-Well Injection," in Surface Disposal in Geologic Basins— A Study of Reservoir Strata, ed. John E. Galley, Tulsa, Oklahoma: American Association of Petroleum Geologists, Memoir 10, 1968, p. 15. Reprinted by permission.)

Both the chemical and physical properties of the fluid to be injected are important, insofar as they must not destroy the suitability of the rock as a reservoir. Because they might plug the rock so that it can receive no more fluid, suspended solids, gas in the fluid, products of the reaction between the injected fluid and the fluid in the rock, or even the rock itself are all potentially dangerous. To compensate for reduced flow under these conditions, more pressure is exerted at the well head; this can result in hydraulic fracturing of the formation, reducing the effectiveness of the confining strata. When this happens the site may have to be permanently abandoned.

Economics   The cost of injection as a method of disposal includes drilling the well, operating the well, and, often, pre-injection treatment of the waste to make it suitable for injection. Geologic factors are a major determinant of the construction and operation costs of the well. Factors such as well depth and the nature of the rock being drilled greatly influence the cost of construction, while such geologic parameters as fluid pressure in the injection zone, thickness of the reservoir rock, permeability, and porosity determine, in large measure, the cost of operating the well. Even the pre-injection treatment of the waste may be affected by geologic factors, including the mineralogy of the reservoir rock and the chemistry of the interstitial fluids in the rock.

Legal aspects   Apparently no state prohibits by law the disposal of noxious fluids by deep-well injection, but several do not allow, as a matter of state policy, the use of such wells. Other states regulate what can be disposed of—in some cases, only potable water or oil-field brines—in deep wells. Several states have no regulation of deep wells at all.

One legal problem that will probably arise in the future in connection with deep-well disposal is that of subsurface trespass by the noxious wastes. The oil industry has already confronted this problem in connection with the disposal of oil-field brines. This problem will undoubtedly arise in connection with other fluids, too, unless the lessons learned by the petroleum industry are carefully applied.

Deep-disposal wells have failed in various ways. Although wells are cased so that they will contain the noxious fluids until they reach the disposal aquifer, in at least one instance the casing above the disposal level was corroded, causing the wastes to spill into an underground freshwater reservoir. Because the wastes are pumped into the containing beds under pressure which is sometimes very high, the formation is occasionally unable to contain the pressures involved over a long period, and a rupture may occur which extends into or through the confining impervious beds. As a result, in at least one case the noxious fluids reached the surface and erupted as a

20-foot geyser. Injection of fluids under pressure has been blamed for causing a series of earthquakes which shook Denver, Colorado, between 1962 and 1967 (see Chapter 10).

Underground Disposal

In 1954 the United States Congress amended the Atomic Energy Act to allow nuclear fission to be used in the production of electric power, as well as for other uses. At that time the Atomic Energy Commission (AEC) began to study all the ramifications implicit in the safe and practical use of nuclear fission for nonmilitary purposes, among these waste-handling. Early in this process, the National Academy of Sciences and the National Research Council were asked to establish a committee charged with reviewing the problem of radioactive waste disposal and with making recommendations for specific courses of action. The committee's recommendations, presented in 1957 after two years of deliberation, suggested that disposal in geologic formations was the best solution. Such a solution would isolate the wastes from man and his environment over a long period of time without requiring his long-range supervision of them. Because of the nature of these wastes, this last factor is extremely important. For about the first 1000 years after burial the wastes undergo a great deal of thermal activity, which generates much heat. A more distant problem is that having to do with Plutonium-239, which has a half-life of 24,000 years. If we use the "rule of thumb" that ten half-lives are necessary before an isotope ceases to be harmful to the biosphere, we can see that it will be approximately a quarter of a million years before this waste decays to the level at which it is no longer dangerous to man.

As now envisioned, disposal in geologic formations will utilize preexisting caverns or caverns excavated specifically for the disposal of particular kinds of waste products. So far, only nuclear wastes have been considered for large-scale disposal of this kind.

> *Because no form of these radioactive wastes (including refactory oxides and glasses) has yet been found sufficiently stable for long-term storage in the unprotected environment, the principle emphasis has been placed on finding safe, stable, natural locations for their storage that are remote from man's environment.**

An Evaluation of the Concept of Storing Radioactive Wastes in Bedrock Below the Savannah River Plant Site, Publication ISBN 0-039-02035-2, Committee on Radioactive Waste Management, National Academy of Sciences–National Research Council, Washington, D. C., 1972, p. 60. Reprinted by permission.

With 37 nuclear power plants operable, 57 under construction, and 89 planned as of September 30, 1973, the problem of what to do with the high-level wastes generated by these installations is becoming of increasing concern, although it is not yet critical. Two recent and somewhat extensive studies of the feasibility of storing nuclear wastes underground involve disposal in bedded salt and bedrock.

Disposal in bedded salt   One of the NAS-NRC Committee's conclusions was that bedded salt deposits, of all geologic formations, offered the most attractive possibilities for the disposal of high-level radioactive waste. Among its unique properties, and of prime importance, is salt's ability to flow under reasonably low pressure, thus enabling it to self-seal in the event that fractures are formed during an earthquake. This is important because of the long time required until the isotopes included in the waste decay to a level which is not hazardous to man. The half-life of these isotopes ranges from several hundred years for some to as much as one-half million years for others, and an earthquake occurring during that time could rupture the rocks, allowing the radioactive material to contaminate the surrounding area.

In addition to its ability to flow under low pressure, salt is strong enough to withstand the excavation of large cavities without collapsing. This means that men and equipment can have more work space within the facility. Salt deposits are usually dry and are not associated with underground sources of usable water. One of the fears connected with the use of natural formations for the storage of nuclear waste is that the waste will somehow undetectedly mingle with underground water and thereby contaminate a major water supply. The melting point of salt (around 1400° F, which is nearly double the temperature of most radioactive wastes) would seem to be an additional favorable factor. Since bedded salt underlies slightly less than half a million square miles in the United States, it offers a seemingly unlimited storage facility.

During the four-year period from 1963 to 1967, the Atomic Energy Commission conducted experiments in a 300-foot-thick bed of salt in an abandoned salt mine (Fig. 13-6) situated 1000 feet below the town of Lyons, Kansas, and owned by the Carey Salt Company. This set of experiments, called Project Salt Vault, was designed to give the AEC information about the effect of heat and radiation on salt and related geologic materials.

Based on the information provided by these experiments, it was determined that a site chosen as a permanent repository should have the following characteristics:

1. The salt bed must be at least 200 feet thick.
2. It must be at least 500 feet below the surface.

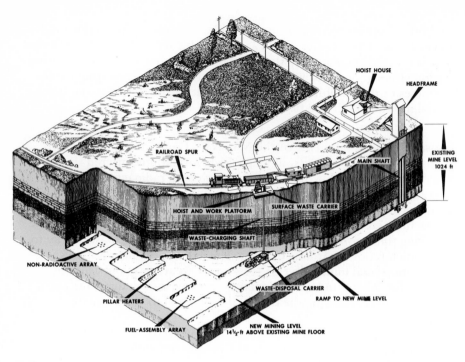

13-6
*Schematic block diagram of the Carey Salt Company Mine in Lyons, Kansas, site of the
AEC's Project Salt Vault experiments. (Courtesy U.S. Atomic Energy Commission)*

3. Because of operational and cost factors, the maximum depth must
   not exceed 2000 feet.
4. The salt beds should have an areal extent of several tens of miles
   beyond the burial site.
5. The area should be tectonically stable.
6. The site should be away from any areas with high population
   density, i.e., towns and cities.
7. The site should have rail and highway transportation available.
8. An existing, but abandoned, mine is desirable but not necessary.

However, some questions, most of which deal with consequences of
releasing the waste into the salt, must still be answered satisfactorily before

13-7
Cannisters containing radioactive waste are lowered down a shaft to this specially designed mover waiting 1000 feet below the surface. It then carries the waste to the experimental area, shown in Fig. 13-8. (Photograph courtesy U.S. Atomic Energy Commission)

this method is committed to large-scale use. The AEC proposes that the waste be brought to the repository as a solid embedded in a ceramic materal which will, in turn, be placed in metal containers. These containers will be lowered into holes, similar to those shown in Fig. 13-8, which have been prepared in the mine floor and which will then be backfilled with crushed salt. After the entire floor area of a room has been used, the room itself will be backfilled with crushed salt. The metal containers are expected to last no longer than six months, while indications are that the ceramic material will, in a few years, deteriorate enough to allow the radioactive particles to migrate into the salt. It is expected that during the thermal stage (the first 1000 years of burial) the crushed salt used to backfill the individual holes and the rooms will refuse to form a solid, coherent chunk of salt having the waste trapped inside. However, as studies indicate

13–8
The seven holes shown here were drilled 12 feet deep into the floor of the mine and lined with stainless steel and were used to contain radioactive fuel assemblies for the experiments. Monitoring cables lead to a control panel at the side of the room. (Photograph courtesy U.S. Atomic Energy Commission)

that salt tends to store high heat when in contact with radioactivity, some workers in the field have suggested that this might result in an explosion under certain conditions. Although recent laboratory work suggests that this will not be the case, additional investigation is presently under way. Ultimately, the pros and cons of the problems that have been raised will be satisfactorily settled, and bedded salt will then be used for the storage of high-level, solid radioactive waste.

Bedrock disposal   The concept of using chambers cut into bedrock as containers for storing high-level wastes in liquid form is under evaluation at two Atomic Energy Commission sites. The first exploratory drilling of the bedrock underlying the AEC's Savannah River Plant near Aiken, South Carolina, was done by the Corps of Engineers in 1961, and the evaluation of the surface-based exploration was completed in 1971. At the AEC's Hanford Reservation in the state of Washington, the first continuous rock cores of the bedrock underlying that site were taken in mid-1971. We can expect that this evaluation will not be completed for some time. The suggestion that bedrock might serve as a suitable disposal site was made by the United States Geological Survey in 1951.

The proposed bedrock disposal site at the Savannah River location would have a 100,000,000-gallon storage capacity which would be provided by six tunnels, each of which would be 3400 feet long, 26 feet wide, and 28 feet high. A transverse access tunnel would connect all of the six storage tunnels, and each storage tunnel would have at least one access shaft, 10 to 15 feet in diameter, and one service shaft to carry the fill and vent pipes. After the storage tunnels have been quarried to the desired dimensions, concrete bulkheads will isolate the tunnel from its access shaft. Waste would then be pumped from the 18 underground carbon-steel tanks (with capacities ranging from 750,000 to 1,300,000 gallons) presently being used to store the wastes into the storage tunnels. After a suitable monitoring period, the service shaft used to fill and monitor the tunnel would be sealed.

The bedrock underlying the Savannah River Plant site (see Fig. 13–9) consists of pre-Mesozoic metamorphic rocks, Triassic sedimentary rocks, and the Tuscaloosa Formation of Cretaceous age. The Tuscaloosa Formation, a prolific source of fresh water, is separated from the metamorphic rocks and Triassic sediments by a layer of clay which would act as a barrier to prevent the migration of radioactive ions into the freshwater source above. In order to demonstrate conclusively which of these formations would provide the best and safest waste-storage unit, it has been recommended that an exploratory shaft be sunk and that exploratory tunnels be driven into the rock.

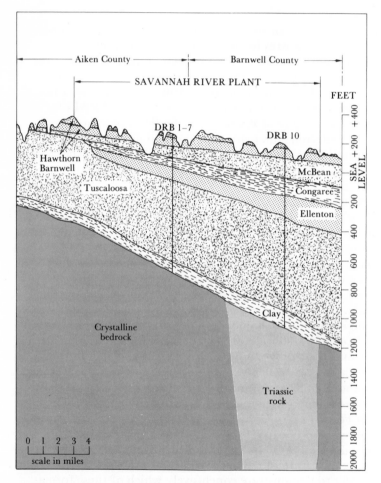

13-9
*Geologic cross-section of the beds underlying the Savannah River Plant. (*An Evaluation of the Concept of Storing Radioactive Wastes in Bedrock Below the Savannah River Plant Site, *Publication ISBN 0-309-02035-2, Committee on Radioactive Waste Management, National Research Council, Washington, D.C., 1972, p. 15. Reprinted by permission.)*

SANITARY LANDFILLS

Every year more and more people inhabit the earth, and every year these people produce more and more trash and garbage. The proper disposal of trash from households and businesses has become a major concern because many communities are running short of space in which to put their waste and because of damaging effects to the air and water environment. At one time, trash was disposed of by burning or dumping it in the oceans or in a distant, isolated ravine. "Out of sight, out of mind" was the philosophy behind trash disposal. Now, because of air and water pollution, safety, the problem of rat and insect control, and esthetic values, we are attempting to find other ways of disposing of trash. Recycling of some materials is possible, but it is expensive and cannot possibly reclaim all waste. Incineration is sometimes practiced, but not all material can be, or is, safely disposed of by this means. Most of the ordinary trash produced by households and many businesses must be disposed of in other ways, the most important of which is the sanitary landfill. The sanitary landfill is the successor to the old city dump, but differs from it in many ways: no burning takes place, the waste is compacted and buried, water contamination is avoided, and the area is fenced or screened off.

The selection and maintenance of a landfill site involves problems of economics, health, engineering, geology, and land-use planning. Geological problems are primarily those of the earth materials present and the hydrologic situation. The topography — the "lay of the land" — must be such that both space is available for the disposal of waste and drainage over the area can be controlled to prevent water from entering the waste materials. If other conditions are favorable, valleys or, in certain cases, depressions created by mining, can be filled in. In some instances, ground that is nearly level can be utilized if trenches are dug for the waste. Occasionally, waste is piled up above the normal ground level, producing a hill; an example is DuPage County, Illinois, where a small mountain has been created. When completed, the mountain will be covered with soil, landscaped, and used as an area for skiing and other recreational activities.

Water that seeps through a city dump is bound to leach out any number of pollutants. If the polluted water flows from the dump area into surface streams or infiltrates the soil and becomes part of the ground water, hazardous environmental problems are created. By creating an impermeable barrier around the waste in a sanitary landfill so that water cannot reach it and fluids in the waste cannot escape, the waste is effectively segregated from precipitation, surface flow, and ground water.

Because clay is the most effective earth material for this purpose, an adequate supply of high clay-content soil is sought in picking a site for a landfill. Each day's haul of trash is dumped onto a clay-bearing imperme-

13-10
*Dumping waste at a sanitary landfill. After compaction and covering with soil, another
layer of waste will be placed on top. Ultimately, the landfill will form a ski hill. (Photograph courtesy USDA—Soil Conservation Service)*

able bed of earth, is compacted, and is then covered with the same material.
Thus, a pod of compacted trash that is completely walled off from the environment at the top, bottom, and sides is formed. The next day's haul is
deposited next to the first pod, and this pattern is repeated until the entire
landfill is filled, as shown in Fig. 13-10. Most sanitation standards require
that each day's deposit of waste be covered within about an hour of the
operation's daily closing time, so that no waste is left uncovered overnight
or remains exposed for very long.

Slopes on the landfill are maintained so that only a very minimal
amount of rainfall ever reaches the area and that rainwater which does is
effectively removed through drainage ditches. Fences prevent the trash

13–11
A golf course will occupy this site when the landfill is completed. The homes in the background, which were built with knowledge of this fact, will appreciate in value. (Photograph courtesy Environmental Projection Agency)

from being blown away, sprinkler trucks keep the dust down in dry weather, access roads are graded and maintained to prevent muddy conditions, and rows of rapidly growing bushes and trees are sometimes established around the area to absorb and deaden the sound of machinery and to improve the appearance of the site.

Settling takes place during, and subsequent to, the filling of a site. Materials weather, decay, and compact; gases, such as methane, are discharged, rise, and escape into the air. These are sufficiently dilute that they do not create a problem if allowed to escape freely and continually. Because settling and possible gas accumulation occurs in basements, buildings should not be placed on the site of a former landfill. The area can instead be used for parks, golf courses, playgrounds, parking lots, or agricultural lands. A reasonable land-use plan that includes possibilities for future use should be a prerequisite of the development of a sanitary-landfill operation.

The ideal sanitary landfill is very different from the old town dump. If one takes into account the earth-moving equipment, manpower, fencing, and shrubbery, it is also more expensive. On the other hand, the environmental protection it provides compensates for the additional cost. The benefits of health, safety, wildlife conservation, water conservation, clean air, and efficient land use far outweigh the operational costs.

REFERENCES

Anderson, D. M., *et al. Compaction of Radioactive Solid Waste,* Washington, D. C.: U. S. Atomic Energy Commission, 1970.

Barnes, H. L. and S. B. Ronberger. "Chemical aspects of acid mine drainage," *Water Pollution Control Federation Journal* **40**, 3 (1968):371–384.

Claus, G. and G. J. Halasi-Kin. "Environmental pollution," in *Encyclopedia of Geochemistry and Environmental Sciences,* ed. R. W. Fairbridge, New York: Van Nostrand Reinhold, 1972, pp. 309–337.

Cleaning Our Environment – The Chemical Basis for Action. Washington, D. C.: American Chemical Society, 1969.

Council on Environmental Quality. *Ocean Dumping – A National Policy,* Washington, D. C.: U. S. Government Printing Office, 1970.

de Laguna, W. "Importance of deep permeable disposal formations in location of a large nuclear-fuel reprocessing plant," in *Subsurface Disposal in Geologic Basins – A Study of Reservoir Strata,* ed. John E. Galley, A.A.P.G. Memoir 10, 1968, pp. 21–31.

An Evaluation of the Concept of Storing Radioactive Wastes in Bedrock Below the Savannah River Plant Site. Washington, D. C.: National Academy of Sciences –National Research Council, 1972.

Food and Agriculture Organization. *Fertilizers: An Annual Review of World Production, Consumption, and Trade, 1954–1964,* New York, 1965.

Fox, Charles H. *Radioactive Wastes,* Washington, D. C.: U. S. Atomic Energy Commission, 1969.

Fredericksen, W. and R. Gentile. "Field trip 1, proposed salt mine repository for radioactive wastes," in *Guide to Field Trips,* 1972 Annual Meeting of the Association of Engineering Geologists, Kansas City, October 24-28, 1972, pp. 12–17.

Gross, M. G. "New York metropolitan region – a major sediment source," *Water Resources Research* **6** (1970):927–931.

Hartman, C. D. "Deep-well disposal of steel-mill wastes," *Water Pollution Control Federation Journal* **40**, 1 (1968):95–100.

James, R. D. "Underground dumps: waste disposal wells, once considered safe, now seen as polluters," *Wall Street Journal,* May 21, 1970.

National Academy of Engineering. *Wastes Management Concepts for the Coastal Zone,* Washington, D.C.: National Academy of Sciences, 1970.

National Industrial Pollution Control Council. *Animal Wastes*, Washington, D. C.: U. S. Government Printing Office, 1971.

————. *Waste Disposal in Deep Wells*, Washington, D. C.: U. S. Government Printing Office, 1971.

Queal, C. "Radiation in Grand Junction poses perilous puzzle," *The Denver Post*, April 11, 1971.

"Questioning deep-well disposal," *Science News* 97 (1970):314-315.

Ross, R. D. *Industrial Waste Disposal*, New York: Van Nostrand Reinhold, 1968.

Schneider, W. J. *Hydrologic Implications of Solid-Waste Disposal*, U. S. Geological Survey Circular 601F, 1970.

Sheffer, H. W., E. C. Baker, and G. C. Evans. *Case Studies of Municipal Waste Disposal Systems*, Washington, D. C.: U. S. Department of the Interior, Bureau of Mines, Circular Ic-8498, 1971.

Small, W. E. *Third Pollution—The National Problem of Solid Waste Disposal*, New York: Praeger, 1970.

Snelling, R. N. "Environmental survey of uranium mill tailings pile, Monument Valley, Arizona," *Radiological Health Data and Reports* 11 (1970):511-517.

Summers, C. M. "The conversion of energy," *Scientific American* 224, 3 (1971):149-160.

Ullmann, J. E., ed. *Waste Disposal Problems in Selected Industries*, Hofstra University Yearbook of Business, Series 6, 1, 1969.

"Uranium tailings: no one's responsibility," *Chemical and Engineering News*, V. 49, (October 4, 1971):12-13.

U. S. Department of the Interior. *Surface Mining and Our Environment: A Special Report to the Nation*, Washington, D. C.: U. S. Government Printing Office, 1967.

Ward, J. C., G. A. Margheim, and G. O. G. Lof. *Water Pollution Potential of Spent Oil Shale Residues*, Washington, D. C.: Environmental Protection Agency, 1971.

Warner, Don L. "Subsurface disposal of liquid industrial wastes by deep-well injection," in *Subsurface Disposal in Geologic Basins—A Study of Reservoir Strata, op. cit.*, pp. 11-20.

————. *Survey of Industrial Waste-Injection Wells: Final Report to the U. S. Geological Survey*, 3 vols., NTIS documents AD-756 641, AD-756 642, and AD-756 643, 1972.

(Photograph by George Sheng)

Appendixes

Appendix A   The Common Minerals and Rocks

Table A-1 Some common rock-forming minerals and generalized formulas

Silicates	Oxides	Carbonates
Quartz—SiO_2	Limonite—$Fe_2O_3 \cdot NH_2O$	Calcite—$CaCO_3$
Chert—SiO_2	Hematite—Fe_2O_3	Dolomite—$CaMg(CO_3)_2$
Orthoclase feldspar—	Magnetite—Fe_3O_4	
$\quad$ $KAlSi_3O_8$		
Plagioclase feldspar—		
$\quad$ $NaCaAlSiO_n$*		
Muscovite mica—		
$\quad$ $HK(AlSiO_n)$†		
Biotite mica—		
$\quad$ $HKFeMg(AlSiO_n)$		
Pyroxene group—CaMgFe	**Others**	
$\quad$ $(AlSi)_2O_6$		
Amphibole group—CaMgFe	Halite—NaCl	
$\quad$ $(AlSi)_8O_{22}(OH)_2$	Gypsum—$CaSO_4 \cdot 2H_2O$	
Talc—$HMgSiO_n$		
Chlorite—$HMgSiO_n$		
Serpentine—$HMgSiO_n$		
Olivine—$(FeMg)_2SiO_4$		
Clay group—$H(AlSiO_n)$		

*n=some number denoting the ration of O to Si or Si+Al

†denotes a metal

CLASSIFICATION OF ROCKS

Rock classification is not simple, because depending on the conditions of their origin, rocks vary through a wide spectrum of mineral composition, structure, and texture. Based on their origin, all rocks may be put into one of three great groups: igneous, sedimentary, or metamorphic, each of which is subdivided into other groups.

$\quad$ Figure A-1 shows that all rocks form from pre-existing rocks by the actions of various physical and chemical processes. Melting of any

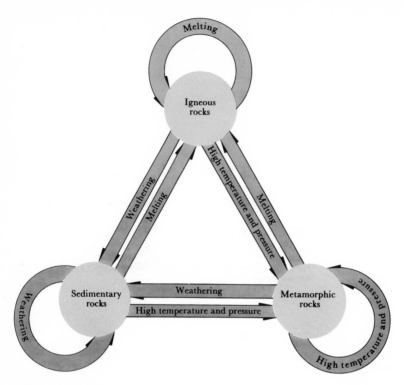

A–1
The rock cycle.

rock produces magma which, upon cooling below its freezing point, forms *igneous* rocks. Weathering destroys rocks, producing sediment or chemicals in solution which, when deposited and cemented together or precipitated from solution, forms *sedimentary* rocks. The action of high temperature and pressure, as might occur when rocks are deeply buried, alters the grain size and arrangement or mineral composition, forming *metamorphic* rocks.

Igneous Rocks

Igneous rocks form when molten rock material, called *magma*, freezes. The lava that pours out of a volcano during eruption is derived from magma; when it cools, it forms igneous rock. However, igneous rocks of volcanic origin, or extrusive igneous rocks, represent only a small portion of this rock group. By far the largest portion is comprised of rocks that have cooled from magmas deep within the crust, called plutonic or

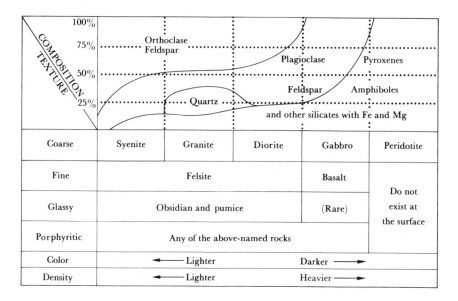

COMPOSITION / TEXTURE					
Coarse	Syenite	Granite	Diorite	Gabbro	Peridotite
Fine	Felsite			Basalt	Do not exist at the surface
Glassy	Obsidian and pumice			(Rare)	
Porphyritic	Any of the above-named rocks				
Color	←— Lighter		Darker —→		
Density	←— Lighter		Heavier —→		

A–2
Simplified classification of igneous rocks. Fragmental volcanic rocks, which form from material shattered in volcanic explosions and settle in the immediate vicinity of the volcanoes, do not fit the basis on which this chart is built and therefore are not included here.

intrusive igneous rocks, and which were not exposed at the surface until long after they solidified.

Igneous rocks are classified according to two parameters: their mineral composition and their texture (that is, the size of the mineral grains). Figure A–2 is a simplification of the classification of igneous rocks. To use the chart effectively one must actually see and handle the rock specimens; however, one can understand the general classification without engaging in laboratory work.

Texture   This classification uses only four basic textures. The "texture" of a rock describes how coarse or how fine its individual crystals or grains are. When we speak of a rock's texture, for example, we are describing how it looks to us. In most cases it is impossible to distinguish textures in igneous rocks simply by feel. For example, if you were blindfolded and handed a variety of igneous rocks, you probably would be unable to accurately describe the size of the grains making up the specimens.

Coarse textures consist of grains larger than about 1 millimeter in diameter; the fine-textured rocks are comprised of grains smaller than

1 millimeter (a millimeter being about the thickness of a dime). We use the following as a workable definition: if the grains are large enough to identify the individual minerals which constitute the grains, the rock is coarse-textured; if they are too small, it is fine-textured. In general, the slower the rate at which the magma cools, the larger the crystal grains will be.

Rocks having glassy textures simply look like pieces of glass and have no grains at all. The molten material froze before any crystalline structure could begin. Porphyritic rocks show two or more *distinctly* different grain sizes, which are the result of changes in the rate of cooling during the solidification process. If the magma cools very slowly, a coarse-textured rock forms; if it cools more rapidly, a fine-textured rock forms; if it cools very rapidly, a glassy rock results; and if it cools first at one rate and then something causes it to solidify at a faster or slower rate, the result will be a rock having distinctly different grain sizes.

Although the texture of an igneous rock is the result of the rate at which magma cools, many factors influence the rate of cooling.

1. Rate of cooling The slower a magma cools, the longer the crystals have to form; hence, the larger they are. Factors that affect the rate of cooling are:

    a) The depth at which the magma is buried
    b) The size of the magma
    c) The shape of the magma
    d) The composition of the magma

2. Composition Given the same cooling rate, a magma of basaltic composition will form larger grains than will a magma of granitic composition, because the higher silica content of the granitic magma significantly increases its viscosity. This impedes the movement of atoms and molecules which are trying to get together to build a crystal.

3. Presence of mineralizers Bases, e. g., such as H_2O (as super-heated steam), SO_2, CO_2, and NH_3; H_2SO_4; compounds of fluorine; and others are collectively referred to as "mineralizers." Their presence affects texture by increasing the fluidity of the magma and lowering the freezing point of most minerals, which keeps the magma fluid longer.

Composition of igneous rocks Only a few minerals are important components of igneous rocks, and these are the only ones used in rock identification. The percentages of these minerals are shown graphically in the upper portion of Fig. A–2. To determine the composition of any rock, look at what minerals are placed directly above the rock name and estimate (using the percentage scale on the left of that area) what percent of the rock is composed of that mineral. Knowing the names and percentages of the minerals present is essential only for identifying the coarse-textured rocks.

The grains in fine-textured rocks are too small to identify; therefore, their identification is based on color. If, for example, the rock is fine-grained and light in color, it is called *felsite*; fine-grained black rocks are called *basalts*. The coarse-textured rock composed of the same minerals as basalt is known as *gabbro*. Coarse-textured rocks that are light in color include *syenite* and *granite*, while *diorite* has an intermediate shade. Light-colored, fine-textured rocks are collectively called felsite, as it is not possible to distinguish among them. Thus, we know that a felsite has the same composition as syenite, granite, or diorite, but which of these cannot always be easily determined. *Obsidian* and *pumice* have the chemistry which *would have* resulted in a composition similar to that of felsite if they had cooled slowly enough to have formed crystals. Therefore, they are positioned below felsite.

Two important factors that influence the composition of igneous rocks are the depth at which the magma originated and the magmatic differentiation. In general, magmas that originate at depths of 10 miles or less are granitic in composition, those originating at depths between 10 and 40 miles are gabbroic, and those originating deeper than 40 miles are peridotitic.

Magmatic differentiation is a process whereby the composition of a magma changes as material is crystallized and removed from the melt. Since the specific gravity of the mineral crystals is usually greater than that of the molten material, the newly formed minerals may sink to the floor of the magma chamber and form a rock having a certain composition. The remainder of the magma will have a different composition and should therefore form a rock having a composition which differs from that of the first. Thus, several different rock compositions can be formed, theoretically, from a single magma.

N. L. Bowen found, through laboratory experiments, that the silicates which crystallize from a magma could be arranged in a series related to their order of crystallization (Fig. A–3). Because each mineral is derived from the preceding mineral as a result of the reaction of the mineral with the remaining liquid magma, and because the minerals can be arranged in a series, the series is called the *Bowen Reaction Series*.

Color and specific gravity   Generally, rocks listed on the left side of the chart are lighter in color, while those on the right side are darker. An exception is obsidian, which is usually black but appears on the left side of the chart. Its dark color is a result of the dissemination of fine grains of iron throughout the glass, but in thin pieces obsidian is clear.

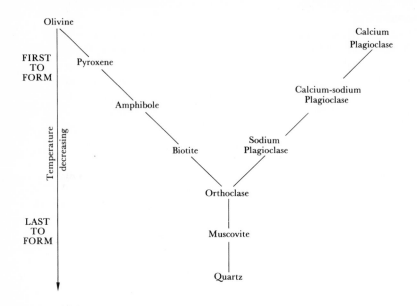

A–3
Bowen's reaction series shows the general sequence by which minerals crystallize from a cooling magma.

Because of their chemical composition, rocks on the left side of the chart have a lower specific gravity than those on the right. More iron and magnesium occur in the rocks on the right side of the chart, while the rocks on the left are composed of relatively lighter elements.

Sedimentary Rocks

Sedimentary rocks are composed of the weathered debris of pre-existing rocks which has been deposited by wind, water, or ice. This debris occurs in two forms: detritus, which consists of discrete particles carried by the agent of deposition and dissolved material carried in solution. The mode in which the material is available dictates the classification of sedimentary rocks: (1) *detrital*, or *clastic*, rocks and (2) *chemical* rocks, comprised of material that has precipitated out of solution (Table A-2).

Since detrital rocks are grouped according to the size of the particles making up the rock (Table A–3), it is convenient to divide them into three groups: (1) *conglomerate*, made up of particles whose size is 2 mm or more; (2) *sandstone*, made up of particles which are between 2 mm and

Table A-2 Classification of sedimentary rocks

DETRITAL		CHEMICAL	
Particle size	Name	Chemical (mineral) composition	Name
Gravel (2mm)	Conglomerate	Calcite	Limestone
Sand (2mm–1/16mm)	Sandstone	Dolomite Chert	Dolomite Chert
Silt and clay (1/16mm and smaller)	Shale (siltstone, mudstone)	Halite Gypsum	Rock salt Gypsum

Table A-3 Modified Wentworth scale of particle sizes for clastic sediments

Size		Fragment name
mm	inches	
256	10	Boulder
64	2.5	Cobble
4	5/32	Pebble
2	5/64	Granule
1/16	.0025	Sand
1/256	.00015	Silt
		Clay

(Adapted from Wentworth, *Journal of Geology* 30, 1922, p. 381)

1/16 mm in size; and (3) *shale*, which is composed of grains which are less than 1/16 mm in size. Such terms as mudstone, siltstone, claystone, or mudrock are associated with material in the size range we call shale. We include these terms within our definition of *shale* because the materials in them are all smaller than 1/16 mm.

Although classification of detrital rocks is not based on composition, some generalizations can be made about the mineralogy of detrital rocks. Sandstone may be composed of any mineral (of the appropriate size), but *most* sandstones are composed of quartz grains which have been cemented together. Even though shale may consist of any mineral whose grains are within the appropriate size range, most shale is made of clay minerals (see Chapter 4).

Table A-4 Simplified classification of metamorphic rocks based on texture and composition

Structure	Monomineralic rocks	Multimineralic rocks
Foliated		Slate
		Schist
		Gneiss
Nonfoliated	Quartzite	Hornfels
	Marble	Skarn

Chemical sedimentary rocks are classified entirely by composition, and grain size is not a factor. Rocks composed mostly of calcite are called *limestone*, rocks composed mainly of the mineral dolomite are called *dolomite*, and rocks made up of silica are called *chert* or *flint*. Less abundant, but nevertheless important, are gypsum, rock salt, and potash, the latter composed of potassium salts of which sylvite is a common example.

Metamorphic Rocks

The term *metamorphosis* means a "change in form." Metamorphic rocks, then, are those rocks whose forms have been altered. Although all rocks are changed in some way, the term "metamorphic" is reserved for those rocks which have been altered by either high temperature and high pressure or by one of the two. The requisite temperatures are, of course, lower than those necessary to melt the rock, for if those temperatures were achieved we would be dealing not with metamorphism, but with melting and igneous rocks. The melting temperature of a rock establishes the upper limit of its potential thermal metamorphism.

Although the classification table of metamorphic rocks (Table A-4) is abbreviated, the group as a whole is the most complex of the rock groups. *Any* existing rock type (igneous, sedimentary, or metamorphic) can undergo metamorphism—which makes for great variation at the start. This, in addition to the fact that the same rock can have completely different end results, depending on the heat and pressure inputs, produces a highly varied system. If chemical additions and subtractions occur during the metamorphic process, the complexity increases still further.

The major subdivision in the classification of metamorphic rocks is based on the presence or absence of *foliation*, which is an approximate parallelism or alignment of mineral grains in a rock caused by the preferential orientation on the part of the mineral grains. Foliation is possible if the rock contains mineral grains, at least one dimension of which is relatively longer than the rest. These can then be oriented in response to the application of differential pressure. Minerals such as the micas, which tend to be broad, and the feldspars, pyroxenes, and amphiboles, which form elongate crystals, produce foliation when pressure is applied. Such minerals as quartz and calcite usually are not visibly oriented; nonfoliated metamorphic rocks are generally composed of these minerals.

Within the foliated subdivision, the rocks are grouped according to the character of the foliation. A rock having coarse, or thick, foliation (the result of the presence of thick, stubby minerals, such as the feldspars, pyroxenes, or amphiboles) is called *gneiss*. Minerals in gneiss are often arranged in alternate bands, each consisting mainly of one mineral. A rock having fine, or thin, foliation (the result of the presence of thin, flat minerals such as the micas or chlorite) is called *schist*. A rock that shows a very fine, but less distinct, foliation and was originally shale or mudstone is called *slate*. The fine size of the grains nearly always precludes identification of the minerals in slate.

The composition of nonfoliated rocks is used as the basis for classification. Two common rocks that occur in this category are *marble*, which consists of calcium or magnesium carbonate and is the result of metamorphosing limestone or dolomite, and *quartzite*, which is composed of silica and is the result of metamorphosing quartz sandstone. *Hornfels* is a thermally metamorphosed shale, and *skarn* is the result of the metamorphosis of impure, silica-bearing limestone or dolomite.

Appendix B   Physical Characteristics of the Earth

Table B-1 Dimensions of the earth

Size	Equatorial radius: 3963.5 miles; 6,378.388 kilometers Polar radius: 3950.2 miles; 6,356.912 kilometers Difference: 13.3 miles; 21.476 kilometers Circumference at equator: 24,900 miles Circumference at 60° latitude=1/2 equatorial circumference (12,450 miles)

Internal
dimensions
	Core: from 1800 mi (2900 km) depth
	          to 3960 mi (6371 km) (center)
	
	Mantle: from about 20 mi (30 km) depth
	            to 1800 mi (2900 km)
	
	Crust: continental 15–30 mi (25–50 km) thick
	          oceanic 4 mi (6 km) thick

Shape	Oblate ellipsoid (the shape of a spheroid slightly flattened at the poles; flattening due to rotation)
Weight and density	Weight: 6,600,000,000,000,000,000,000 (6.6×10^{21}) tons Density: 350 lbs per cubic foot Specific gravity: 5.6
Area	Land: 57,487,200 mi^2; 148,892,000 km^2 Ocean: 139,404,900 mi^2; 361,059,000 km^2 Total: 196,892,100 mi^2; 509,951,000 km^2
Relief and elevation	Highest point: Mt. Everest +29,028 feet Lowest point: Mariana Trench –36,204 feet Total relief: 65,232 feet Average elevation of continents: +2000 feet Average elevation of ocean basins: –14,000 feet

STRENGTH

The strength of crustal material can be thought of in terms of both absolute and relative strength. Strength is defined as the force per unit area which is necessary to cause a body to break at room temperature and pressure. In order to break granite, for example, a compressional force of over 20,000 pounds per square inch, on the average, is required. The average strength necessary to break limestone is about 15,500 pounds per square inch, and marble requires a force of over 14,000 pounds per square inch. To us, who are so small and weak compared to these values, earth materials seem very strong. However, let us look at these strengths on a scale relative to the earth's size. To do so in understandable terms necessitates our reducing everything to a human scale. This "scaling down" applies not only to the linear dimensions of length, width, and depth but also to the strength of the material. This was done most graphically in an article written by M. King Hubbert over 25 years ago. If you are thoughtful, you will be able to use the concept illustrated in this passage to help you understand many geologic situations.

> *To render our problem specific, let us consider the quarrying operation illustrated in [Fig. B–1] . . . . Here we suppose that we are able to quarry as a single block with a thickness roughly one-fifth of its width the entire state of Texas, and that we have a quarry crane capable of hoisting it. Let us suppose further that this block is composed of the strongest of rock, and moreover, that it is monolithic and flawless. . . Since the original assignment to this case is manifestly impossible of execution, suppose we imagine the block to be reduced to such a size that it can be lifted conveniently. What must the properties of this reduced block be in order that it should respond to lifting in the same manner as the original* if the latter could be lifted?*
>
>   *In this case gravity will be unchanged and it will be convenient to keep the density the same, so we are left with the result that we must reduce the strength of the materials by the same amount as the length reduction. The state of Texas is about 1200 kilometers wide and a convenient size for the reduction would be a reduction of 5×10^{-7}. Assuming the original to be composed of very strong rock, the crushing strength would be of the order of 30,000 lbs/in^2. The strength of the properly reduced version should then be*

$$30{,}000 \times 5 \times 10^{-7} = 0.015 \ lb/in^2$$

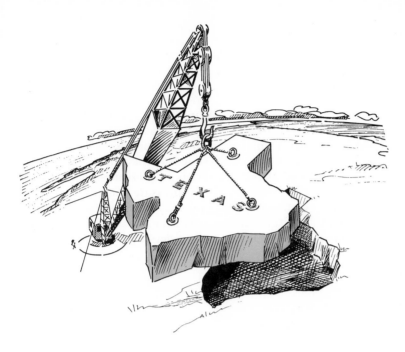

B-1

The strength of Texas. (Reprinted by permission of the American Association of Petroleum Geologists from M. King Hubbert, "The Strength of the Earth," AAPG Bulletin 29, 11, November 1945, p. 1634.)

It is difficult to envisage a solid of this weakness. A crushing strength of 0.015 lb/in^2 is the same as 1 gm/cm^2, which for a density of 3 gms/cm^3 would be the pressure at the base of a column one-third of a centimeter high. Any column higher than this would collapse under its own weight. Yet the size of the reduced block would be such that its thickness would be about 15 cm (6 in) and its total weight about 180 lbs. The pressure at its base would be about 45 gm/cm^2, or 45 times the crushing strength of the materials.

Consequently, if we tried to lift such a block in the manner indicated in [Fig. B-1], the eyebolts would pull out; if we should support it on a pair of sawhorses, its middle would collapse; were we to place it upon a horizontal table, its sides would fall off. In fact, to lift it at all would require the use of a scoop shovel. That this is not an unreasonable result can easily be verified *by direct calculation upon the original block. For it, too, the pressure at its base would exceed the crushing strength of its assumed material by a factor of 45. The inescapable conclusion, therefore, is that the good state of Texas is utterly incapable of self-support!**

This example makes it obvious that we must use a new reference framework when we begin to consider earth problems on an earth scale. Our normal human framework cannot deal with the problems we encounter. Another dimension that must be added to the concept of strength is time. The response of rock materials to pressure varies according to the length of time over which the pressure is applied. Long-term (thousands or millions of years) pressures may result in *fatigue* in rocks. Folding, instead of faulting, may occur. The precise behavior of materials in the time dimension is not well understood.

---

* M. King Hubbert, "Strength of the Earth," *AAPG Bulletin* 29, 11 (November 1945), pp. 1635–1638. Reprinted by permission of the American Association of Petroleum Geologists.

Appendix C   Geologic Time

INTRODUCTION

In the eighteenth and nineteenth centuries, when man began to unravel
the history of the earth, it became increasingly apparent that the earth
is much older than anyone had previously realized. As the events in the
long history of the earth were sorted out, it was recognized that there
were periods of time during which certain events took place. These time
periods were given names, their chronological order was established, and
thus a *geologic time scale* came into being. The rocks representing
those times comprise the geologic column (Table C–1).

The more recent the time, the more we know about the earth's
history. The record of the earth lies in its rocks and in the fossils
contained in them. Almost nothing is known of the first four billion
years, because the rocks of those times have been variously remelted,
metamorphosed, eroded, or buried beyond our reach. It is also signif-

Table C–1 The geologic time scale

Millions of years before present	Era	Period	Epoch (only Cenozoic epochs are listed)
.01 1	Cenozoic	Quaternary	Holocene (=Recent) Pleistocene
		Tertiary	Pliocene Miocene Oligocene Eocene Paleocene
65 136 190 225	Mesozoic	Cretaceous Jurassic Triassic	
280 345 395 430 500 570	Paleozoic	Permian Pennsylvanian Mississippian Devonian Silurian Ordovician Cambrian	
4600		Precambrian (subdivisions, not listed here, have not been entirely determined)	

icant to note that these rocks contain virtually no fossils. The presence
of fossils in rocks greatly assists attempts to reconstruct the past, since
the abundance and kinds of life give clues about past environments. The
record of life in the earth is shown in Fig. C-1.

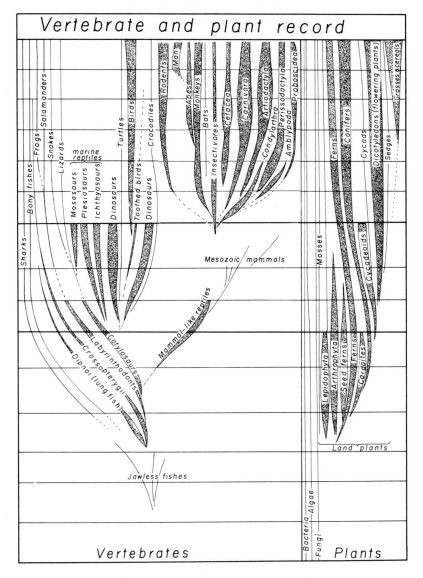

C-1
The range of the chief vertebrate and plant groups. (Carl O. Dunbar and Karl M. Waage,
Historical Geology, *3rd ed., New York: Wiley, 1969, p. 17. Reprinted by permission.)*

Table C-2 Landmarks in earth history on a scale of one inch
equals one million years

Event	Millions of years before present	Distance from present in feet and inches
First man and beginning of Great Ice Age	2	0 ft–2 in.
Dinosaur extinction	70	5 ft–10 in.
Uplift of Rocky Mountains	80	6 ft–8 in.
First dinosaurs	225	18 ft–9 in.
First reptiles and the making of great coal deposits	345	28 ft–9 in.
First amphibians	375	31 ft–3 in.
First fish	500	41 ft–8 in.
First abundant preservation of animal life	600	50 ft–0 in.
First photosynthetic organisms	2000	166 ft–8 in.
First primitive organisms (nonphotosynthetic)	3000	250 ft–0 in.
Oldest dated rocks	3300	275 ft–0 in.
Origin of earth	4600	383 ft–4 in.

An idea of the sequence of events in a proper time scale can be obtained by the following experiment: by using a football field for reference, you can plot some landmarks of earth history (Table C-2). Any scale can be used, but in order to be able to record the most recent events, e.g., the Ice Age, a scale of one million years to the inch, half-inch, or centimeter should be used. Using a scale in which one inch is equivalent to one million years, four and a half billion years would require an area longer than the length of a football field. If we stand on one goal line and look back in time, we would see that the first organisms (probably nothing we would recognize) originated at about the far goal line. We would have to advance to the 15-yard line nearest us before we could find a well-preserved fossil record. Life emerging from the sea in the form of amphibians would be plotted about ten yards in front of us. The great dinosaurs appeared and became extinct within the last five yards. Man does not appear until the last two inches!

NAMING OF GEOLOGIC TIME DIVISIONS

It is common practice to name formations of rock after areas or some prominent, permanent feature in the area in which the rocks were first studied. Many geologic periods were named in this way. Others were named for processes or events which more or less characterize the period.

Four *eras*—Precambrian, Paleozoic, Mesozoic, and Cenozoic—comprise geologic time. The Precambrian received this name because it includes all rock formations that formed prior to the Cambrian Period. The beginning of the Cambrian is generally designated as the advent of the abundant preservation of fossils. The Precambrian is composed mostly of unfossiliferous igneous and metamorphic rocks. The Paleozoic (Greek *palaios*: ancient; Greek *zoe*: life) was named because it contains fossil evidence of the more ancient life forms, mostly extinct, of those periods which show abundant life. The Mesozoic (Greek *mesos*: middle; Greek *zoe*: life) contains fossil remains of those organisms which appear to have comprised an intermediate stage between the primitive life in the Paleozoic and the more advanced forms of the present. Many of the organisms which are preserved as fossils in Mesozoic rocks are extinct. The Cenozoic (Greek *kainos*: recent; Greek *zoe*: life) contains the more recent life forms, which are in most cases the most advanced forms. The fossils found in Cenozoic rocks are usually still represented by living forms. Indeed, many of the Cenozoic forms are indistinguishable from modern forms.

Each era can be subdivided into *periods.*

Cambrian   Rocks of this period were first studied in an area in Wales, called *Cambria* by the Romans.

Ordovician   These rocks were first described in an area in Wales which had been populated by the Ordovices, the Roman name for this ancient Celtic tribe.

Silurian   Rocks of this period were first studied in an area in Wales which had been populated by the Silures, an ancient Celtic tribe conquered by the Romans in about A.D. 80.

Devonian   This period was named for rocks first studied in Devonshire, England.

Carboniferous   This name consists of the Latin roots: *carbo,* meaning coal, and *ferre,* to bear, that is, the "coal-bearing" rocks. These rocks were first named in England after the upper strata which contain the famous coal beds of Europe and North America. The Carboniferous rocks, or more exactly the coal in them, supported the industrial revolution in England. In North America this period is represented by the combined Pennsylvanian and Mississippian periods. The *Mississippian* is named for lower Carboniferous rocks which are best exposed in the upper Mississippi River valley. The *Pennsylvanian* represents the upper part of the Carboniferous, the coal-bearing strata, which fist became well known from extensive study in the coal-mining regions of Pennsylvania. These two terms are North America's only contributions to the geologic time scale.

Permian   Rocks exposed in Perm, a former province in eastern Russia, gave this period its name.

Triassic   The name comes from the Latin *trias,* which means a division into three parts. The reason for this particular name was that studies of three geologic units in western Germany, which were at the time considered to be the German equivalent of a "period," were shown to be the result of conditions in a single geologic period. The three units were declared to be subdivisions of the same period; hence, the name Triassic.

Jurassic   This period is named for outcrops that occur in the Jura Mountains of Switzerland.

Cretaceous   The name is derived from the Latin word *creta,* meaning chalk. This period was named for the rocks along the English Channel, where this age was first defined. Although more limestone and shale than chalk were deposited, more chalk was deposited in this period than in any other period in geologic history.

Tertiary   This term is derived from a classification made by an Italian geologist, Giovanni Arduino, in the mid-1700s. He recognized three kinds of mountains, consisting of three different kinds of rocks, and named them *Primitive, Secondary,* and *Tertiary.* Tertiary, which comes from the Latin word *Tertius,* meaning third, is a term still used today.

Quaternary   In the early and middle 1800s deposits that were obviously younger than those of the Tertiary, were recognized in some areas. These were given the name *Quaternary,* from the Latin word *Quattuor,* meaning four, to indicate the position of these deposits above the *Tertiary,* or third order of rocks.

The periods are subdivided into epochs, the best known of which are the epochs of the Tertiary and Quaternary periods. Division of these periods into epochs is based on the percentage of the fossil fauna represented by living forms of life. Whereas the epochs in the lower Tertiary may have only 1–10 percent of their fossil fauna represented by living forms, the epochs in the upper part have as much as 65 percent representation. The meanings of the epoch terms of the Cenozoic are listed in Table C–3.

Table C–3 The Cenozoic epochs and derivation of their names

Period	Epoch	Meaning
Quaternary	Holocene	Gr. *holos* (whole) + *kainos* (recent)
Tertiary	Pleistocene	Gr. *pleistos* (most) + *kainos* (recent)
	Pliocene	Gr. *pleion* (more) + *kainos* (recent)
	Miocene	Gr. *meion* (less) + *kainos* (recent)
	Oligocene	Gr. *oligos* (slight) + *kainos* (recent)
	Eocene	Gr. *eos* (dawn) + *kainos* (recent)
	Paleocene	Gr. *palaica* (ancient) + *kainos* (recent)

GEOLOGIC DATING

Two methods are available to us in dating any event: *relative dating* and *absolute dating*. When using relative dating, we simply say that one event took place before or after another event, without quantifying the length of time between the two events. With absolute dating, an exact number of years or other time unit, plus or minus an allowance for error, is given for the length of time between the event and the present.

Relative dating   Until very recently no reliable means of absolute dating was known, and therefore all geologic work, including the geologic time scale, was the result of relative dating (Table C–4). A number of geologic principles are involved in relative dating. These are not shown in our tables.

The *Law of Superposition* states that in a normal sequence of bedded sedimentary rocks, the lower bed is older than the upper, as the lower must have been there before the upper bed could have been deposited. The *Law of Original Horizontality* states that sedimentary beds were originally deposited horizontally, or nearly so, regardless of their present position. The *Law of Cross-Cutting Relationships* states that beds cut by dikes or faults are older than the event that formed the dike or fault. The *Principle of Faunal Succession* is important in geology, because it has allowed us to comprehend the bulk of the geologic puzzle. This principle states that fossil groups, both plant and animal, succeed one another in an orderly and predictable fashion and that any period of geologic time can be recognized by the fossil assemblages it contains. This principle works because of the law of superposition and because of organic evolution.

Although these principles are techniques of only relative dating, paleontologists and stratigraphers have been able to infer some approximate absolute periods of time between events. A "feeling" for the time involved in organic evolution, coupled with hints given by some lake sediments, in-

Table C–4 Summary of early attempts to date the earth

Method used	Age determined (in years)
Rate at which sediments are deposited; based on 3.5 inches per century as determined by finding a 3200-year-old statue under nine feet of Nile River mud	13,500
Counting seasonal layers in sediments; Green River Shale in Wyoming used	6,500,000
Calculation, based on assumption that the earth started as a molten mass, of the length of time necessary for the earth to cool to present temperature	20,000,000 to 40,000,000
Calculation, based on present-day values for certain assumptions, of the length of time necessary for the oceans to reach their present saltiness	100,000,000

crements of growth in the hard parts of some animals (such as bone or shell), growth rings in trees, and so forth, enabled these scientists to construct a geologic time scale before any method of absolute dating of long periods of geologic time was available. These dates are now being verified by radioactive dating methods and in most instances are proving to be valid.

Radiometric dating   In 1896 Antoine Henri Becquerel, a French physicist whose father and grandfather had been physicists and had pioneered in photochemistry and electrochemistry, respectively, announced that certain natural ores spontaneously generated a strange kind of energy that affected his photographic plates. Two years later Pierre and Marie Curie announced the discovery of radium. Becquerel and the Curies shared the 1903 Nobel Prize in physics for their accomplishments.

Becquerel's announcement heralded the discovery of radioactivity and ushered in a new age. Among the most important uses of this discovery was the development of a new technique for measuring the age of the earth and for dating, on an absolute scale, the events that had occurred since the earth's beginning. In 1904 Ernest Rutherford, a British physicist, suggested that geologic time could be measured by finding the ratio of uranium to helium in a given sample. This was theoretically possible, but there were two practical difficulties to overcome: the first was to prevent the helium from escaping as it was generated by the process; the second was the lack of an analytic technique to measure the decay product. Less than a year after Rutherford made his suggestion, Bertram Boltwood, an American physicist, proposed using the ratio of uranium to lead. Although this index was far more suitable than that proposed by Rutherford, the lack of an analytic technique was still a major obstacle. It was not until about 1939 that the mass spectrograph was used to determine the isotopic ratios in several leads generated by the radioactive process, and this, for the first time, made it possible to date geological material.

A number of theoretical considerations form the basis for using this technique for age determination (Table C-5). First, all radioactive elements ultimately decay into new products. Although the time at which any given atom of a radioactive element will disintegrate is totally unpredictable and is not governed by any physical laws that are presently known, the reaction of aggregates of atoms of these elements is subject to rules of mathematics and is predictable within statistical limits. Although it is not possible to predict when the last atom will disintegrate, it is possible to tell accurately what length of time is necessary for half of the atoms to disintegrate. Although this length of time, known as the *half-life* of the element, varies from element to element from a few millionths of a second to billions of years, each radioactive element has its own characteristic half-life. No circumstances are known that can change the length of the half-life.

In this process the original material is called the parent material, and the end result is called the daughter product. If one can determine the

Table C-5 Radioactive elements used in geologic dating
and their daughter products

Element	Daughter	Half-life (in years)
Uranium 238	Lead 206	4,510,000,000
Uranium 235	Lead 207	713,000,000
Thorium 232	Lead 208	14,100,000,000
Potassium 40	Argon 40	1,300,000,000
Rubidium 87	Strontium 87	47,000,000,000
Carbon 14	Nitrogen 14	5,730

ratio of parent material to daughter product, and if one knows the rate of disintegration (half-life), it is simple to tell how long the process has been operating. It is obvious that certain conditions could interfere with our obtaining a correct answer from our nuclear "clock." If the material has not been contained in a closed system, our results will be erroneous. That is, if any amount of either the parent or the daughter material is allowed to escape, the ratio between the two will be changed and therefore will cause us to miscalculate the time involved. Also, if either the parent or daughter material is enriched through an addition of material from outside the system, the ratio will be changed, and the lapsed time will be calculated incorrectly.

Index